Public Administration and Information Technology

Volume 11

Series Editor
Christopher G. Reddick
San Antonio, Texas, USA

Public Administration and Information Technology publishes authored and edited books that examine the application of information systems to common issues and problems in public administration. This series examines both the successes of Information and Communication Technology (ICT) adoption and some of the most important challenges to implementation. The books published in this series will address all areas of public administration, through the use of information technology adoption in the public and nonprofit sectors, and in the private sector where important lessons can be learned for public managers and policy analysts. New and emerging technologies that will have a lasting impact on public administration will be featured in this series. Both developed and developing countries will be examined in this series. The series particularly welcomes books that cover international dimensions of public administration and information technology. The research in this series will be able to bridge both theory and practice to provide relevance to public managers. The series will cover all aspects of e-governance/e-government research, and new and emerging trends and issues in this research. The series publishes edited books, monographs, upper-level textbooks and research handbooks. For authors interested in submitting a book proposal, please email the editor. Prof. Christopher Reddick, The University Texas at San Antonio, USA, chris.reddick@utsa.edu

Editorial Advisory Board members

More information about this series at http://www.springer.com/series/10796

J. Ramon Gil-Garcia • Theresa A. Pardo
Taewoo Nam
Editors

Smarter as the New Urban Agenda

A Comprehensive View of the 21st Century City

Springer

Editors
J. Ramon Gil-Garcia
University at Albany
State University of New York
Albany
New York
USA

Centro de Investigacion y Docencia
Economicas (CIDE)
Mexico
Distrito Federal
Mexico

Theresa A. Pardo
Center for Technology in Government
University at Albany
State University of New York
Albany
New York
USA

Taewoo Nam
Myongji University
Seoul
Korea, Republic of (South Korea)

Public Administration and Information Technology
ISBN 978-3-319-17619-2 ISBN 978-3-319-17620-8 (eBook)
DOI 10.1007/978-3-319-17620-8

Library of Congress Control Number: 2015946567

Springer Cham Heidelberg New York Dordrecht London

Printed on acid-free paper

Springer International Publishing AG Switzerland is part of Springer Science+Business Media (www.springer.com)

To my Amores Nadia, Dante, and Julieta, with all my love

J. Ramon Gil-Garcia

To my dear husband Manuel, with love and gratitude

Theresa A. Pardo

To my lovely wife Dayoung and baby

Taewoo Nam

Foreword

The 20-chapter book before you is a highly important academic contribution to advancing the understanding of smartness as a distinctive and essential characteristic of urban society in the twenty-first century. In this capacity the book is considered a must read for academics in the field of smart city and smart government studies. It is also of the highest value to practitioners in both urban public administration and urban communities.

The agenda of urban "smartness" evolved for more than a decade; as a result, at the midpoint of the second decade of the twenty-first century, a comprehensive and well-grounded understanding of the term "smartness" in the context of cities and municipalities emerged, which this book duly documents.

So, in a nutshell what does smartness as an urban agenda for the twenty-first century mean and stand for, and why is it so important?

According to the United Nations' WHO, by mid-2014, 54 % of the world population lived in urban rather than rural spaces, which is a 20 % increase from the year 1960.[1] By 2050, the urban share in the world population has been projected to surpass the 70 % mark.[2] So, in less than a century, the global ratio between urban and rural populations will be more than reversed; and today, some urban centers already host populations within their city limits (not even counting the population in the entire metropolitan area around them) that are multiple in sizes compared with quite a few sovereign nation states around the world. Urban centers and their metropolitan areas have become the dominant hubs of economic, social, and cultural activities in the early twenty-first century. Also, these hubs compete on a global scale against each other for resources, talent, investments, influence, and wealth.

The rapid urbanization of the globe presents both great challenges and great opportunities for urban society as a whole as well as for urban government and urban self-governance as two important organizing elements of the urban society. While the concentration of human life in dense urban centers provides numerous economies of scale and efficiencies (centralized services, short-range service distribu-

[1] See http://www.who.int/gho/urban_health/situation_trends/urban_population_growth_text/en/—accessed 10/17/2014).

2 See http://www.fastcodesign.com/1669244/by-2050-70-of-the-worlds-population-will-be-urban-is-that-a-good-thing—accessed 10/17/2014).

tion, proximity of connected logistic entities, short-distance transport of goods and people, low per-capita land use, and leverage of energy resources, to name a few), it also comes with a price tag (for example, in terms of crowding and its various side effects, pollution, higher crime rates, and infrastructural vulnerabilities).

A dense urban space is a highly dynamic and complex system, which at the least is self-referential, if not self-organizing,[3] that is, it cannot be steered like a car, and for its governance it requires sophisticated interventions and precise triggering at certain leverage points to be effectively influenced in ways that help it evolve towards and maintain desirable directions as defined by its very stakeholders.[4] A twenty-first-century urban space is expected to offer its citizens and business stakeholders high standards along the six dimensions of (1) quality of life (also referred to as livability or overall attractiveness), (2) democratic governance and institutions, (3) safety, (4) security, (5) rich economic opportunity and competitiveness supported by sophisticated and effective infrastructures of all kinds (individual and public transports, education, communication, information, health care, retail, research and innovation, public utilities, recreation, entertainment, culinary services, the arts, and cultural institutions, among others), and last, but not least, (6) a healthy and intact natural environment.[5]

Neighborhoods in twenty-first-century urban centers will increasingly play semi-autonomous roles in their governance, since they cannot be effectively managed from a single remote central point or in a stringent hierarchical fashion. At the same time, these semi-autonomous neighborhoods need to stay connected and well integrated in the greater urban whole. This interplay requires new ways of interaction and procedures, which need to be developed.

Furthermore, urban centers and their metropolitan areas compete globally for resources, talent, and investments, and the attractiveness of an urban space hinges upon its capacities to provide a balanced and sustainable mix of the six dimensions outlined above. Obviously, these dynamic, larger, denser, and multilayered urban centers need new and smarter approaches to governance, public service provision, and public administration than were available and practiced in the past.

This leads me to the concept of smartness as a twenty-first-century urban agenda. A smart urban space (also referred to as a smart city) is one that is able to create and maintain a strong attractiveness, safety and security, abundant economic opportunity, sophisticated and effective infrastructures of all kinds, and a healthy natural environment based on a model of smart democratic governance. Elements of smart governance encompass balanced innovation-, competitiveness-, and sustainability-oriented norms and policies that foster smart practices by using and sharing high-quality actionable

[3] P. M. Hejl, "Towards a theory of social systems: self-organization and self-reference, self-reference and syn-reference," in Self-organization and management of social systems: insights, promises, doubts, and questions, H. Ulrich and G. Probst, Eds. Berlin; New York: Springer-Verlag, 1984, pp. 60–78.

[4] J. W. Forrester, Urban dynamics. Cambridge, Mass.: MIT Press, 1969.

[5] S. AlAwadhi, A. Aldama, H. Chourabi, J. R. Gil-Garcia, S. Leung, S. Mellouli, T. Nam, T. Pardo, H. J. Scholl, and S. Walker, "Building Understanding of Smart City Initiatives.," in Electronic Government. vol. 7443, H. J. Scholl, M. Janssen, M. A. Wimmer, C. E. Moe, and L. S. Flak, Eds. Berlin, Heidelberg: Springer Berlin Heidelberg, 2012, pp. 40–53.

information facilitated by state-of-the-art information and communication and other technologies along with top-notch human capital and skills as well as other resources.[6]

Smart government, hence, can be defined as the skilled and effective orchestration of the elements of smart governance and their interplay towards areas of administrative focus such as smart budgeting, continuous administrative modernization, security and safety, continuous infrastructure overhaul and upgrading, carbonless and clean individual mobility, participation, transparency, and collaboration, and data science-based information generation and dissemination. Over time, the areas of focus of smart government may change in the twenty-first century; however, the basic elements of smart governance will remain relatively constant.

The reader may have noticed that unlike most discourses on smart cities, smart governance, and smart government, I have so far not mentioned information and communication technologies (ICTs) or other breakthrough technologies in any prominent way. This is, of course, neither negligence nor coincidence. Modern ICTs and other technologies have facilitated the evolution of a truly global and densely connected economy in the course of only a few decades. They have facilitated the rapid creation of new markets and industries as well as the complete demolition of traditional markets and industries. Generally, they have helped obliterate many, if not most, traditional ways of doing business of whatever kind.

While ICTs and other breakthrough technologies do not "drive" change by themselves, by employing, deploying, and using novel ICTs in new ways towards new ends and new purposes, knowledgeable, purposeful, and intentioned human actors indeed do drive change. For decades ICTs have been touted as "enablers" of change, organizational, behavioral, and other; however, if not used and deployed properly, they have proven to be ineffective and were found capable of even powerfully stalling progress and change. So, whether or not novel ICTs can help bring about change towards desired new ends and new purposes critically depends upon the thorough understanding of a given new technology's potential, a vision for its novel uses, and a resolve and capacity to make this vision a reality on part of purposeful and intentioned human actors. In other words, novel ICTs and other breakthrough technologies need to be in the hands and under the purview of smart and savvy human actors to be deployed in the ways that they can enact their full potential and make a real difference. Then, these technologies can, in fact, act as grandiose facilitators and as massive levers for change of all sorts.

Smartness, hence, is an inherent human capacity. When referring to smart cities, smart governance, and smart government, we implicitly understand that novel technologies (ICTs and others) are necessary, albeit not sufficient engines for making an urban space smarter. It requires smart people who share the vision about and then create the future of the urban space they live in.

The ongoing process of urbanization of the globe requires extraordinary human savviness and smartness in decision-making in order to cope with the looming chal-

[6] H. J. Scholl and M. C. Scholl, "Smart Governance: A Roadmap for Research and Practice," in Proceedings of the 9th iConference, M. Kindling and E. Greifeneder, Eds. Berlin, Germany: Illinois Digital Environment for Access to Learning and Scholarship (IDEALS), 2014, pp. 163–176.

lenges and to seize the emerging opportunities. In other words, smartness is not a nice-to-have attribute, but rather a necessity.

Ideally, smart (democratic) governance paves the path for smart government, which helps instigate the evolution of a smart urban space. But the smart urban agenda of the twenty-first century also has to engage, mobilize, and contribute the various urban communities and stakeholders toward cocreating the smart urban space they want to share and live in. In this vein, smart governance, smart government, and smart city are literary vehicles for promoting the discourse about our future directions of human life on this planet, which will predominantly provide an urban experience. How this experience will shape up is for us to find and smartly decide.

This book makes an important academic contribution to the discourse on smartness in the context of urban environments, governance, and government. It has three parts, which are dedicated to theories, concepts, and methodologies of smartness in the urban context (part I), case studies from around the world (part II), and citizen participation in building smart cities (part III). It represents a wide range and great diversity in terms of regions covered as well as themes in the various chapters. For example, smart-government or smart-participation initiatives in Barcelona/Spain, Istanbul/Turkey, Mexico, Milan/Italy, Moscow/Russia, New Taipei City/Taiwan, Norwegian cities, Rio de Janeiro, Porto Alegre, Curitiba and Campinas (all Brazil), and Stockholm, Göteborg, Malmö, Jönköping, and Umeå (all Sweden) were analyzed and presented among others. Themes ranged from studying and developing conceptual models for smart city-related research over detailed single and multiple case studies to recommendations and lessons learned from the cases and the literature.

The book provides global coverage and a balanced approach in representing the various scholarly approaches to smart city, smart governance, and smart government research. The list of renowned contributors to this volume is impressive, and the contributions are authoritative. Therefore, with great pleasure, I recommend to you the reading of this volume on "Smarter as the New Urban Agenda: A Comprehensive View of the 21st Century City," which in my view represents a milestone in current research on the subject.

Seattle, November 2014

Hans J (Jochen) Scholl PhD
Chair of IFIP WG 8.5 (Information Systems and Public Administration)
Chair of the HICSS EGOV Track
Former President of the Digital Government Society
Associate Editor of Government Information Quarterly

Acknowledgments

This book is the result of the dedicated efforts of many wonderful people who worked together and apart to bring this volume to fruition. We take this opportunity to express our sincere and deep regards and appreciation to all of these people who helped and supported us in the conception and completion of this work. Thank you to all the authors for sharing your knowledge through this editorial project and your interest in smart cities and smart governments. This volume would not have been possible without the hard work and collegiality of authors, reviewers, the editorial advisory board, the series editor, Christopher Reddick, and the staff at Springer. We are grateful for the guidance provided by members of the Editorial Advisory Board, who, regardless to the great distances and time differences, were always willing to share their insights on ways to improve the book. A special thank you to all the reviewers who not only gave their time and effort, but also very useful and constructive comments that enhanced the book's overall quality and contribution to the field.

From the staff at Springer, we would like to mention the dedication and commitment of Kevin Halligan during the entire process. We always received support and useful guidance from him. We want also to thank Ana Catarrivas, whose dedication and tireless effort have been instrumental for the completion of this book. We are also grateful to *Centro de Investigación y Docencia Económicas (CIDE)* and the Center for Technology in Government, University at Albany from which we have received strong institutional support and great encouragement and motivation from colleagues and friends.

Finally, we send love and gratitude to our families. They have tirelessly encouraged and wholeheartedly supported our academic adventures.

J. Ramon Gil-Garcia
Theresa A. Pardo
Taewoo Nam

Contents

Contributors

Mortaza S. Bargh Ministry of Justice & Security, WODC, The Hague, EH, The Netherlands

Lasse Berntzen Department of Business and Management, Buskerud and Vestfold University College, Drammen, Kongsberg, Norway

Dario Cavenago Department of Sociology and Social Research, University of Milano-Bicocca, Milano, Italy

Wojciech Cellary Department of Information Technology, Poznan University of Economics, Poznań, Poland

Sunil Choenni Ministry of Justice & Security, WODC, The Hague, EH, The Netherlands

Edward Curry Insight Centre for Data Analytics, National University of Ireland, Galway, Galway DERI Building, IDA Business Park, Lower Dangan, Republic of Ireland

Zamira Dzhusupova United Nations Department of Economic and Social Affairs (UNDESA), Division for Public Administration and Development Management (DPADM), E-Government Branch, NY, USA

Enrico Ferro Istituto Superiore Mario Boella, Torino, Italy

Roberto Garcia Alonso Bogotá, Colombia

Mila Gascó Institute of Public Governance and Management, ESADE-Ramon Llull University, Barcelona, Spain

J. Ramon Gil-Garcia University at Albany, State University of New York , Albany, NY, USA & Centro de Investigacion y Docencia Economicas (CIDE), Mexico City, Mexico

Ahmet Guler Lock Haven University, Department of Criminal Justice, USA

Karim Hamza Institute for European Studies (IES), Vrije Universiteit Brussel, Cairo, Egypt

Marius Rohde Johannessen Department of Business and Management, Buskerud and Vestfold University College, Drammen, Kongsberg, Norway

Artem Kostyrko National Research University "Higher School of Economics", Moscow, Russia

Sedat Kula Kars Emniyet Mudurlugu, Turkish National Police, Ortakapi Mah., Kars, Turkey

Sebastian Lippez-De Castro Bogotá, Colombia

Xosé María Mahou-Lago Facultade de Ciencias Sociais e da Comunicación, University of Vigo, Pontevedra, Spain

Hamid Mcheick University of Quebec in Chicoutimi, Chicoutimi, QC, Canada

H. Patricia McKenna AmbientEase, Victoria, BC, Canada

Albert Meijer School of Governance, Utrecht University, Utrecht, Netherlands

Ronald F. Meijer Ministry of Justice & Security, WODC, The Hague, EH, The Netherlands

Paulo Roberto de Mello Miranda Curitiba/PR, Brazil

Sehl Mellouli Laval University, Québec, QC, Canada

Karam Mustapha University of Quebec in Chicoutimi, Chicoutimi, QC, Canada

Taewoo Nam Myongji University, Seoul, Korea, Republic of (South Korea)

Juan Carlos Núñez Armas Universidad Autónoma del Estado de México, Toluca, MEX, Mexico

Adegboyega Ojo Insight Centre for Data Analytics, National University of Ireland, Galway, Galway DERI Building, IDA Business Park, Lower Dangan, Republic of Ireland

Michele Osella Istituto Superiore Mario Boella, Torino, Italy

Theresa A. Pardo Center for Technology in Government, University at Albany, State University of New York, Albany, NY, USA

Elisa Pautasso Istituto Superiore Mario Boella, Torino, Italy

Christophe Premat Södertälje, Sweden

José Maria Pugas Filho Curitiba/PR, Brazil

Manuel Pedro Rodríguez Bolívar Department of Accounting and Finance, Faculty of Business Studies, University of Granada, Granada, Spain

Carmelita Roepan Rotterdam, The Netherlands

Jarogniew Rykowski Department of Information Technology, Poznan University of Economics, Poznań, Poland

Rodrigo Sandoval-Almazán Universidad Autónoma del Estado de México, Toluca, MEX, Mexico

Evgeny Styrin National Research University "Higher School of Economics", Moscow, Russia

Benedetta Trivellato Department of Sociology and Social Research, University of Milano-Bicocca, Milano, Italy

Enrique José Varela-Álvarez Facultade de Ciencias Sociais e da Comunicación, University of Vigo, Pontevedra, Spain

Maria Alexandra Viegas Cortez da Cunha Curitiba/PR, Brazil

Kevin Y. Wang College of Communication, Butler University, Indianapolis, IN, USA

Chao-Ming Wu Research, Development and Evaluation Commission, New Taipei City Government, New Taipei City, Taiwan

Editors Biography

J. Ramon Gil-Garcia is an associate professor of Public Administration and Policy and the research director of the Center for Technology in Government, University at Albany, State University of New York (SUNY). Dr. Gil-Garcia is a member of the Mexican National System of Researchers and of the Mexican Academy of Sciences. In 2009, he was considered the most prolific author in the field of digital government research worldwide and in 2013 he was selected for the Research Award, which is "the highest distinction given annually by the Mexican Academy of Sciences to outstanding young researchers." Dr. Gil-Garcia is the author or coauthor of articles in prestigious international journals in public administration, information systems, and digital government and some of his publications are among the most cited in the field of digital government research worldwide. His research interests include collaborative electronic government, smart cities and smart governments, digital divide policies, new public management, public policy evaluation, and multi-method research approaches.

Theresa A. Pardo, is the director of the Center for Technology in Government at the University at Albany, SUNY. She holds research associate professor appointments in public administration and policy at the Rockefeller College of Public Affairs and Policy and in informatics at the College of Computing and Information. Professor Pardo is a codeveloper of the top-ranked Government Information Strategy and Management concentration at the University of Albany and is president of the Digital Government Society, a global multidisciplinary organization of scholars and practitioners interested in the development and impacts of digital government.

Taewoo Nam has been working for the Department of Public Administration at Myongji University, Korea as an assistant professor since 2013. He earned PhD degree in public administration from the University at Albany, State University of New York. He is interested in citizen participation, intergovernmental relationships, city and local administration, and digital government. His publications appear in various academic journals in the field of public administration and information science including *Government Information Quarterly, International Journal of Public Administration, International Review of Administrative Sciences,*

International Review of Public Administration, Social Science Computer Review, Journal of Urban Technology, Journal of Information Technology and Politics, and Information Polity. As an advisor in the performance evaluation board of the Korean national government, he helps the government evaluate the performance of national informatization initiatives and programs.

A Comprehensive View of the 21st Century City: Smartness as Technologies and Innovation in Urban Contexts

J. Ramon Gil-Garcia, Theresa A. Pardo and Taewoo Nam

Abstract This chapter introduces important concepts such as smart cities, innovation, technologies, and "smartness" in government. It argues that a concept of smart city should be more than the use of information technologies in local governments. The study and analysis of cities today requires a multidimensional and multifaceted concept and, therefore, multiple components. Some of these components are (1) public services, (2) city administration and management, (3) policies and other institutional arrangements, (4) governance and collaboration, (5) human capital and creativity, (6) knowledge economy and pro-business environment, (7) built environment and city infrastructure, (8) natural environment and ecological sustainability, (9) ICT and other technologies, and (10) data and information. In addition, the authors refer to smartness as public sector innovation in urban contexts. A comprehensive view of a smart city should encompass government innovation in management and policy as well as technology and must acknowledge how the context of a city shapes the data and the technological, organizational, and policy aspects of a specific initiative. The chapter also briefly describes the content and the target audience of this book.

Keywords Smart city · Information technologies · Innovation · Smart government · Urban context · Digital government · Electronic government · Smartness · Data · Organizational change · Policy

---

J. R. Gil-Garcia (✉)
University at Albany, State University of New York, 187 Wolf Road, Suite 301, Albany, NY 12205, USA
e-mail: jgil-garcia@ctg.albany.edu

Centro de Investigacion y Docencia Economicas (CIDE), Mexico, Distrito Federal, Mexico

T. A. Pardo
Center for Technology in Government, University at Albany, State University of New York, Albany, NY, USA
e-mail: tpardo@ctg.albany.edu

T. Nam
Myongji University, Seoul, Korea, Republic of (South Korea)
email: namtaewoo@gmail.com

J. R. Gil-Garcia et al. (eds.), *Smarter as the New Urban Agenda A Comprehensive View of the 21st Century City,* Public Administration and Information Technology 11,
DOI 10.1007/978-3-319-17620-8_1

1 Introduction

As of 2008, more than half of the world's population lived in cities. The trajectory of further urbanization is unprecedented[1] and by the year 2050 it is expected that 70 % of the world population will live in urban areas, with most urban growth occurring in less developed countries.[2] Such substantial and rapid changes are putting cities all over the world in a state of flux as they seek new approaches to both old and new problems that exhibit increasingly complex dynamics and require new and innovative solutions.

Becoming a "smarter city" is being pursued around the world as part of new urban strategic agendas aimed at addressing these issues and threats. Cities are working to respond to their changing reality and to become "smarter" through a new agenda focused on improving convenience, facilitating mobility, creating process efficiencies, conserving energy, improving the quality of air and water, identifying problems and fixing them quickly, recovering rapidly from disasters, collecting data to make better decisions, deploying resources effectively, and sharing data to enable collaboration across entities and domains (Caragliu et al. 2011; Schaffers et al. 2011). Although many cities are benefitting from such agendas, some are struggling under the weight of them (Gil-Garcia et al. 2014). Many of these cities suffer from a multitude of threats to their efforts, including aging infrastructure, scarce resources, and competing priorities, among others. Regardless of this complexity, the adoption of smart city solutions, while not pervasive yet, is according to Hase (2014) is inevitable. "Few cities have put the Internet of Things to large-scale use, with the continuing growth of cities and calls for sustainability, the adoption of smart city solutions will eventually be inevitable" (Hase 2014).

In spite of such global recognition of need and increasing commitment to and investments in making cities smarter, there is a lack of clarity and consensus around what a smart city is and what its components actually are. There are many labels representing the "smart city" and related phenomena—among them, digital city, urban innovation, intelligent city, creative city, knowledge city, and information city, to mention a few. Some of these labels draw attention to the technological aspects, while others to the development of human capital or physical infrastructure. In many cases, the use of information and communication technologies (ICTs) appears central to defining a smart city. However, increasingly scholars and practitioners are recognizing that smart cities are not built simply through harnessing the potentials of ICTs. Nam and Pardo (2014), for example, argue that technology needs to be understood as a means to enable social, environmental, economic, and cultural progress in cities. Hollands (2008) as well calls for a conceptualization of the idea "smart city" to include most of the important aspects, beyond technology, and all their conceptual amplitude.

[1] World Health Organization's Global Health Observatory www.who.int/gho/en/.

[2] Population Reference Bureau's homepage www.prb.org.

As an introduction to this book, this chapter provides background information and introduces the reader to the subject matter. With this book we argue that a comprehensive view of the twenty-first century city is necessary to better understand the array of phenomena now called “smartness” in cities and other urban environments. This chapter contributes to the maturation of this research area by proposing an integrative and more comprehensive conceptualization of a smart city. The chapter is organized in seven sections, including the foregoing introduction, which provides some background and explains why research on smart cities is important and timely. Section 2 discusses various definitions of a smart city and argues that smartness in cities should be conceptualized considering multiple dimensions and a range of diverse components. Section 3 presents a view of smartness as a creative combination of the use of emergent technologies and innovation in the public sector. Based on the previous sections, Sect. 4 proposes a comprehensive view of a smart city and how this view, we argue, will support more robust understanding of smart city initiatives in the present and future. Section 5 briefly describes each of the chapters included in this book outlining also the structure and main sections of the book. Section 6 suggests the main audience of the book and how different readers might benefit from the theories, empirical data, and cases it contains. Finally, in Sect. 7, the editors highlight some unique characteristics and contributions of this volume and suggest ideas for future research on smart cities.

2 Conceptualizing Smart Cities

While for several authors the conceptualization of a smart city is clearly based on the use of ICTs as the only, or at least the most, important component of smartness, authors from different disciplines are increasingly proposing definitions that go beyond the use of information technologies (Chourabi et al. 2012). The editors of this book argue that a smart city is indeed a multidimensional and multifaceted concept and, therefore, smart cities should be studied and analyzed on the basis of multiple components. Some of these components are (Chourabi et al. 2012; Gil-Garcia 2012; Gil-Garcia et al. 2014; Nam and Pardo 2014): (1) public services; (2) city administration and management; (3) policies and other institutional arrangements; (4) governance, engagement, and collaboration; (5) human capital and creativity; (6) knowledge economy and pro-business environment; (7) built environment and city infrastructure; (8) natural environment and ecological sustainability; (9) ICT and other technologies; and (10) data and information. We briefly describe each of these components in the following paragraphs.

Public services refer to the efficient and effective production and delivery of critical city services (Folz and Abdelrazek 2009; Hollands 2008) such as public safety, transportation, health and social services, emergency services, culture, tourism, recreation, and so on. City administration and management, for example, emphasizes the importance of organizational capacity, leadership, and design strategies to become intelligent (Santinha and de Castro 2010). Some examples are e-governance, performance management, funding, staffing, leadership, vision, policy instruments, and policy learning. Policies and institutional arrangements (e.g., laws, regulations, norms, and others) influence the way people act in organizational settings (Scott 2008). Institutional arrangements should be considered key components of any city government initiative or program, including smart city efforts. Governance, engagement, and collaboration refer to structures and efforts that go beyond city government and include other social actors, for example, engagement in collective decision-making efforts to enhance transparency and accountability of government. Very few studies address governance in smart cities (Johnston and Hansen 2011; Paskaleva 2009), but some are found dealing more broadly with urban innovation. Three main elements fall into this category: (1) e-governance; (2) engagement of stakeholders, citizens, and communities; and (3) networks, partnerships, and collaboration. Concepts such as wiki government and crowdsourcing in government emerge (Bertot et al. 2010a, 2010b, 2012, 2010c; Musso et al. 2011, 2006; Noveck 2009).

The human capital and creativity component represents human resource-related elements such as education, training, culture, arts, and creative economy and industry (Bartlett 2005). People, education, learning, high-skilled workforce and knowledge are of central importance to cities (Landry 2000; Nam and Pardo 2011a), since a creative and diverse culture is an important element to attract smarter people to a city and, therefore, become a more efficient and sustainable city (Švob-Đokiæ 2007). Similarly, knowledge economy and pro-business environment refers to the economic component of a smart city. A knowledge economy primarily involves research and development, technology transfer, and technological innovation as a hotbed for innovative industries (Edvinsson 2006; Komninos 2002, 2009; van Winden 2008, 2010; Yigitcanlar et al. 2008a, b).

Another component is built environment and city infrastructure. It includes city infrastructures such as roads, bridges, tunnels, buildings (residential, business, and recreational), pipelines, electrical and communication lines, and so forth, in terms of economic sustainability and management challenges (Kaklauskas et al. 2009; Kim 2010; Lin 2007; Moser 2010; van Heur 2010; Wang 2011). Natural environment and ecological sustainability refer to the vision of cities becoming clean and green (Odendaal 2003) referring to the ecological implications of urban growth and development (Hollands 2008, p. 310). Cities are increasingly faced with scarcer resources, necessitating the establishment of large-scale monitoring systems, assessments and evaluations (Gabrys 2007; Yovanof and Hazapis 2009).

Previous literature seems to focus on the inclusion of ICT and other technologies as the core component. In fact, for some, smartening an entire city can be seen as a massive IT effort (Helal 2011, p. 30). In addition, some of the actions and efforts identified within this component could be considered central to a smart city project: broadband, wireless, virtual, and ubiquitous technologies and infrastructures offer benefits to city dwellers with mobile lifestyles. Many cities are beginning to claim that they are smart because they employ ICTs in their operations (Caragliu et al. 2011; Hollands 2008). Technology could be seen as a component that enables the development and progress in other smart city components becoming a means and not an end in itself (Eger 2009, p. 48). Another element to consider is data and information. Using data and information is central to making cities smart. Data management capacity, information processing, and information sharing through ICTs are considered key to partnerships and interorganizational communications between multiple entities related to smart city initiatives (Odendaal 2003). Also, data are important in terms of security, availability, accuracy, integrity, and shareability through common data standards, architectures, protocols and practices.

These components of a smart city can be grouped in the following dimensions (Chourabi et al. 2012; Gil-Garcia 2012; Gil-Garcia et al. 2014; Nam and Pardo 2014): technology, infrastructure, services, systems integration, community, and environment. Considering all these aspects clearly reflects that the smart city as a phenomenon refers to much more than the use of information technologies and as such, a more comprehensive view is also needed. Such a comprehensive conceptualization of the smart city will become a resource for researchers and practitioners. In this sense, instead of a dichotomy between "being smart" or "not being smart," a smart city should be seen as a continuum in which local government officials, citizens, and other stakeholders could think about the initiatives that attempt to make the city a better place to live. In order to do this, researchers and practitioners must take a holistic approach where the initial conceptualization of smart city includes, at its base, technology, management, and policy components (Nam and Pardo 2014).

3 Smartness as Public Sector Innovation in Urban Contexts

While many see smart city initiatives as laboratories for technological advancements, increasingly researchers and practitioners alike see a smart city as a broad, integrated approach to enhancing the efficiency of city operations, improving the quality of life for citizens, and growing the local economy (AlAwadhi et al. 2012; Allwinkle and Cruickshank 2011; Caragliu et al. 2011; Chourabi et al. 2012; Giffinger et al. 2007; Harrison et al. 2010; Hollands 2008; Washburn et al. 2010; Winters 2011). For instance, a number of recent studies point to the risks behind the use of sensors and data analytics software in cities and how data-driven transformation can lead cities to lose control over innovation projects (Newcombe 2014; Marshall 2014). Alternatively, some city governments are focusing on urban innovation that has very little to do with emergent technologies. Our view of smart cities is aligned with the more current, broader, and integrated view of smart cities rather than the prevalent perspectives in which the role of emerging technologies is highlighted as the most important element of smartness in the urban context. There is no doubt technologies play a decisive and pivotal role in making cities smarter, but how cities utilize technologies for certain purposes and how cities determine the role of technologies depends, in our view, on many additional important components such as city management capability and urban policy dynamics.

In our view, smartness is a matter of public sector innovation and the role of city governments is essential in leading smart city initiatives that reflect a creative mix of emerging technologies and innovation (Gil-Garcia et al. 2014). As the meanings of smart include being percipient, astute, shrewd, and quick; a smart government uses emerging technologies and various innovation strategies "to gain a good understanding of communities and constituencies (being percipient), accurately assess situations or people (being astute), show sharp powers of judgment (being shrewd), and then make decisions and respond quickly or effectively (being quick)" (Gil-Garcia et al. 2014, p. 11). In order to make cities smarter, a smart government requires having a forward-thinking and forward-looking approach to the use and integration of information, technology, and innovation not only in the activities of governing, including internal operations, but also public services and citizen engagement.

A smart city needs a smarter government and smarter governance. A smart government utilizes "sophisticated information technologies to interconnect and integrate information, processes, institutions, and physical infrastructure to better serve citizens and communities" (Gil-Garcia 2012, p. 274). It has the following core elements (Scholl and Scholl 2014, p. 166): openness in decision-making, open information sharing and use, stakeholder participation and collaboration, and improving government operations and services, all through the use of intelligent technologies as they act as a facilitator of innovation, sustainability, competitiveness, and livability. Gil-Garcia (2012) argued that "sensors, virtualization, geographic information technologies, social media applications, and other elements could function like a brain to manage the resources and capabilities of government, but also the participation of social actors, the physical infrastructure, and the machines and equipment using that infrastructure" (pp. 274–275).

Willke (2007) considered smart governance as "an abbreviation for the ensemble of principles, factors, and capacities that constitute a form of governance able to cope with the conditions and exigencies of the knowledge society" (p. 165). Overall, the efforts of making cities smarter include both technological and social activities related to government workings and governance arrangements. The activities involved in creatively investing in emergent technologies are coupled with innovative strategies to achieve more agile and resilient government structures and governance infrastructures (Gil-Garcia et al. 2014, p. 11). Following this logic and rationale, this book acknowledges the key role of technologies, but at the same time takes a closer look at how the use of emerging technologies interplays with other social and organizational factors such as governance and citizen participation in order to generate innovative smart initiatives in cities around the world.

4 A Comprehensive View of a Smart City

A comprehensive view of a smart city should encompass government innovation in management and policy as well as technology and must acknowledge how the context of a city shapes the data, technological, organizational and policy aspects of a specific initiative (Nam and Pardo 2011; Gil-Garcia et al. 2014). Urban performance relies a lot on a city's endowment of hard (physical capital) infrastructure and soft (human and social capital) infrastructure (Boulton et al. 2011; Caragliu et al. 2011). Urban competitiveness, for example, in the form or economic development, an increasingly important concern for city leaders, lies in how cities manage innovative efforts and design relevant policies and practices. Little research on cities discusses innovation in management and policy while the literature on technology innovation is abundant; little is known about inevitable risks from innovation, strategies to innovate while avoiding risks, and contexts underlying innovation and risks (Nam and Pardo 2011). Being smart is not an end state, but rather can be an enabling condition that may lead to other desirable social, economic, or environmental outcomes. With the goal of furthering understanding about how to create such enabling conditions, the book provides new insights on how governments and other actors in the urban context are implementing smart governance practices to cope with complex and uncertain contexts while building requisite capacity to achieve resilience, sustainability, and livability.

Who and what leads a smart city movement in a city merits further discussion. The imperative of urban innovation is not merely in the hands of city governments and/or specific civic groups alone; rather, city residents—ordinary citizens—also have interest in and in many cases, participate in, making their cities smarter (Paskaleva 2009, 2011). Governance and collaboration thus become a crucial axis of smarter cities, since they reflect how public value can be generated with the participation of citizens and other social actors. Eger (2009) argued that a smart city becomes a laboratory for collaboration. Collaboration (interorganizational, inter-sectoral, and citizen–government) is one of the important success factors for smart city initiatives (AlAwadhi et al. 2012; Bakici et al. 2013; Eger 2009; Paskaleva 2011). Deakin and Al Waer (2011) found that innovative and creative partnerships and community-based activities lead the transition from intelligent to smart cities. Many ongoing initiatives for smart cities in practice are rooted in collaborative approaches and grassroots democracy based on communities. Similarly, elements of smart government include coordination between economic and social policy, improvement in intragovernmental coordination in the social sphere, decentralization, increased participation, and renewal of organizational structures (Gil-Garcia et al. 2014, p. 12). Some of the chapters in this book address how collaborative governance and citizen participation help cities innovate and generate value. Various cases from around the globe highlight multiple aspects and variables related to governance and participation. So, this book attempts to present this comprehensive view on smartness and innovation by speaking of different cities from different regions around the world and tells their stories and the lessons they have learned.

Attention to the risks and concerns of smart city initiatives, as well as the opportunities and benefits, contributes to our claim of a comprehensive and balanced view of smart cities. The argument that Gil-Garcia (2012) made regarding a potential smart State could be restated in terms of a smart city: Given that a smart city is technically possible and, to a certain extent, politically feasible, a critical question is whether we, as a society, want smart cities, and if so, in what way do we want them to be smart. What is the social desirability of a situation in which all local government organizations are working together and there is a single integrated system containing all information about citizens and businesses? Is this desirable? If so, why? What concerns does such a city-owned information resource raise? Comprehensive knowledge of smart cities requires an understanding of risks and concerns resulting from innovative efforts to make cities smarter. A comprehensive and multidimensional view of a smart city can help to better understand the risks and benefits, and inform a plan for making a city smarter that reflects what the city government and citizens of that city, really want and need.

5 Smartness and Urban Environments Around the World

Overall this book includes 20 chapters and represents a diversity of theoretical approaches, research methods, and geographical location. The chapters are presented in three parts. The first part entitled "Theory, Concepts and Methodologies" contains eight chapters. These chapters rely on theory and literature reviews to revise and describe the concept of smart cities, the research landscape surround it, and evaluation frameworks. Together, they provide a valuable group of ideas, concepts, and efforts for the study and understanding of smart cities. Chapter 2 by Ojo, Dzhusupova, and Curry, entitled Exploring the Nature of the Smart Cities Research Landscape introduces the field of study by analyzing the existing literature that focuses on innovation and smart cities. The authors argue that smart cities are only just emerging as a research domain, given the number of publications, books, and other scholarly articles on smart cities indexed in Google Scholar and Elsevier's Scopus—an abstract and citation database. However, significant literature is available on related topics such as intelligent city, digital city, and intelligent community based on search results research repositories such as Elsevier's Scopus, ACM Digital Library, and Google Scholar. This chapter maps the research work in the smart city domain, based on the available scholarly publications. The aim is to synthesize an emerging understanding of the smart city concept, determine major research themes, types, and gaps in the current research landscape.

The second chapter in this part is by Rodriguez Bolivar, entitled "Characterizing the Role of Governments in Smart Cities: A Literature Review." This chapter creates a fundamental link between smart city projects and the expected role of the government and the public sector in the success of such initiatives. Rodriguez describes how the growth of smart cities is forcing governments to think about the need to advance in the implementation of ICTs for the improvement of the citizenry's participation in the decision-making processes, the increasing efficiency of public services, and the improvement of transparency and accountability. In this regard, governments in smart cities are called to play a key role in promoting and managing these cities. Based on prior research, the chapter seeks to analyze the role of governments in smart cities, trying to identify different patterns of management styles in these cities. Also, differences between theoretical and empirical studies about the role of governments in smart cities are identified proposing new research themes for the future.

"Smart City Governance: A Local Emergent Perspective" by Albert Meijer presents an emergent perspective on smart city governance. The author argues that smart city governance is about using new technologies to develop innovative governance arrangements. Cities all around the world are struggling to find smart solutions to wicked problems and looking to learn new strategies from successful techno-governance practices in other cities. In the chapter, Meijer argues that learning about successes of smart city governance is important but lessons need to be contextualized: approaches that work in one city may fail in another. This chapter presents the local cooperative knowledge potential and the nature of the problem domain as key contextual factors and develops a model for studying and assessing smart city governance in context.

Expanding on the literature on smart cities, McKenna explores the learning dimension of the smart city and the potential for innovation through use of an early-stage social radio tool. Based on the experience with the tool, this study aims to provide an understanding of: (a) how participatory and collaborative engagement can be fostered and (b) the awareness aspect of emerging technologies. The chapter entitled "Rethinking Learning in the Smart City: Innovating Through Involvement, Inclusivity, and Interactivities with Emerging Technologies" makes several contributions to the e-government literature by providing: (a) insight into the value of under-design approaches to understanding and assessing tools at the early development stages for e-government, transformational government, and lean government; (b) a framework for rethinking and innovating the learning city; and (c) an expanded way of looking at and working with learning and innovation in the smart city that may have implications for other types of e-government relationships (e.g., G2C, G2B, and G2G).

In their chapter, "Ad Hoc BYOD Information Services in Public Places of Smart Cities," Rykowski and Cellary focus on public information broadcasting as an important component of a smart city. To broadcast information, they propose to install miniature and very cheap access points in public places in a city. These access points will be capable of interacting with mobile personal devices, according to BYOD principles, in an ad hoc and anonymous mode, by means of Bluetooth connections. According to the idea "right information at the right place and the right time," the proposed system provides information well suited for the place of its access, as well as minimizes the efforts related to parameterized access and personalized filtering. The broadcasted information is automatically adjusted to the specificity of a given time/place, limiting the need for further processing at the user's side. The proposed system seeks to supplement and extend the already existing intelligent transportation systems.

In "Toward a Methodological Approach to Assess Public Value in Smart Cities," Osella, Ferro, and Pautasso propose a novel framework aimed at measuring performances of smart cities. The methodological approach underlying the framework has its roots in an in-depth analysis of the smart city paradigm conducted from the perspective of urban governance. In this context, the notion of public value is seen as a backdrop for exploring the various ways in which value for society can be created in a smart city. Through a multidisciplinary synthesis of various strands of literature related to smart cities the authors seek to pave the way to a framework meant to evaluate the "smartness" of a city through the lenses of economic, social, and environmental performances, in line with the "triple sustainability" principle. The Italian city of Turin is used as a case study for testing the proposed assessment tool.

In their chapter, "Smart Cities and Resilience Plans: A Multi-Agent Based Simulation for Extreme Event Rescuing," Mustapha, Mcheick, and Mellouli take on the issue of resilience and propose a new framework based on multi-agent systems designed to help cities build simulation scenarios for rescuing citizens in the case of an EE. The authors draw on the view that smart cities must draw on the use of new ICTs to improve services that cities provide to their citizens. The resilience of a city is one of the services that it can provide to its citizens. Resilience is defined as its capacity to continue working normally by serving citizens when extreme events (EEs) occur. The main contribution of the framework, put forward by the authors, is a set of models at different levels of abstraction to reflect the organizational structure and policies within the simulation, which involves the integration of truly dynamic dimensions of this organization. The framework proposes methods to go from one model to another (conceptual to simulation) and be applied to different domains and EEs, such as, smart cities, earthquake, and building fire.

Closing this section, Karim Hamza argues that the current smart city frameworks and models do not meet the needs of cities in developing countries seeking to become smarter. The weak integration of social, economic, and political action found in such cities and the lack of a holistic and integrated approach to sustainable city development limits the utility of such tools. In addition, most developing countries lack the means for implementation, including proper infrastructures, funding, sufficient economic growth, and political stability. Moreover, the challenges that can prevent the success of such a concept, such as poverty, inequality, and cultural barriers are significant. In many developing world cities, Hamza points out that the most difficult obstacles are continuous growth of slums due to unplanned immigration from rural areas to cities. His chapter entitled "Smart City Implementation Framework for Developing Countries: The Case of Egypt" recommends a "Strategic Implementation Framework for Smart City" tailored for developing countries such as Egypt.

Part II of the book, entitled "Smart Cities Around the World," comprises six chapters describing recent innovation projects in cities around the world, including Italy, Mexico, Russia, Spain, Sweden, Taiwan, and Turkey. Each chapter tells a story of a city's effort to better serve citizens through a smart city initiative. For instance, in their chapter entitled "How Do Southern European Cities Foster Innovation? Lessons from the Experience of the Smart City Approaches of Barcelona and Milan," Gasco, Trivellato, and Cavenago address the need for a smart city strategy to guide efforts to prevent and manage challenges and to guide implementation toward a successful outcome. By drawing on a comparison between Barcelona (Spain) and Milan (Italy), this chapter explores similarities and differences in the ways these two Southern European cities, both the second largest in their respective country, are building their smart city agendas. The ultimate aim of the chapter is to identify the main features of the two still developing approaches, which appear to be influenced by the increasing integration of smart dimensions and initiatives in cities' strategic agendas, and the related opportunities and challenges.

Focusing on Europe, in his chapter, "Smart Cities in a Digital Nation: Are Swedish Cities Enough Innovative?" Premat investigates the relationship between a country's desire to be smarter and the smart city efforts of cities in that country. Sweden, as the focus of the chapter, is known to be a well-connected society with a strong will to transform into a digital nation with specific attention to the unique and important role of cities. Sweden's main goal, as outlined by Premat, is to achieve sustainable development with well-balanced policies. Framed by such national efforts Sweden's cities began to map a digital agenda where e-business, universities, smart transportation, and green and cultural policies are combined. The aim of this chapter is to analyze a few municipal digital agendas (Stockholm, Göteborg, Malmö, Jönköping, and Umeå) to describe the Swedish model of smart cities with the goal of understanding whether the national digital agenda can be seen as linked to a strong development of smart cities. The chapter explores the differences between the three main metropolitan areas and the other smart cities in Sweden and discusses whether there is real digital growth that can support the development of smart cities in Sweden.

The chapter "Implementing Smart Services in Moscow: The Integrated Mobile Platform" by Styrin and Kostyrko argues that the city of Moscow is a leading implementer of ICTs in public services. The high level of ICT and mobile penetration among Muscovites creates a strong demand for mobile and electronic services. This chapter explores a case study involving the creation and development of Moscow's Integrated Mobile Platform (IMP). The case study illustrates ICT usage policy and the Moscow government's priorities in terms of delivering and providing access to mobile public services. The case study takes a framework approach to mobile platform development and is also based on the lean government concept. Key success factors in IMP development as well as challenges involved in the collaboration and coordination of various IMP stakeholders are examined. The governance decision-making process and regulatory framework for IMP management are examined as well.

Likewise, in another analysis referring to mobile technology, Kula and Guler's chapter presents a case study about the application of mobile electronic system integration (MOBESE) in Istanbul, Turkey. MOBESE has been in use to fight crime in more effective and efficient ways and to manage traffic since 2005. From the perspective of smart city and innovation, the chapter entitled "Smart Public Safety: Application of Mobile Electronic System Integration (MOBESE) in Istanbul" explains the process of MOBESE and its applications for crime prevention, crime fighting, and traffic management. The chapter reviews smart city research from an information science perspective and surveillance research from a criminal justice perspective. Then, drawing on the literature from these two perspectives, the case is used to highlight the potential of such systems and the need for a multidisciplinary view. Finally, its relevance and contribution to the literature is discussed.

With their chapter "Building an Intelligent Government, Intelligent City, and Intelligent Citizenry Through ICTs: Smart City Innovations in New Taipei City,

Taiwan," Wang and Wu shift the focus to Asia and present a case study on smart city initiatives in New Taipei City (NTPC) in Taiwan. Using the integrative framework proposed by Chourabi et al. (2012) as an analytical lens, the authors discuss the characteristics and scope of NTPC's innovations in delivering services and managing resources using ICTs and examine the policy and the human, social, and cultural contexts that shape the adoption and development of these smart technologies. A SWOT analysis (strengths, weaknesses, opportunities, and threats) was conducted in responding to the research question, contributing to the identification of lessons learned, from this particular case, which may have broader relevance for other smart city initiatives.

The last chapter in this part by Sandoval-Almazán and Núñez Armas tells the story about the use of social media in several counties in Mexico. The goal of the chapter entitled "Social Media Experiences at County Level: The Case of the State of Mexico" is to create new understanding about the link between social media and smart cities. In particular, they seek to understand how citizen's use of information technologies to communicate, collaborate, and interact can be leveraged in helping make a city smarter. In order to understand this linkage, the authors brought together two different but complementary frameworks related to social media and smart cities and use this to explore the relationship with CIO's of eight cities in the State of Mexico. Findings reveal that social media tools are very much a part of smart city practices in these cities and are changing departments, systems, and relationships among internal agencies. The study also shows there are important gaps to be closed if social media tools are going to fully address issues of inclusion, empowerment, and information quality.

The third set of chapters addresses an important component of our vision of smart cities; the role of the citizenry. This third and final part of the book contains five chapters that further our understanding of this component by examining the link between smart city initiatives and the citizenry in the form of studies of e-participation, e-governance, and the active engagement of citizens.

In the first chapter of Part III, Berntzen and Rohde-Johannessen examine the role of citizen participation in the development of smart cities through literature and example cases from Norwegian cities. Their chapter entitled "The Role of Citizen Participation in Municipal Smart City Projects: Lessons Learned from Norway" presents an overview of technologies used for participation, including their strengths and weaknesses, discussing how different types of projects should be handled differently in the decision-making process. The authors examine three different categories of participation: citizen competence and experience, data collection through citizens' use of technology, and participation as democratic value. Finally, they offer suggestions for technologies to collect citizen input in each of the three categories and present recommendations for how practitioners can set up citizen participation projects in smart city initiatives.

In their chapter entitled "eParticipation in Smart Cities of Developing Countries: Research-Based Practical Recommendations," Miranda, Cunha, and Pugas argue that cities are facing major challenges and the expansion of citizen participation has been considered a valuable strategy for meeting these challenges. The authors discuss some of the enabling and inhibiting elements for the creation of virtual spaces for participation by governments in a developing country of Latin America—Brazil. A list of enabling and inhibiting elements grouped in five dimensions was produced from a survey with Brazilian experts, including public officials, academic researchers, and consultants involved in the implementation of e-democracy projects. To contribute to the smart city initiatives in developing countries, interviews with the CIOs of four major cities in Brazil—Rio de Janeiro, Porto Alegre, Curitiba, and Campinas were conducted. The authors found that enablers to the smart citiy initiatives in these cities were concentrated in the sociocultural and technological dimensions, while the inhibitors are mainly in the political and governance dimensions.

In "Technology Helps, People Make: A Smart City Governance Framework Grounded in Deliberative Democracy," García Alonso and Líppez-De Castro explore an important dimension of the full development of smart cities—the human dimension. Based on the theoretical model of Habermasian deliberative democracy, the chapter proposes a revision of the Chourabi et al. (2012) analytical framework for smart city initiatives, in which people are as important as technology. This chapter discusses the role played by ICTs, particularly e-government, in walking the road toward smarter cities. In sharp contrast to perspectives which overestimate the role played by technology, the authors argue that simply using ICTs such as those provided by e-government tools, m-government applications, the integration of information systems, and even the capabilities of Web 2.0, is not enough for a city to become smarter. In particular, they argue that the success of a smart city initiative depends upon the capability of integrating people and communities with the advantages of ICTs, within a comprehensive smart city governance framework.

The chapter "Privacy and Security in Smart Data Collection by Citizens" by Choenni, Bargh, Roepan, and Meijer explores how exploiting information and communication infrastructures makes a city or government better and the potential risks such use creates. Large quantities of data are generated from these infrastructures and infusing these data into the physical infrastructure of a city or government may lead to better services to citizens. Collecting and processing of such data, however, may result in privacy and security issues that should be faced appropriately to create a sustainable approach for smart cities and governments. The authors focus on data collection through crowdsourcing with smart devices and identify the corresponding security and privacy issues in the context of enabling smart cities and governments. They categorize these issues in four classes. For each class, they identify a number of threats as well as solution strategies.

In the final chapter of the book Mahou and Álvarez present their study entitled "Innovation and Opportunities for Citizen Participation in Spanish Smart Cities." With this chapter they argue that smart cities are a social, political, administrative, and technological phenomenon. The growth and consolidation of smart cities depends, they posit, on sociotechnical conditions that implement 2.0 virtual platforms in specific sociopolitical and organizational contexts. To determine whether Spanish smart city web portals facilitate fluid interaction between local administrations and citizens, the authors designed a study using heuristic test techniques to analyze web portal usability in 20 Spanish smart cities. The authors identify the type and development level of electronic participation features on selected municipal websites and confirmed that smart cities must articulate an integrated framework in which citizen participation emerges as an absolute requirement. From their results, the authors concluded that while the Spanish smart city web portals they studied are informative they do not generate virtual environments favoring fluid interaction between local administrations and citizens.

6 Who Should Read This Book?

Academics and professionals who want to improve their understanding of smart city initiatives, in particular, of such initiatives being carried out and studied in very different political, economic, and cultural contexts are the target audience of this book. More specifically, there are two primary audiences for the book. First, researchers and students in the digital government field and various disciplines, for example, public administration, political science, communication, information science, administrative sciences and management, sociology, computer science, information technology and urban planning, among others, seeking a comprehensive account of topics related to cities engaged in "smart city" initiatives. Second, government officials and public managers seeking empirically based, practical recommendations, and context-specific lessons that contain insights and guidance for the development, management, and evaluation of smart city initiatives. For all of them this book could provide some of the necessary knowledge, analytical frameworks, and interesting examples from cities around the world.

7 Unique Contributions of Smartness as the New Urban Agenda for the Twenty-First Century City

This book makes a significant and unique contribution to what is known about the twenty-first century city and how city governments are working to become "smarter." The book creates this comprehensive view by providing a coherent collection of concepts and cases for understanding twenty-first century city governments pursuing innovation agendas organized around the idea of becoming "smarter." The book goes beyond what is currently available in the field by addressing innovative methodologies for the analysis and evaluation of smart city initiatives. Through the compilation of high-quality chapters covering cases, concepts, methodologies, experiences, and practical recommendations focused on cities and city governments worldwide and their efforts to become smarter through the use of technology and related organizational and policy innovations, the book uniquely supports the efforts of those seeking to understand and lead twenty-first century cities.

The range of policy domains addressed and the wide variety of cities covered in the chapters contributes to the comprehensiveness of the book. Authors from Europe to the Americas to the Middle East, Russia, and Asia consider the nature and evolution of smart city research and evolving implementation frameworks. Case studies are used to further understanding about how the concepts of smart city are being enacted around the world. Taken collectively, for example, the book shares what is being learned in Quebec, Moscow, and Stockholm together with insights from New Taipei City, and provides new understanding on how both long standing and newer cities are pursuing smarter city objectives.

Chapters include attention to issues of strategy development, privacy and security, service delivery innovations, and indicator development and use. Attention to these issues in the context of a city provides insight that can help guide cities in framework development and execution of related strategic plans. Some chapters take a national and subnational look while others do comparisons between regions. For example, one chapter makes a contribution both to the development of comparative methods and provides new insight about the differences between cities in Latin America and the European Union. Another chapter provides a national look by looking inside Sweden to explore the development of a national model for linking the cities of that country. Chapters provide readers with lenses to look into the efforts of city-level public safety programs, emergency management, integrated mobile services, and participation. Several chapters take some of the early conceptualizing work done by leading smart city scholars and expand and enhance models with a new understanding of the role of citizens in making cities smarter. Three chapters round out the contribution of the book by providing case studies on place-based information services, policing, and resilience.

Another unique contribution of the book is the comingling of examinations of how cities are innovating in the provision of specific services together with chapters which address the more social side of smart cities. One chapter for example, explores the impact of social media on county governments; another examines the creation of enabling spaces in the cities of Brazil. Two more look at the roles of citizens themselves in the development of such capabilities in cities. Whether through an examination of issues, locations, policy innovation, or services, among others, this book provides the scholar and practitioner with a unique compilation of the latest in smart city research produced by some of the world's leading scholars in smart cities, doing work focused on some of the most interesting and innovative cities of the world. The new student of the world's cities, the experienced scholar, and the scholarly practitioner are just a few of the audiences who will benefit from this volume and the contribution it makes to a very current and complex set of questions about the cities of the world and how and why they are becoming smarter.

Furthering the argument for a comprehensive view of smart cities is one of the main contributions of this book. Indexes and indicators for ranking smart cities seem to agree that there is real value in having smart economy, smart environmental practices, smart governance, smart living, smart mobility, and smart people (Giffinger et al. 2007; Harrison et al. 2010; Washburn et al. 2010). And while these tools may help cities track their performance with specific actions developed for specific needs they provide little guidance for cities as they seek a comprehensive strategy for becoming "smarter." While some practical tools used to rank cities in terms of "smartness" propose a seemingly comprehensive set of components for evaluating urban conditions, the tools do not suggest how cities can go forward to achieve such high standards. Through an examination of cases of cities that sought to become smarter, this book provides some guidance to cities as they commit their energies and resources to efforts toward their aspirational goals.

The geographic diversity of the cases presented represents yet another contribution of this book. The cases, taken together provide a unique view of smart cities through the inclusion of cities from the developed and the developing world, as well as cases of mature and emerging smart cities. The coverage of case studies in this book includes many urban areas in Asia, Africa, America, and Europe and suggests both general lessons and context-dependent unique findings. A merit of the book is not only the exploration and delivery of good stories (benchmarking practices) but the concrete introduction and systematic analysis of how cities create their own unique stories and build innovative capabilities (processes and dynamics).

The overall contribution of the book is found in the enhanced theories and new frameworks for smart cities it presents. The book broadly encompasses conceptual developments, theoretical frameworks, methodological approaches, empirical research, comparative studies, a geographically dispersed and domain-diverse set of case studies of cities, practical recommendations, and policy suggestions. Understanding and adapting these frameworks, models, and stories to ensure that cities around the world and those who live and work in them benefit is the next step for researchers and practitioners on their critical journey to making the world's cities smarter. The comprehensive perspective used in the book helps bridge the gap between sound research and practical expertise in the area of smarter cities and innovation in policy, management, and technology aspects and serves as a guide for such efforts going forward.

References

AlAwadhi, S., Aldama-Nalda, A., Chourabi, H., Gil-Garcia, J. R., Leung, S., Mellouli, S., et al. (2012). Building understanding of smart city initiatives. In *Proceedings of EGOV 2012* (pp. 40–53).

Allwinkle, S., & Cruickshank, P. (2011). Creating smart-er cities: An overview. *Journal of Urban Technology, 18*(2), 1–16.

Bakici, T., Almirall, E., & Wareham, J. (2013). A smart city initiative: The case of Barcelona. *Journal of the Knowledge Economy, 4*(2), 135–148.

Bartlett, L. (2005). Smart city: Social entrepreneurship and community engagement in a rural regional city. In *Proceedings of the International Conference on Engaging Communities*, Brisbane, Australia, August 14–17.

Bertot, J. C., Jaeger, P. T., & Grimes, J. M. (2010a). Crowd-sourcing transparency: ICTs, social media, and government transparency initiatives. In *Proceedings of the 11th Annual International Conference on Digital Government Research* (pp. 51–58). Puebla, Mexico, May 17–20.

Bertot, J. C., Jaeger, P. T., & Grimes, J. M. (2010b). Using ICTs to create a culture of transparency: Egovernment and social media as openness and anti-corruption tools for societies. *Government Information Quarterly, 27*(3), 264–271.
Bertot, J. C., Jaeger, P. T., Munson, S., & Glaisyer, T. (2010c). Social media technology and government transparency. *Computer, 43*(11), 53–59.
Bertot, J. C., Jaeger, P. T., & Grimes, J. M. (2012). Promoting transparency and accountability through ICTs, social media, and collaborative e-government. *Transforming Government: People, Process and Policy, 6*(1), 78–91.
Boulton, A., Brunn, S. D., & Devriendt, L. (2011). Cyberinfrastructures and "smart" world cities: Physical, human, and soft infrastructures. In P. Taylor, B. Derudder, M. Hoyler, & F. Witlox (Eds.), *International handbook of globalization and world cities*. Cheltenham: Edward Elgar.
Caragliu, A., Del Bo, C., & Nijkamp, P. (2011). Smart cities in Europe. *Journal of Urban Technology, 18*(2), 65–82.
Chourabi, H., Nam, T., Walker, S., Gil-Garcia, J. R., Mellouli, S., Nahon, K., et al. (2012). Understanding smart cities: An integrative framework. In *Proceedings of the 45th Hawaii International Conference on System Sciences* (pp. 2289–2297), January 4–7, Maui, Hawaii, USA.
Deakin, M., & Al Waer, H. (2011). From intelligent to smart cities. *Intelligent Buildings International, 3*(3), 140–152.
Edvinsson, L. (2006). Aspects on the city as a knowledge tool. *Journal of Knowledge Management, 10*(5), 6–13.
Eger, J. M. (2009). Smart growth, smart cities, and the crisis at the pump a worldwide phenomenon. *I-Ways, 32*(1), 47–53.
Folz, D. H., & Abdelrazek, R. (2009). Professional management and service levels in small U.S. communities. *The American Review of Public Administration, 39*(5), 553–569.
Gabrys, J. (2007). Automatic sensation: Environmental sensors in the digital city. *Senses & Society, 2*(2), 189–200.
Giffinger, R., Fertner, C., Kramar, H., Kalasek, R., Pichler-Milanović, N., & Meijers, E. (2007). *Smart cities: Ranking of European medium-sized cities*. Vienna: Centre of Regional Science (SRF), Vienna University of Technology.
Gil-Garcia, J. R. (2012). Towards a smart state? Inter-agency collaboration, information integration, and beyond. *Information Polity, 17*(3–4), 269–280.
Gil-Garcia, J. R., Helbig, N., & Ojo, A. (2014). Being smart: Emerging technologies and innovation in the public sector. *Government Information Quarterly, 31*(S1), 11–18.
Harrison, C., Eckman, B., Hamilton, R., Hartswick, P., Kalagnanam, J., Paraszczak, J., et al. (2010). Foundations for smarter cities. *IBM Journal of Research and Development, 54*(4), 1–16.
Hase, J. (2014). Smart city integration: Why open IT architectures are key (industry perspective). Deutsche Telekom, Government Technology. http://www.govtech.com/data/Smart-City-integration-Why-open-IT-architectures-are-key.html. Accessed 10 Oct 2014.
Helal, S. (2011). IT footprinting—Groundwork for future smart cities. *Computer, 44*(6), 30–31.
Hollands, R. G. (2008). Will the real smart city please stand up? Intelligent, progressive or entrepreneurial. *City, 12*(3), 303–320.
Johnston, E. W., & Hansen, D. L. (2011). Design lessons for smart governance infrastructures. In A. Balutis, T. F. Buss, & D. Ink (Eds.), *Transforming American Governance: Rebooting the public square* (pp. 197–212). Armonk: M.E. Sharpe.
Kaklauskas, A., Zavadskas, E. K., & Šaparauskas, J. (2009). Conceptual modeling of sustainable development. *Technological & Economic Development of Economy, 15*(1), 154–177.
Kim, C. (2010). Place promotion and symbolic characterization of New Songdo City, South Korea. *Cities, 27*(1), 13–19.
Komninos, N. (2002). *Intelligent cities*. London: Spon Press.
Komninos, N. (2009). Intelligent cities: Towards interactive and global innovation environments. *International Journal of Innovation and Regional Development, 1*(4), 337–355.
Landry, C. (2000). *The creative city: A toolkit for urban innovation*. London: Earthscan.

Lin, Z. (2007). From megastructure to megalopolis: Formation and transformation of mega-projects in Tokyo Bay. *Journal of Urban Design, 12*(1), 73–92.
Marshall, A. (2014). Are cities losing control over 'smart' initiatives? (opinion), *Government Technology*, published March 4, 2014. http://www.govtech.com/data/Are-Cities-Losing-Control-Over-Smart-Initiatives-Opinion.html. Accessed 10 Oct 2014.
Moser, S. (2010). Putrajaya: Malaysia's new federal administrative capital. *Cities, 27*(4), 285–297.
Musso, J. A., Weare, C., Oztas, N., & Loges, W. E. (2006). Neighborhood governance reform and networks of community power in Los Angeles. *The American Review of Public Administration, 36*(1), 79–97.
Musso, J., Weare, C., Bryer, T., & Cooper, T. L. (2011). Toward "strong democracy" in global cities? Social capital building, theory-driven reform, and the Los Angeles neighborhood council experience. *Public Administration Review, 71*(1), 102–111.
Nam, T., & Pardo, T. A. (2011). Smart city as urban innovation: Focusing on management, policy, and context. In *Proceedings of the 5th International Conference on Theory and Practice of Electronic Governance*, September 26–28, Tallinn, Estonia.
Nam, T., & Pardo, T. A. (2011a). Conceptualizing smart city with dimensions of technology, people, and institutions. In *Proceedings of the 12th Annual International Conference on Digital Government Research*, DG.O 2011, College Park, MD, USA, June 12–15, 2011; 01/2011.
Nam, T., & Pardo, T. A. (2014). The changing face of a city government: A case of Philly311. *Government Information Quarterly, 31*(Supplement), S1–S9.
Newcombe, T. (2014). Can we trust smart cities? *Government Technology*, published June 2. http://www.govtech.com/data/Can-We-Trust-Smart-Cities.html?utm_source=related&utm_medium=direct&utm_campaign=Can-We-Trust-Smart-Cities. Accessed 10 Oct 2014.
Noveck, B. S. (2009). *Wiki Government: How technology can make Government better, democracy stronger, and citizens more powerful*. Washington, DC: Brookings Institution Press.
Odendaal, N. (2003). Information and communication technology and local governance: Understanding the difference between cities in developed and emerging economies. *Computers, Environment and Urban Systems, 27*(6), 585–607.
Paskaleva, K. A. (2009). Enabling the smart city: The progress of city e-governance in Europe. *International Journal of Innovation and Regional Development, 1*(4), 405–422.
Paskaleva, K. A. (2011). The smart city: A nexus for open innovation? *Intelligent Buildings International, 3*(3), 153–171.
Santinha, G., & de Castro, A. E. (2010). Creating more intelligent cities: The role of ICT in promoting territorial governance. *Journal of Urban Technology, 17*(2), 77–98.
Schaffers, H., Komninos, N., Pallot, M., Trousse, B., Nilsson, M., & Oliveira, A. (2011). Smart cities and the future internet: Towards cooperation frameworks for open innovation. In *The future internet* (pp. 431–446). Heidelberg: Springer.
Scholl, H. J., & Scholl, M. C. (2014). Smart governance: A roadmap for research and practice. *iConference 2014* (pp. 163–176). http://dx.doi.org/10.9776/14060.
Scott, W. R. (2008). *Institutions and organizations: Ideas and interests* (3rd ed.). Thousand Oaks: Sage.
Švob-Dokia, N. (Ed.). (2007). *The creative city: Crossing visions and new realities in the region*. Zagreb: Institute for International Relations. http://www.culturelink.org/publics/joint/cultid08/Svob-Djokic_Creative_City.pdf. Accessed 10 Oct 2014.
Van Heur, B. (2010). The built environment of higher education and research: Architecture and the expectation of innovation. *Geography Compass, 4*(12), 1713–1724.
Van Winden, W. (2008). Urban governance in the knowledge-based economy: Challenges for different city types. *Innovation: Management, Policy & Practice, 10*(2–3), 197–210.
Van Winden, W. (2010). Knowledge and the European city. *Journal of Economic & Social Geography, 101*(1), 100–106.
Wang, S. W. (2011). The evolution of housing renewal in Shanghai, 1990–2010: A 'socially conscious' entrepreneurial city? *International Journal of Housing Policy, 11*(1), 51–69.
Washburn, D., Sindhu, U., Balaouras, S., Dines, R. A., Hayes, N. M., & Nelson, L. E. (2010). *Helping CIOs understand "Smart City" initiatives: Defining the smart city, its drivers, and the role of the CIO*. Cambridge: Forrester Research.

Willke, H. (2007). *Smart governance: Governing the global knowledge society*. Frankfurt a. M.: Campus.
Winters, J. V. (2011). Why are smart cities growing? Who moves and who stays. *Journal of Regional Science, 51*(2), 253–270.
Yigitcanlar, T., O'Connor, K., & Westerman, C. (2008a). The making of knowledge cities: Melbourne's knowledge-based urban development experience. *Cities, 25*(2), 63–72.
Yigitcanlar, T., Velibeyoglu, K., & Martinez-Fernandez, C. (2008b). Rising knowledge cities: The role of urban knowledge precincts. *Journal of Knowledge Management, 12*(5), 8–20.
Yovanof, G. S., & Hazapis, G. N. (2009). An architectural framework and enabling wireless technologies for digital cities & intelligent urban environments. *Wireless Personal Communications, 49*(3), 445–463.

J. Ramon Gil-Garcia is an associate professor of public administration and policy and the research director of the Center for Technology in Government, University at Albany, State University of New York (SUNY). He is also a professor (on leave) at "Centro de Investigacion y Docencia Economicas (CIDE)" in Mexico City. Dr. Gil-Garcia is a member of the Mexican National System of Researchers and of the Mexican Academy of Sciences. In 2009, he was considered the most prolific author in the field of digital government research worldwide and in 2013 he was selected for the Research Award, which is "the highest distinction given annually by the Mexican Academy of Sciences to outstanding young researchers." Dr. Gil-Garcia is the author or coauthor of articles in prestigious international journals in public administration, information systems, and digital government and some of his publications are among the most cited in the field of digital government research worldwide. His research interests include collaborative electronic government, smart cities and smart governments, digital divide policies, new public management, public policy evaluation, and multi-method research approaches.

Theresa A. Pardo, is the director of the Center for Technology in Government at the University at Albany, SUNY. She holds research associate professor appointments in public administration and policy at the Rockefeller College of Public Affairs and Policy and in informatics at the College of Computing and Information. Professor Pardo is a codeveloper of the top-ranked Government Information Strategy and Management concentration at the University of Albany and is president of the Digital Government Society, a global multidisciplinary organization of scholars and practitioners interested in the development and impacts of digital government.

Taewoo Nam has been working for the Department of Public Administration at Myongji University, Korea as an assistant professor since 2013. He earned PhD degree in public administration from the University at Albany, State University of New York. He is interested in citizen participation, intergovernmental relationships, city and local administration, and digital government. His publications appear in various academic journals in the field of public administration and information science including *Government Information Quarterly, International Journal of Public Administration, International Review of Administrative Sciences, International Review of Public Administration, Social Science Computer Review, Journal of Urban Technology, Journal of Information Technology and Politics, and Information Polity.* As an advisor in the performance evaluation board of the Korean national government, he helps the government evaluate the performance of national informatization initiatives and programs.

Part I
Smart Cities Concepts and Methodologies

Exploring the Nature of the Smart Cities Research Landscape

Adegboyega Ojo, Zamira Dzhusupova and Edward Curry

Abstract As a research domain, Smart Cities is only emerging. This is evident from the number of publications, books, and other scholarly articles on smart cities indexed in Google scholar and Elsevier's Scopus—an abstract and citation database. However, significant literature is available on related topics like intelligent city, digital city, and intelligent community based on search results research repositories such as Elsevier's Scopus, ACM Digital Library, and Google Scholar. This chapter maps the research work in the smart cities' domain, based on the available scholarly publications. The aim is to synthesize an emerging understanding of the smart city concept, determine major research themes, types, and gaps in the current research landscape.

Keywords Smart cities · Intelligent cities · Smart city research · Research mapping · Bibliometrics · Scientometrics

1 Introduction

The unprecedented level of urbanization and consequent growth in size and numbers of cities in different parts of the world present both challenges and opportunities. On the one hand, phenomenal growth (~tenfold) in the urban population

A. Ojo (✉) · E. Curry
Insight Centre for Data Analytics, National University of Ireland, Galway DERI Building, IDA Business Park, Lower Dangan, Galway, Republic of Ireland
e-mail: adegboyega.ojo@insight-centre.org

Z. Dzhusupova
United Nations Department of Economic and Social Affairs (UNDESA), Division for Public Administration and Development Management (DPADM), E-Government Branch, P. O. Box 10017, NY, USA
e-mail: dzhusupova@un.org; zdjusupova@gmail.com

E. Curry
e-mail: ed.curry@insight-centre.org

J. R. Gil-Garcia et al. (eds.), *Smarter as the New Urban Agenda A Comprehensive View of the 21st Century City*, Public Administration and Information Technology 11,
DOI 10.1007/978-3-319-17620-8_2

(from 250 million at the beginning of the 20th century to 2.8 billion at the beginning of the 21st century, with an expected rise to about 9 billion in 2050 (DiChristina 2011)) challenges the traditional approaches to city management and urban lifestyle. For instance, the traditional approach to the management of transportation systems, water resources, waste, energy, and natural environment in the cities must be fundamentally rethought to cope in a sustainable manner with the pressure induced by growth in demand for these resources. On the other hand, given that cities constitute the social nexus (Ratti and Townsend 2011) of the larger society providing the much needed intellectual and social capital for growth, and that bigger cities appear to be able to do more with less (Bettencourt and West 2011), city growth may offer opportunities for more optimal city management and innovation. The collaborative creativity resulting from proximity and constant interchange of ideas offered by cities has helped many cities, particularly in the developing world to get out of poverty and integrate with the wider world economy (Glaeser 2011).

In responding to challenges and opportunities of rapid urbanization and city growth, many governments at different levels—international, regional, national, and local—have initiated some programs on digital and intelligent cities and lately smart cities. Digital, intelligent, and smart cities are related concepts all involving the information and communication technology (ICT)-enabled transformation of the city, city management, as well as city inhabitants and actors. While the difference between the three concepts remain fuzzy (Hollands 2008), differences have been noted in the changing focus and consequently the required capabilities in implementing initiatives associated with the three concepts.

Smart cities as urban innovation and transformation initiatives aim to harness physical infrastructures, ICT, knowledge resources, and social infrastructure for economic regeneration, social cohesion, better city administration, and infrastructure management (Ojo et al. 2014). A distinguishing feature of the smart city concept is the centrality of people or the welfare of its residents in its essence. Specifically, smart cities are concerned with the transformation of life and work of city inhabitants (Hollands 2008). Smart cities also focus on harnessing human collaboration for generating ideas which are considered as the currency of the current age (Ratti and Townsend 2011). This extended scope and focus on integration of different aspects of city—administration, resource management, lifestyle, mobility, etc.—makes the smart cities' research more challenging and ambitious with respect to previous research on intelligent and digital cities which focused primarily on the technology dimension (e.g., ICT infrastructure and services) and its transformational effect on other dimensions of the city.

The research in urban transformation is fairly mature with over three decades of work, whereas the research in smart cities, intelligent cities, and related areas is relatively new. However, given the close link between smart cities and major issues of interest to policymakers such as sustainability and technology innovation in city governments, research interests and outputs have been growing at a more rapid rate lately. Currently there are over 800 papers in Scopus with "smart cities" in their titles, abstract, or keywords and over 7000 scholarly resources indexed in Google Scholar. We believe that this level of research outputs in the domain is sufficient to explore the emerging and future trends in smart city research.

This chapter examines the smart cities' research domain by analyzing scholarly publications on the subject matter based on data available on Elsevier's Scopus database—the largest abstract and citation database of peer-reviewed literature: scientific journals, books, and conference proceedings. The study complements existing research landscaping studies in the e-government domain (Grönlund 2004; Heeks and Bailur 2007; Yildiz 2007). The chapter is primarily targeted at researchers. Findings may also be of interest to practitioners to guide funding policies for research in the smart cities' domain.

2 Conceptual Framework

This section provides the conceptual underpinning for the study definitions of core concepts of a smart city. The term smart city (or smart cities) has been adopted by different governments, consulting organizations (IBM 2013), and research groups. Despite the wide use of the term, its meaning remains fuzzy (Caragliu et al. 2009; Nam and Pardo 2011). Smart city according to Giffinger et al. (2007) is "A city performing in a forward-looking way in economy, people, governance, mobility, environment, and living, built on the smart combination of endowments and activities of self-decisive independent and aware citizens." This definition is based on the traditional, regional, and neoclassical theories of urban growth and development. In particular, the axes are based—respectively—on the theories of regional competitiveness, transport and ICT economics, natural resources, human and social capital, quality of life, and participation of societies in cities. Based on Giffinger's definition, Caragliu et al. (2009) offer a similar definition of the concept as follows—"We believe a city to be smart when investments in human and social capital and traditional (transport) and modern (ICT) communication infrastructure fuel sustainable economic growth and a high quality of life, with a wise management of natural resources, through participatory governance."

Smart cities are expected to dramatically improve their citizens' quality of life, encourage business to invest, and create a sustainable urban environment (Vasseur and Dunkels 2010). Interestingly, while the term *smart city* literarily implies an outcome or a result, most consider the term as an "activator" of change through exploring relevant open innovation processes (Paskaleva 2011). Other conceptualizations, such as Nam and Pardo (2011) consider smart city as urban innovation involving technological, organizational, and policy innovations. Finally, smart city could be understood as a certain intellectual ability that addresses several innovative socio-technical and socioeconomic aspects of growth (Zygiaris 2012).

In Hollands (2008), three characteristic elements of the smart city concept were identified to include: (1) utilization of networked infrastructures to improve economic and political efficiency and enable social, cultural, and urban development infrastructures including ICT; (2) business-led urban development; and (3) social and environmental sustainability. Social sustainability implies social cohesion and a sense of belonging, whereas environmental sustainability refers to the ecological and "green" implications of urban growth and development. In Komninos (2011), the concept of spatial intelligence of cities is presented as a composite capability enabling communities within the city to harness the intellectual capital, institutions, and material infrastructure in dealing with problems and challenges. Spatial intelligence is composed of three types of intelligence: (1) the inventiveness, creativity, and intellectual capital of the city; (2) the collective intelligence of the city's institutions and social capital; (3) the artificial intelligence of public and citywide smart infrastructure, virtual environments, and intelligent agents. These three types of intelligence involve all dimensions of the city and maps to three types of spaces—physical, institutional, and digital. The "physical space" corresponds to the inventiveness and creativity of the city; the "institutional space" includes the social capital and collective intelligence of the city population; and "digital space" contains the artificial intelligence embedded into the physical environment, including public broadband communication infrastructure and digital technologies.

Focusing on the digital space, the following infrastructure networks for smart cities were identified in Vasseur and Dunkels (2010). Some of these networks are related to transport, public safety and security, public services, utilities, and social networking. In the physical space, skills and human capitals are considered as arguably the most important elements. For instance, it is argued that the greatest competitive advantages of the cities are qualities that attract the best and brightest from the world to a city (Bloomberg 2011). This is supported by the fact that educated cities grow more quickly than the less educated ones, since skilled cities are economically more productive and better at adapting to economic shocks (Glaeser and Saiz 2003).

We summarize the different elements of the definitions of the smart city concept below in Table 1. Further discussions on the conceptualizations and definitions of the smart city are provided in Hollands (2008), Caragliu et al. (2009), Chourabi et al. (2012), and Nam and Pardo (2011).

Table 1 Elements of "smart city" definitions

No	Description	Reference
Nature	Is a (1) forward-looking city in the areas of economy, people, governance, mobility, environment, and lifestyle; (2) form of urban innovation; and (3) intellectual capital profile of a city	Giffinger et al. (2007), Nam and Pardo (2011), Zygiaris (2012)
Essence	Means (1) information access, bridging digital divide, lifelong learning, social inclusion, and economic development; sustainable economic growth and urban development, higher quality of life; and wise management of natural resources; (2) innovative socio-technical and socioeconomic growth of a city	Hollands (2008), Vasseur and Dunkels (2010), Zygiaris (2012)
Approach	Involves (1) investments in human and social capital; (2) investment in traditional (transport) and modern (ICT) communication infrastructure; (3) promoting participatory governance and engagement of citizens; (4) technological, organizational, and policy innovation	Caragliu et al. (2009), Nam and Pardo (2011)

3 Methodology

3.1 Research Objectives and Questions

This study aims to capture the emerging understanding of the "smart city" concept, examining the nature of the smart city research and concluding on the overall research maturity and indications on areas where future research efforts could be targeted. Specific objectives for the study include:

1. Strengthening the conceptual foundations of smart cities' research by: (a) developing an analytical definition for the "smart city" concept by integrating existing definitions in literature; (b) establishing conceptual similarities between "smart cities" and related concepts like intelligent cities, ubiquitous cities, digital cities, e-cities, etc.; and (c) determining the major dimensions of the smart city concept.
2. Determining the trend in "smart cities" research by identifying the major research themes and types in available smart cities' publications and noting how these themes change over time.
3. Eliciting the research gaps by identifying research issues and questions from publications providing critical perspective, critique and lessons from planning, pilot development, and full-scale deployment of smart cities-related initiatives.

Guided by these objectives, the study answers the following questions:

R1. How can the smart city concept be defined and what are the major dimensions of the concept?

R2. Is there a discernible conceptual distinction among the three related concepts—smart city, intelligent city, and digital city? To what extent can previous studies in digital and intelligent cities fundamentally contribute to smart city research?

R3. What trend can be observed in terms of theme, nature, and approach of research carried out in the smart city domain?

R4. What are the areas of "smart city" research that are relatively uncovered and to what extent are the governance aspects of the smart cities studied?

3.2 Research Method

The research method adopted in the study combines research mapping and visualization technique with content analysis of scholarly publications used in scientometric or bibliometric studies. The main source of data was journal articles and conference papers related to smart cities or intelligent cities provided in the Scopus database—the largest abstract and citation database of peer-reviewed literature and quality web sources. Our decision to include publications on intelligent cities as a part of the publications on studies of smart cities is based on the results of preliminary analysis of the definitions of the three related concepts (digital, intelligent, and smart city) presented in Sect. 2. *Our analysis showed that the concept of intelligent city is significantly closer to smart city when compared with the digital city concept.*

By taking publication as the unit of analysis, each publication is mapped to a number of dimensions—defining of key terms for the publication, the research theme addressed in the publication, the nature of research documented in the publication, the overall approach adopted in the research, and the aspects of the smart city concept addressed in the research. Microsoft Excel was used as an analysis tool, while "VOSviewer" and "Pajek"—Program for Large Network Analysis—were used for mapping and visualizing the research landscape. The details of our mapping and analysis are discussed in the subsections below.

Selecting Publications The source of data for our research is the Elsevier's Scopus Abstract and Citation database. We considered all publications in the database with the terms—"smart city" or "intelligent city"—in their titles, abstracts, or keyword lists. The same query was run a number of times between the period October 2011 and March 2012, to track new articles. Given that over 70 % of the available publications were published within the past 5 years, all available 209 publications were considered initially in analysis. After reviewing each publication for relevance and removing publications without abstracts, we were left with 170 publications. The 170 publications were exported from the Scopus to Microsoft Excel for further analysis.

Generating the Publication Map The mapping process entailed determining the nature of research contained in the publication, research approach adopted, aspects of the smart city concept addressed in the research, and the neoclassical dimensions of the smart city concept related to the publication. For all four dimensions, our strategy was to start off with an initial set of possible values or labels and extend the categories as they emerge from the publications. The initial set of values defined for the four dimensions are highlighted below:

1) *Nature of research*—describes the nature of research in the publication. A scheme combing the traditional classification of research in social science with the design-oriented research in computing and engineering was adopted. Publications were classified as (Heeks and Bailur 2007; Grönlund 2004): polemic, debate, position, conceptualization, theoretical, theory development, theory testing, survey, design, or simply descriptive.
2) *Research approach*—describes the methodological aspect of research publications including (Grönlund 2004): single and multiple case study approach, comparative analysis, empirical analysis, action research, modeling and simulation, experimentation were considered as options. When no specific approach is reported or implied in the abstract, the research approach was simply considered to be analytical.
3) *Research themes and subthemes*—the initial set of themes were related to the classical policy cycle for any major initiative, in addition specific themes related to improving current understanding of smart cities, policy and strategy, models and frameworks, technology, governance, organizational processes in smart city initiatives or projects.

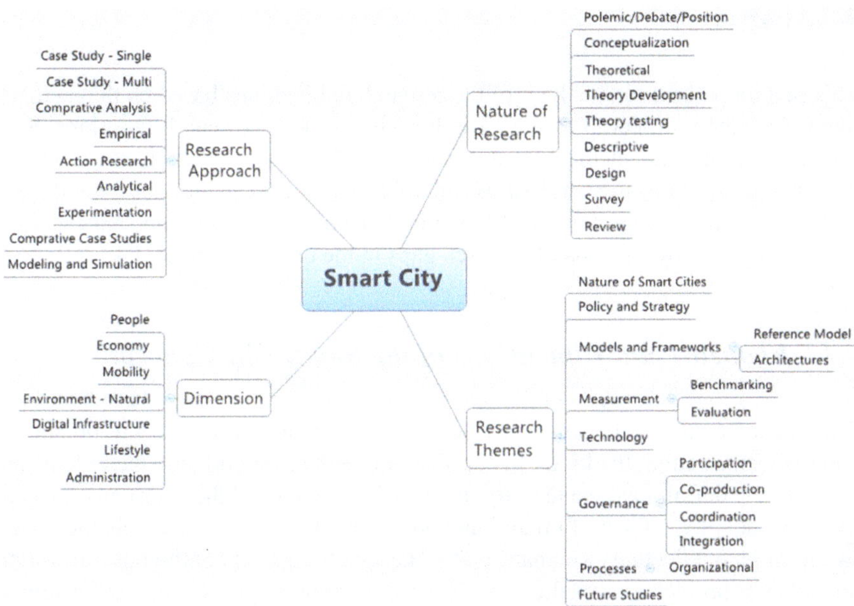

Fig. 1 Taxonomy for analysis dimensions

4) *Smart city dimension*—the initial set of dimensions specified for the smart city concept include people, economy, mobility, natural environment, ICT infrastructure, lifestyle, and public administration (or public governance).

Analyzing the Publication Map After mapping each publication along the four dimensions described above, pivot tables were generated to summarize the publications along each of these dimensions to produce corresponding tables and graphs showing frequency counts and trends over years. The second type of analysis involved organizing and visualizing the research domain based on the titles, keywords, generated research themes and subthemes, as well as abstracts of the publications. The typical workflow for domain visualization was adopted using the VOSviewer and Pajek tools. See Fig. 1 and Table 2 for the taxonomy and parameters for the analysis, respectively.

Table 2 Parameters for analysis

Parameter	Value
Unit of analysis	Individual publication
Measure	Counts of attributes—keywords, title, themes, and subthemes
Similarity measure	Co-term based using VOSviewer mapping and clustering algorithm
Visualization	Pajek "Draw" algorithm

4 Analysis

This section presents an analysis of the research publications based on the methodology described in Sect. 3 as the basis for answering the research questions. Section 4.1 examines the elements and dimensions of the smart city concept, Sect. 4.2 attempts to determine the similarities and differences between the smart, intelligent, and digital city concept. Section 4.3 explores the observed trends in smart city research, followed by analysis of research gaps in the domain in Sect. 4.4.

4.1 Elements and Dimensions of the Smart City Concept

This section analyzes the smart city definitions provided in Sect. 2.3 to obtain the nature of smart cities, the kind of goals they are built to support, and their elements.

From the definitions, we identify three basic attributes of the smart city concept as an "actual city, a form of urban innovation." First, smart cities are characterized by the "high intellectual or human capital" needed to support continuous innovation and address problems or challenges. Second, in terms of goals, smart cities aim at social inclusion, significantly improved quality of life, and economic development. Third, smart city policies also target the development of human capital through lifelong learning, optimal management of natural resources, and sustainable urban development in general.

Aspects of the smart city concept include: participatory governance, human capital development, ICT infrastructure development, technological innovation, organizational innovation, policy innovation, integration of city endowments, and developing active self-decisive citizenry. Details are provided in Table 3 below.

4.2 Conceptual Analysis of Smart Cities and Related Concepts

In this section, we attempt to identify the similarities and differences between the smart city concept and intelligent and digital cities.

Digital cities by their nature are considered to be some form of an extension—a virtual extension of a real city. In a weak sense, it is synonymous to a label assigned to cities with good ICT infrastructure or that widely employ ICT applications. Similar to smart cities, digital cities are targeted at democratic participation, economic development, and social cohesion. Despite the wide applicability of digital city concept, its elements are limited to mainly ICT infrastructure or digital networks and software applications.

Intelligent City as a Concept Appears to Lie Between the Smart City and Digital City Concept Intelligent city is conceived as a real city endowed with collaborative, learning, and innovation environments or spaces. The intelligent city concept is also considered as a transformational instrument in urban development. Core purposes

Table 3 Analysis of "smart cities" definitions in literature

No	Description	Summary
Nature	Is a forward-looking city in economy, people, governance, mobility, environment, and living	Forward-looking city
	Is urban innovation	Urban innovation
	Is intellectual capital of a city	Intellectual capital
Purpose	Means to information access	Information access
	Means to lifelong learning	Lifelong learning
	Means to bridging the digital divide	Bridging digital divide
	Means to social inclusion	Social inclusion (2)
	Means to economic development	Quality of life (2)
	Means to high quality of life	Economic development (4)
	Means to wise management of natural resources	Natural resources management
	Means to dramatically improve their citizens' quality of life	Sustainable urban development
	Means to encourage business to invest	
	Means to creating a sustainable urban environment	
	Means to sustainable economic growth	
	Means to address innovative socio-technical and socioeconomic aspects of growth	
Elements	Includes participatory governance	Participatory governance
	Involves investments in human and social capital	Investment in human capital
	Involves investment in traditional (transport) and modern (ICT) communication infrastructure	Investment in social capital
	Includes technological, organizational, and policy innovation	Investment in communication infrastructure
	Involves smart combination of endowments	Technological innovation
	Involves activities of self-decisive independent and aware citizens	Organizational innovation
		Policy innovation
		Integration of endowments
		Activities of self-decisive citizens

for intelligent city initiatives center on transformation of lifestyle, work, and recreational activities in addition to sustainability of the associated cities. Similar to the smart city concept, an important objective of intelligent cities is developing the problem-solving capabilities of the cities.

Aspects of the intelligent city include the ICT infrastructure development, development of intelligence infrastructure, and services, in addition to building institutional leadership and organizational capacity of the cities. We summarize the similarities and differences among these concepts in the Tables 4 and 5 below.

Based on the analysis here, we note that the three terms—smart-, intelligent-, and digital city—are closely related concepts particularly in terms of their purpose.

Table 4 Similarity of smart city concept to intelligent- and digital city concepts

Similarities	Intelligent city	Digital city
Nature	*Like smart city, intelligent city concept:*	*Like smart city, digital cities are characterized by:*
	Serves as metaphor for real cities	High urban ICT adoption and usage
	Emphasizes on urban innovation	
	Is transformational	
Purpose	*Like in the smart city concept, intelligent cities also:*	*Like smart cities, digital cities support:*
	Focus on the urban sustainability	Socioeconomic development and other aspects of urban life
	Focus on the impact on different aspects of urban life	
Elements	*Like smart cities, intelligent cities also:*	*Like smart cities, digital cities also:*
	Involve ICT infrastructure development	Involve ICT infrastructure development
	Involve creativity and social capital development	

Table 5 Difference between smart city concept and digital- and intelligent city concepts

Difference	Intelligent city	Digital city
Nature	Smart city concept places relative emphasis on human aspect of urbanism, whereas intelligent focuses more on the technical supporting infrastructure and services	Digital cities appear to be virtual extensions of the real cities, whereas smart cities are conceptualized as "real cities"
Purpose	The scope of intelligent cities' initiatives is relatively narrower than that of the smart cities	The scope of digital cities' initiatives is even more limited or focused when compared with smart cities
Elements	Smart cities emphasize integration among elements, whereas elements of intelligent cities are more specialized and standalone	Digital cities are characterized largely by ICT infrastructure and services, whereas smart cities involve activities in all major aspects of the city development

When considering the constituent elements of the concepts, digital city is restricted to ICT infrastructure and services. The intelligent city concept in addition to the basic ICT infrastructure and services includes other specialized technical elements such as "intelligence" infrastructure to support acquisition of information (instrumentation) and learning. The intelligent city also includes elements that involve the development of innovation.

Smart city elements include those of the intelligent city in addition to socio-organizational and institutional infrastructure to support the specific policy domains and governance mechanisms for integrating the solutions to concerns in the different domain.

4.3 Trends in Smart City Research

This section presents the trends in smart city research based on the analyzed publications. The information on research themes is presented in Sect. 4.3.1 followed by the analysis of trends in the nature of smart city research in Sect. 4.3.2. The research approach adopted in smart city research is presented in Sect. 4.3.3. The section is concluded with discussion on research gaps in Sect. 4.3.4.

4.3.1 Research Themes

This section on smart city research themes provides information on the broad research areas and specific themes, and recurring smart city research topics extracted from titles, keywords, abstracts, and themes of publications.

Research areas and themes of publications were mapped into five broad research areas: (1) smart city attributes; (2) smart city implementation; (3) smart city policy domains; (4) management and governance of smart cities; and (5) foundations of smart cities. The first research area on city attributes involves investigation of one or more attributes of a smart city such as its population, size, employment rate, etc. and relationship among these attributes. The second research area on smart city implementation includes work aiming to provide models, tools, and guidelines for developing smart cities. This category of work covers discussions on experiences and success factors for smart cities. The third research area on smart city policy domains includes research on different policy domains such as transportation, urban infrastructure management, energy, and health. Management and governance constituting the fourth research area includes works addressing how smart cities' initiatives could be planned, managed, and governed. This category also covers the regulations and standard issues in smart cities. The fifth category of research on foundation of smart cities aims at providing a better domain understanding. For example, work in this category includes conceptualizations of smart cities and studies to understand the evolution and the future of smart cities (see Table 6).

Analysis of the thematic category mappings in Fig. 2 shows that about 43 % of the available publications focus on different policy domains. The research on implementation aspects of smart cities accounts for about 20 % of the publications, whereas the research on management and governance accounts for about 19 % of the publications. Studies on the foundational aspects of smart cities include about 19 % of the publications, whereas publication volume on specific smart city attributes is very low—about 1 % at the moment.

Regarding concrete themes, the top four most common research themes are: (1) technology—about 29 % of the publications; (2) nature of smart cities—roughly 17 % of the publications; (3) model and frameworks—about 13 % of the publications; and (4) policy and strategy—roughly 8 % of the publications. Figure 3 below provides the details of this distribution.

Table 6 Thematic categories for smart cities' research

Category	Themes
City attribute	Population growth
Implementation	Models and frameworks
	Success factors
	Information sharing
Policy domains	Intelligent building
	Urban infrastructure management
	Sustainable mobility
	Intelligent transportation
	Energy and technology
	Human capital and employment
	Education and municipal governance
	ICT infrastructure, applications, and services
Management and governance	Governance
	Integrated urban planning
	Measurement
	Organization
	Policy and strategy
	Standards and regulations
Foundations	Future studies
	Nature of smart cities
	Trends in e-cities

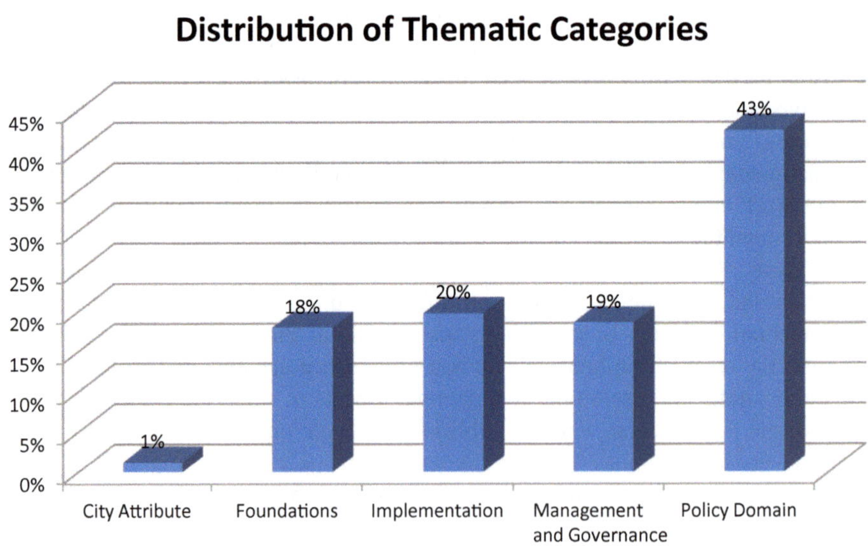

Fig. 2 Distribution of major smart city research areas

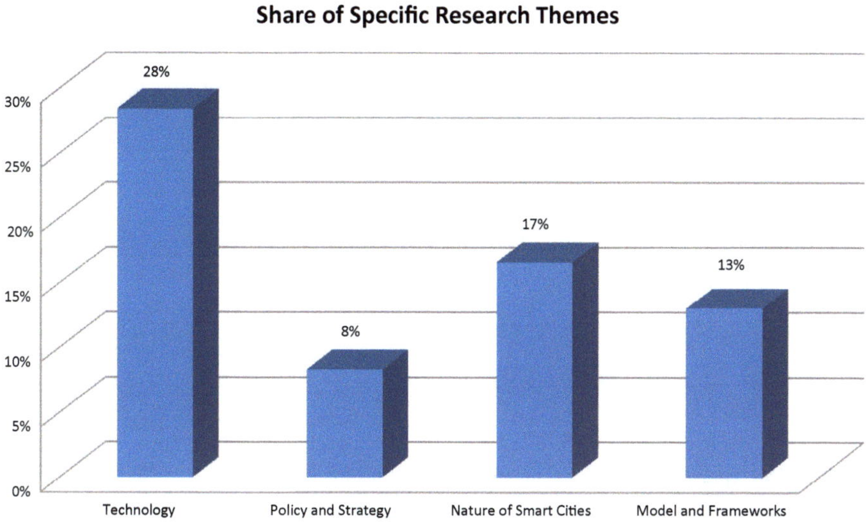

Fig. 3 Share of most common smart city research themes

In addition to information on research areas and themes, recurring topics appearing in the titles, keywords, and abstracts are discussed next. Two kinds of analyses were carried out to produce this information from the input text. The first is the clustering of the key terms representing the text, and the second is the mapping of these key terms into a two-dimensional surface, in such a way that similar concepts are located close to one another. The VOSviewer tool was used for both visualization and mapping. The resulting maps could be visualized using several applications in addition to the VOSviewer. The Pajek tool was used for visualizing the maps generated from VOSviewer. Figures 4 and 5 are examples of the maps generated

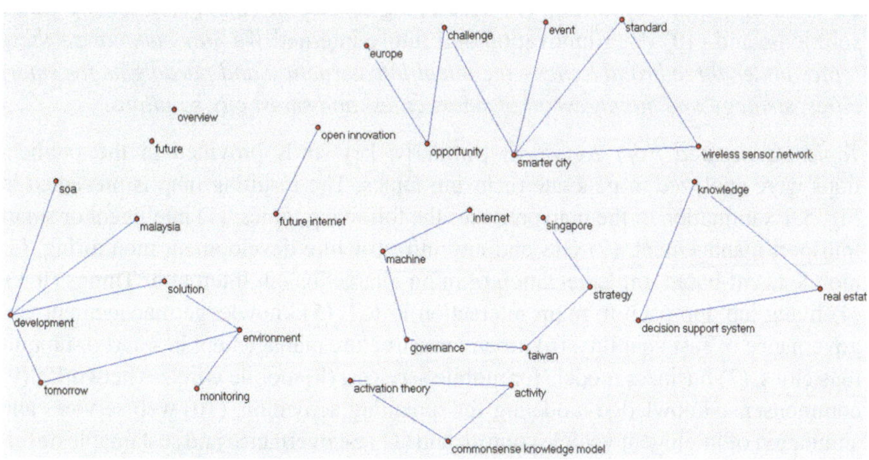

Fig. 4 Most representative terms generated from titles of publications

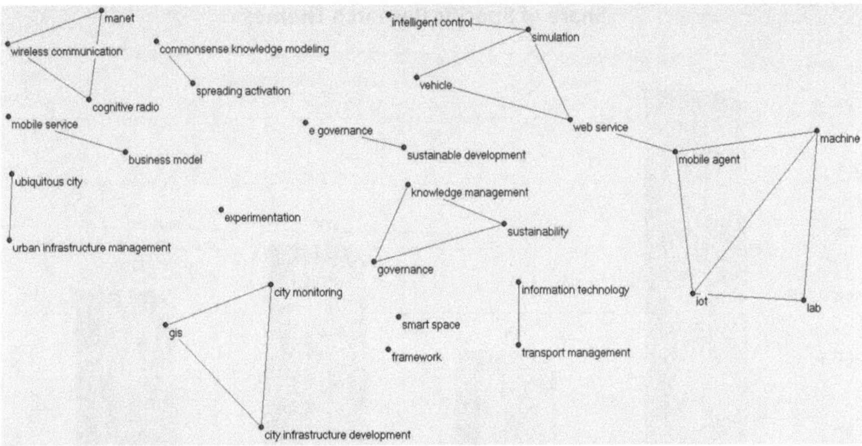

Fig. 5 Most representative terms generated from publication keywords

using Pajek. Information provided on the generated maps is discussed below. *Note that "topics" as used below in the various text analytics tasks connote semantically related (or co-occurring) terms over a number of smart city publications. Therefore, these topics highlight "inherent structures" in the body of smart city literature.*

Topics Generated from Publication Titles The first set of analysis involved titles of publications. The map resulting from this analysis is provided in Fig. 3. Some of the representative topics generated from the titles of publications include: (1) wireless sensor network standards in smart cities; (2) event-driven design in smart cities; (3) challenges in smart cities; (4) decision support system as knowledge management solution for Real Estate; (5) smart city strategies, examples from Taiwan and Singapore; (6) governance of machine-to-machine interaction; (7) supporting activities through commonsense knowledge modeling and activation theory; (8) solutions for environmental monitoring; (9) developing service-oriented, architecture-based solutions; and (10) open innovation and future internet. *We may summarize these topics under three broad topics—technical infrastructure and paradigms for smart cities, strategy and management of smart cities, and smart city solutions.*

Topics Generated from Keywords Similarly, keywords provided in the publications were analyzed to generate recurring topics. The resulting map is presented in Fig. 5. Examination of the map produces the following topics: (1) intelligent or smart transport management, (2) GIS and city infrastructure development monitoring, (3) mobile agent-based implementation of m2m interactions in Internet of Things (IoT), (4) living lab approach to m2m interaction in IoT, (5) knowledge management and governance of sustainability, (6) urban infrastructure management in smart or ubiquitous cities, (7) business models for mobile services, (8) mobile wireless networks, (9) commonsense knowledge modeling for spreading activation, (10) web services and simulation of intelligent vehicle control, and (11) e-governance and sustainable development. *From these topics, four broad categories of topics can be identified—smart*

Table 7 Parameters for analysis

Source of information	High-level topics detected from text analysis
Title of publication	(1) Technical infrastructure and paradigms for smart cities, (2) strategy and management of smart cities, and (3) smart city solutions
Keywords	(1) Smart city solutions, (2) technical infrastructure and paradigm for smart cities, (3) management and governance of smart cities, and (4) business aspects of smart cities

city solutions, technical infrastructure and paradigm for smart cities, management and governance of smart cities, and business aspects of smart cities.

These text-based analyses provide the complementary information on the underlying cognitive structure of the smart city domain based on the available publications. We consolidate the resulting high-level topics from the two analyses in Table 7 below.

Coverage of Smart City Dimensions in Research Publications An important perspective to understanding smart city concept is through its dimensions (please see Fig. 1). The majority of the publications focus on only one specific dimension—about 71 %. Research involving one or more dimensions (e.g., education and governance) constitutes about 21 % of the publications, whereas publications addressing all dimensions as a whole are about 7 % (see Figs. 6 and 7). In terms of relative coverage of specific dimensions, most of the publication in smart cities is on ICT infrastructure (or technology). The next area of focus is governance, followed by people.

With respect to governance, topics found in publication set include reform, integration, policy and strategy, measurement, standards and regulations, public engagement, and partnership. Within governance-related publications, the most common governance topic is policy and strategy (29 %), followed by measurement (24 %) and standards and regulations (18 %) (see Fig. 8).

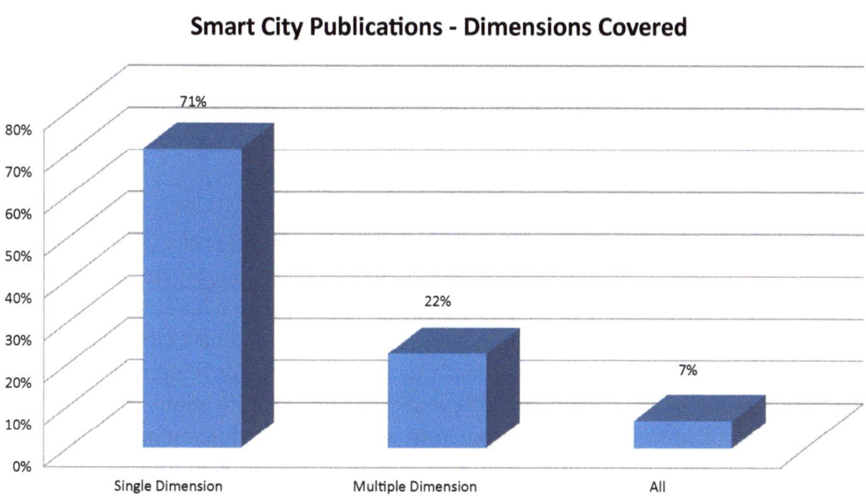

Fig. 6 Distribution of dimensions of smart city research

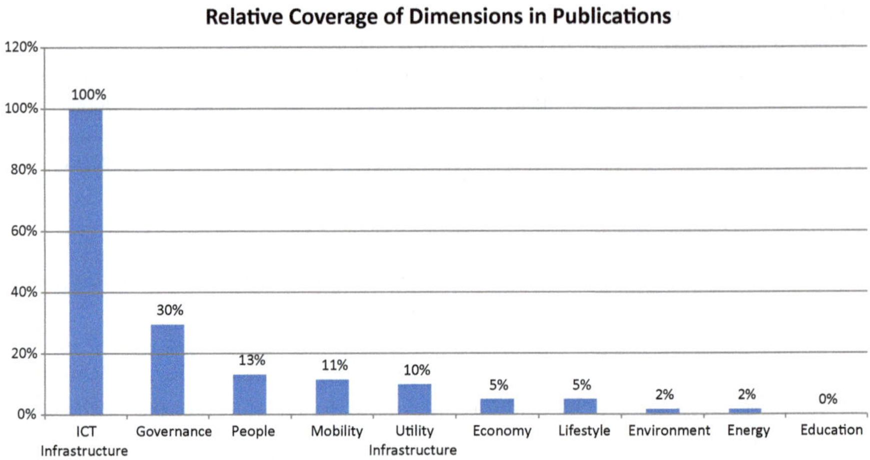

Fig. 7 Distribution of themes of smart city research

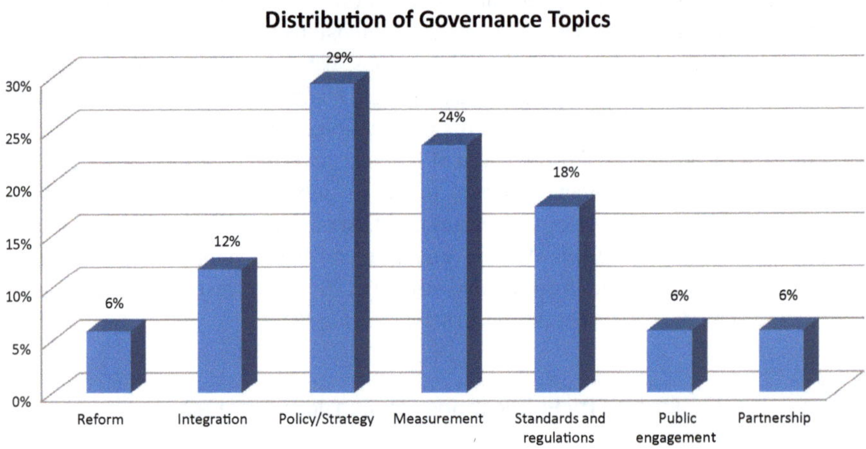

Fig. 8 Distribution of governance topics

4.3.2 Nature of Research

This section reviews the nature of research characterizing smart city domain. About 41 % of the publications contained research works that were simply descriptive with no specific orientation. Publications describing design research made up about 32 % of the total publications. Research on conceptualizations of smart cities constitutes 9 % of the whole publications. Research works on theory development and testing about smart cities account for 5 % and 1 % of the publications, respectively. The numbers of survey, theoretical, polemic, and position publications are very few (virtually nonexistent). Figure 9 provides details of the distribution of smart city research based on their nature.

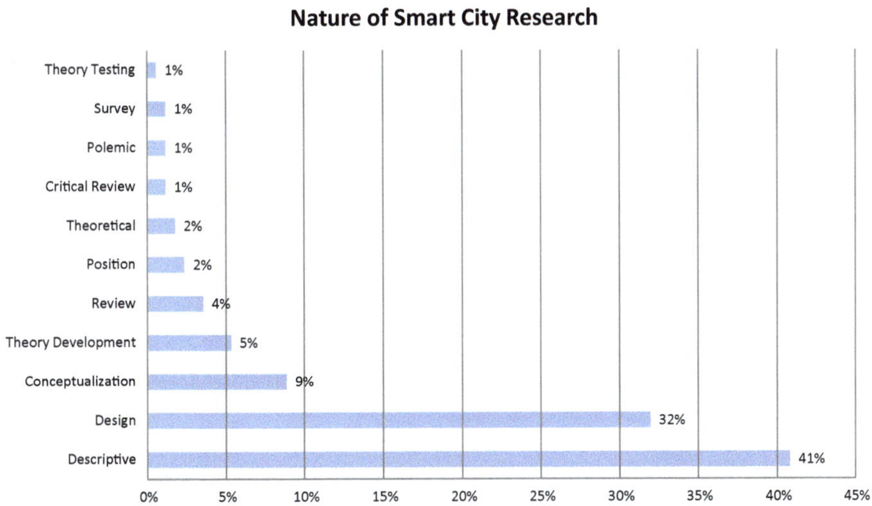

Fig. 9 Distribution of nature of smart city research

In terms of the trends, close to 60 % of the total publications on smart cities were published in the 2011 alone and since 2010 there has been 200 % increase in annual smart city publication volume. Considering the specific research types, research on design and conceptualization has doubled from 2010 to 2011, whereas general descriptive research increased more than sixfold from 2010 to 2011.

From these facts, we gather that about 40 % of the research publications on smart cities are simply descriptive, with no discernible research type or philosophy. About 50 % of the remaining research works with discernible orientation are design oriented. This could be attributed to the fact that computer science and engineering are currently by far the subject areas contributing most to smart cities' research. After design research, works on smart city conceptualization are next indicating ongoing efforts to better understand the concept. The relatively few numbers of theory-related publications and the rapid growth could be attributed to relative young nature of the research domain.

4.3.3 Research Approach

In this section, we summarize the smart city research domain based on the research approach adopted, see Fig. 10. Results show that about 46 % of the research publications have no clear or discernible research approach but offer logical analysis of the problems and solutions (i.e., analytical approach). Research work involving development of one form of technical artifact or another as solution to a problem (i.e., development approach) constitutes about 23 % of the publications. Research based

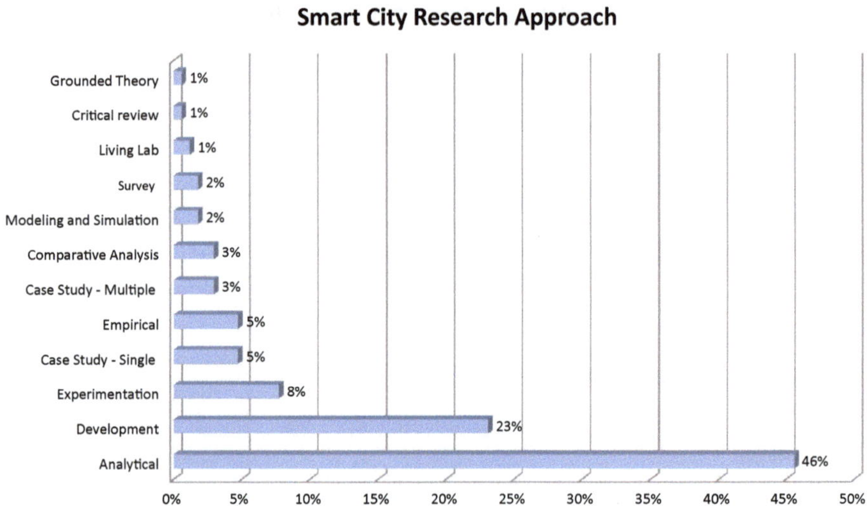

Fig. 10 Distribution of smart city research approach

on experimentation and case studies equally accounts for about 8 % of the research publications. Approaches including modeling and simulation, survey, grounded theory, living lab, and empirical research have also been employed in the smart city research albeit to very limited level. In terms of trends, the use of case studies (multi-case studies in particular) grew significantly from one publication in 2010 to nine in 2011, with respect to other methods such as development. In addition, experimentation and comparative analysis in smart city research grew by 300 % and 200 %, respectively.

4.4 Gaps in Smart City Research

This section summarizes the research gaps arising from the analysis of the research themes, type, and approach of the available smart city publications in Sect. 4.1 and the fundamental research issues obtained from the smart city literature in Sect. 4.2.

4.4.1 Gaps Arising from Research Themes, Type, and Approach

We arrive at the research gaps by identifying research areas, themes, types, and approaches that are relatively underrepresented in the set of smart city publications analyzed in this study (see Table 8). These various thematic and paradigmatic topics could be combined to generate concrete research scenarios that potentially contribute to the smart city research and practice domain. *Given the relative maturity of the domain, interdisciplinary studies on concrete smart city cases including success*

Table 8 Smart cities' research aspects with relatively low numbers of publications

Aspect	Topic
Research areas	Research on smart city characteristics
	Foundations of smart cities
	Management and governance
	Lessons and experiences on smart city implementation
Research themes	Urban infrastructure management
	Smart city success factors
	Information sharing and service integration in smart cities
	Measurement of smart cities
	Human capital and employment
	Sustainable mobility and intelligent transport
	Standards and regulatory framework for smart cities
	Policy and decision optimization across smart city domain
Dimensions	Education
	Energy
	Lifestyle or smart living
	Smart economy
	Utility infrastructure
	Mobility
	People
Nature of research	Theory development and testing
	Review smart cities research and practice
	Survey of smart cities initiatives
	Position papers on aspects of smart cities
Research approach	Modeling and simulation
	Living lab
	Empirical
	Critical review
	Comparative analysis
	Case studies

factors and challenges require attention to build theoretical foundations. From the analyzed publications, another major observation on research gap is the relative disconnect between smart cities' research and the traditional, more mature studies in urban informatics. *In summary, critical research efforts to better understand different aspects of existing smart cities are very few if at all available.*

4.5 Examples of Research Challenges from Literature

This section presents examples of research challenges discussed in some of the selected research publications. These examples serve as concrete instances of problems in the smart city research areas listed in Sect. 4.4.1. Four categories of issues are described here: (1) fuzziness or conceptual ambiguity of the term "smart city," often conflated with terms like intelligent and digital city; (2) dialectics of suburban policy—arguments on the fallacy of suburban policy as a greener option to pro-concentration and city growth policy; (3) top-down versus bottom-up transitional strategy for smart city development; and (4) participatory urbanism—how can citizens serve as human sensors and source of data. *With respect to the gaps identified in Table* 8, *the first challenge is related to the "Foundations of smart cities" research area. The second and third challenges are related to "Lessons and Experiences on Smart City Implementation" area. The fourth challenge is in the area of "management and governance" with focus on the people dimension.* These four categories of issues are discussed below:

1) *Fuzziness and conceptual ambiguity of "smart city"* (Hollands 2008)—There is a general consensus on the fuzziness of smart city as a concept. For instance, existing literature has associated the smart city concept with IT and creativity and urban entrepreneurship. Research in this would address questions like—Are smart cities high-tech variations of urban entrepreneurialism? How to distill the substantive issues from hype or marketing use of the "smart city" terms? How to have a more critical look at urban labeling such as smart city?
2) *Dialectics of suburban policy* (Glaeser 2012)—This research line involves producing empirical evidence to support either the suburban or city center concentration or pro-urban option. There are arguments giving the well-accepted facts that cities provide economic, health, and educational benefits that accrue from face-to-face social networking; policies to favoring citizens settling in suburbs should be carefully rethought. The thesis is linked to the "super-linear scaling" effect—that socioeconomic properties of the cities increase faster than a direct linear relation to their population.
3) *Building smart cities top-down or by retrofitting* (Biello 2011)—This line of research seeks to determine the better strategy for developing smart cities. The top-down school of thought argues that smartness must be engineered into attributes, such as sustainability, and must be built into infrastructures. Reference models for building smart cities top-down are beginning to emerge. Smart city planners could formulate a planning agenda based on the reference model. However, from the bottom-up perspective, planned smart and eco-cities are fizzling out mainly because of cost.
4) *Participatory urbanism or citizen science* (Paulos et al. 2009)—This research line explores how new "personal instruments" such as mobile phones enable an entirely novel and empowering genre of mobile computing usage called citizen science. The problem centers on how individuals or citizens can become active participants and stakeholders as they publicly collect, share, and remix measurements of their city that matter most to them. A related challenge is how to effectively handle and leverage the social media streams produced by citizens through their mobile devices (Vakali et al. 2014).

5 Findings

This section summarizes the findings of the study with respect to the research questions.

R1—Smart City and Its Dimensions From the definitions, we identify three basic attributes describing the nature of smart cities as "actual cities" that are the results of "urban innovation." Smart cities are characterized by the "high intellectual or human capital" needed to support continuous innovation and address problems or challenges. In terms of goals, smart cities aim at social inclusion, significantly improved quality of life, and economic development. Smart city policies also target the development of human capital through lifelong learning, optimal management of natural resources, and sustainable urban development in general.

R2—De-conflating Smart, Intelligent, and Digital Cities The three terms—smart, intelligent, and digital city—are closely related concepts particularly in terms of their purpose. When considering the constituent elements of the concepts, digital city is restricted to ICT infrastructure and services, whereas the intelligent city concept is an addition to basic ICT infrastructure and services including other specialized technical elements such as "intelligence" infrastructure to support acquisition of information (instrumentation) and learning. The intelligent city also includes elements that involve development of innovation. Smart city elements include those of the intelligent city in addition to socio-organizational and institutional infrastructure to support specific policy domains and governance mechanisms for integrating the solutions to concerns in the different domain. However, the notion of intelligent city is significantly closer to that of the smart city when compared with digital city.

R3.1—Smart City Research Themes Five broad areas of research were obtained from the analysis of the smart city research publications—(1) smart city attributes, (2) smart city implementation, (3) smart city policy domains, (4) management and governance of smart cities, and (5) foundations of smart cities.
Mappings to these research areas show that about 43 % of the available publications focus on different policy domains. Research on implementation aspects of the smart cities accounts for about 20 % of the publications. Research on management and governance accounts for about 19 % of the publications. Studies on foundational aspects of smart cities include about 19 % of publications, whereas publication volume on specific smart city attributes is very low—about 1 % at the moment.
Considering concrete research themes, the top four most common themes are: (1) technology—about 29 % of all publications, (2) nature of smart cities—roughly 17 % of the publications, (3) model and frameworks—about 13 % of the publications, and (4) policy and strategy—roughly 8 % of the publications.

In terms of relative coverage of specific smart city dimensions, by far most of the publications in smart cities are on ICT Infrastructure (or technology). The next area of focus is on governance—about one third of the publications on ICT infrastructure, followed by people.

R3.2—Nature of Smart City Research About 41 % of the publications contained research works that were simply descriptive with no specific orientation. Publications describing design research made up about 32 % of the total publications. Research on conceptualizations of the smart cities constitutes 9 % of the whole publications. Research works on theory development and testing about smart cities account for 5 % and 1 % of the publications, respectively. The numbers of survey, theoretical, polemic, and position publications are very few (virtually nonexistent). In terms of the trends, close to 60 % of the total publications on smart cities were published in 2011 alone and since 2010 there has been 200 % increase in the annual smart city publication volume. Considering specific research types, research on design and conceptualization has doubled from 2010 to 2011, while general descriptive research increased more than sixfold from 2010 to 2011.

R3.3—Approaches to Smart City Research Results show that about 46 % of the research publications have no clear or discernible research approach but offer logical analysis of the problems and solutions (i.e., analytical approach). Research works involving development of one form of technical artifact or another as solution to a problem (i.e., development approach) constitute about 23 % of the publications. Research based on experimentation and case studies equally accounts for about 8 % of the research publications. Approaches including modeling and simulation, survey, grounded theory, living lab, and empirical research have also been employed in the smart city research albeit to very limited level.
In terms of trends, the use of case studies (multi-case studies in particular) grew significantly from one publication in 2010 to nine in 2011, with respect to other methods, such as development. In addition, experimentation and comparative analysis in the smart city research grew by 300 % and 200 %, respectively.

R4.1—Gaps in Smart City Research Given the level of maturity of the domain, interdisciplinary studies based on concrete smart city cases providing more insight to success factors, challenges, and peculiar issues are to enable the development of sound theoretical foundation for the domain. A clear gap resulting from subject area contribution pattern in smart city research is the relative disconnect between smart cities' research and the traditional, more mature studies in urban informatics.

R4.2—Governance of Smart City After ICT infrastructure policy, publications on governance are next in terms of volume of production. Specifically, about 17 % of the publications focusing on specific smart city domains address governance. Governance topics found in the reviewed literature include reform, integration, policy and strategy, measurement, standards and regulation, public engagement, and partnership. Within governance-related publications, the most common governance topic is policy and strategy (29 %), followed by measurement (24 %) and standards and regulations (18 %).

6 Conclusions

The objective of this study was to determine the state of smart cities' research and, in particular, the extent to which governance issues are addressed in the domain. The results show that the research domain is just developing, given that over 75 % of the publications in the domain were produced between 2009 to date and there has been over 200 % increase in the publication volume since 2009. A significant proportion of the works have no specific research orientation, paradigm, or methodology that currently characterizes the smart city research landscape. These facts are symptomatic of relatively new domains. One of the most striking features of smart city research is the clear dominance of computing and engineering as the contributing subject areas. Consequently, most of the research works with a discernible approach were generally design oriented, focusing on developing technical artifacts. Perhaps more surprising is the gap between current smart cities' research and root disciplines, such as urban planning and more recently urban informatics. *We conclude from this study that critical research efforts to better understand different aspects of existing smart cities are very few if at all available.*

Therefore, in addition to the expected growth in smart cities' research, emphasis will likely shift to studies on concrete cases such as those reported in AlAwadhi and Scholl (2013), De et al. (2014), and Lee and Lee (2014) *as the basis for foundational insight into the nature of smart cities and at the same time leveraging existing knowledge in urban planning and city transformation for theoretical grounding.*

An important limitation of the mapping exercise carried out in the study is that the mapping was restricted to the analysis of titles, abstracts, and keywords of the publications. Full texts were only considered or read when the abstracts were unclear enough to determine the necessary information about the publications.

References

AlAwadhi, S., & Scholl, H. J. (Jochen). (2013). Aspirations and realizations: The smart city of Seattle. *Proceedings of the 46th Hawaii international conference on system sciences* (pp. 1695–1703). IEEE. doi:10.1109/HICSS.2013.102.

Bettencourt, L. M. A., & West, G. B. (2011). Bigger cities do more with less. *Scientific American, 305*(3), 44–45.

Biello, D. (2011). How green is my city—Retrofitting is the best way to clean up urban living. *Scientific American, 305*(3), 52–55.

Bloomberg, M. R. (2011). The best and the brightest—New York City's bid to attract science talent could serve as a model for other cities. *Scientific American, 305*(3), 11.

Caragliu, A., Del Bo, C., & Nijkamp, P. (2009). Smart cities in Europe. *3rd Central European conference on regional science—CER 2009*, pp. 45–59.

Chourabi, H., Gil-garcia, J. R., Pardo, T. A., Scholl, H. J., Walker, S., & Nahon, K. (2012). Understanding smart cities: An integrative framework. *2012 45th Hawaii international conference on system sciences*, pp. 2289–2297. doi:10.1109/HICSS.2012.615.

De, L. S., Lheureux-, J., Ramos, C., & Anne, M. (2014). A smart city initiative: A case study of Porto Alegre 156. In *dg.o '14, June 18–21 2014, Aguascalientes, Mexico*, pp. 245–252.

DiChristina, M. (2011). Street-Savvy—Meeting the biggest challenges starts with the city. *Scientific American, 305*(3), 27–29.
Giffinger, R., Fertner, C., Kramar, H., Kalasek, R., Pichler-Milanović, N., & Meijers, E. (2007). *Smart cities ranking of European medium-sized cities* (pp. 1–28). Vienna: Centre of Regional Science.
Glaeser, E. (2011). Engines of innovation. *Scientific American, 305*(3), 36–41.
Glaeser, E. (2012). *Triumph of the city* (p. 352). London: Penguin Books. (Reprint edition January 31, 2012).
Glaeser, E., & Saiz, A. (2003). *The rise of the skilled city* (No. 25). Social science research. Massachusetts: Cambridge.
Grönlund, Å. (2004). State of the art in e-Gov research—A survey. In R. Traunmüller (Ed.), *EGOV 2004, LNCS 3183* (pp. 178–185). Berlin: Springer.
Heeks, R., & Bailur, S. (2007). Analyzing e-government research: Perspectives, philosophies, theories, methods, and practice. *Government Information Quarterly, 24*(2), 243–265. doi:10.1016/j.giq.2006.06.005.
Hollands, R. G. (2008). Will the real smart city please stand up? Intelligent, progressive or entrepreneurial. *City: Analysis of Urban Trrends, Culture, Theory, Policy and Action, 12*, 37–41. doi:10.1080/13604810802479126. (March 2012).
IBM. (2013). Smart cities. http://www.ibm.com/smarterplanet/us/en/smarter_cities/overview/index.html. Accessed 12 Nov 2013.
Komninos, N. (2011). Intelligent cities: Variable geometries intelligent cities: Variable geometries of spatial intelligence. *Intelligent Buildings International, 3*, 37–41. doi:10.1080/17508975.2011.579339. (March 2012).
Lee, J., & Lee, H. (2014). Developing and validating a citizen-centric typology for smart city services. *Government Information Quarterly, 31,* S93–S105. doi:10.1016/j.giq.2014.01.010.
Nam, T., & Pardo, T. A. (2011). Smart city as urban innovation: Focusing on management, policy and context. In E. Estevez & M. Janssen (Eds.), *Proceedings of the 5th international conference on theory and practice of electronic governance (ICEGOV2011)*. Tallinn: ACM Press.
Ojo, A., Curry, E., & Janowski, T. (2014). Designing next generation smart city initiatives—Harnessing findings and lessons from a study of 10 smart city programs. *22nd European conference on information systems (ECIS 2014)* (Vol. 2050, pp. 1–14). Tel Aviv: AIS Library.
Paskaleva, K. A. (2011). The smart city: A nexus for open innovation? *Intelligent Buildings International, 3*, 153–171. doi:10.1080/17508975.2011.586672. (March 2012).
Paulos, E., Honicky, R., & Hooker, B. (2009). Citizen science: Enabling participatory urbanism. In M. Foth (Ed.), *Handbook on research on urban informatics: The practice and promise of real-time cities* (pp. 414–436). Hershey: IGI Global.
Ratti, C., & Townsend, A. (2011). The social nexus. *Scientific American, 305*(3), 30–35.
Vakali, A., Anthopoulos, L., & Krco, S. (2014). Smart cities data streams integration: Experimenting with Internet of things and social data flows. ... *of the 4th International Conference on ...*, 1–5. http://dl.acm.org/citation.cfm?id=2611094. Accessed 25 Aug 2014.
Vasseur, J.-P., & Dunkels, A. (2010). Smart cities and urban networks. In *Interconnecting smart objects with IP—the next Internet* (pp. 335–351). Morgan Kaufmann. doi:10.1016/B978-0-12-375165-200022-3.
Yildiz, M. (2007). E-government research: Reviewing the literature, limitations, and ways forward. *Government Information Quarterly, 24*(3), 646–665.
Zygiaris, S. (2012). Smart city reference model: Assisting planners to conceptualize the building of smart city innovation ecosystems. *Journal of the Knowledge Economy*. doi:10.1007/s13132-012-0089-4.

Adegboyega Ojo is a research fellow and the e-government group leader at the Insight Centre for Data Analytics, National University of Ireland, Galway (NUIG), Republic of Ireland. His research investigates how to harness semantic web and linked data in developing next generation electronic public service infrastructures and how to drive open innovation both in government and governance of smart cities through the adoption and exploitation of big, open, and linked data. Before joining the Insight Centre at NUIG, he worked as an academic program officer and research fellow at the Center for Electronic Governance, United Nations University—International Institute for Software Technology (UNU). At UNU, he supported governments in Africa, Asia, and Latin America; including Macao, Korea, Mongolia, Colombia, Cameroon, and Nigeria. Before his international engagements, he was a senior lecturer in computer science at the University of Lagos, Nigeria, where he earlier earned his doctorate and bachelor degrees in 1998 and 1991, respectively. He is a professional member of ACM and the Computer Professionals of Nigeria.

Zamira Dzhusupova is an independent consultant and expert at UNDP in Kyrgyzstan, has experience as a researcher and practitioner in Electronic Governance (EGOV) and ICT for development. For several years, she was involved as a manager and coordinator of relevant projects supported by the United Nations, World Bank, Governments, and other organizations. From 2008 until 2013, she worked for the Center for Electronic Governance at the United Nations University—International Institute for Software Technology (UNU-IIST) located in Macao SAR, China. At UNU-IIST, she was responsible for managing EGOV projects, education, and training. Prior to joining UNU-IIST, she has worked for UNDP Democratic Governance Programme in Kyrgyzstan. Dr. Dzhusupova was involved in capacity building activities by developing and teaching courses and supervising junior researchers and fellows in UNU-IIST, and for master programme on information systems in public administration at the Academy of Management in Kyrgyzstan. Her current research interests include Open Government issues in developing countries. She holds a PhD in management information systems from International Academy on System Studies in Moscow, University degree (MSc equivalent) in automated management systems from Bauman State Technical University in Moscow, and University degree in finance and credit from State National University in Bishkek, Kyrgyz Republic.

Edward Curry has published over 100 scientific articles in journal, books, and international conferences. He has given invited talks at Berkeley, Stanford, and MIT. In 2010, he was a guest speaker at the MIT Sloan CIO Symposium to an audience of 600+ CIOs and senior IT executives. He leads the technical working groups in the BIG project funded by the European Commission to define a research strategy for the big data economy within Europe. He has a PhD from the National University of Ireland, Galway (www.nuigalway.ie) and serves as an adjunct lecturer within the university.

Characterizing the Role of Governments in Smart Cities: A Literature Review

Manuel Pedro Rodríguez Bolívar

Abstract The growth of smart cities is forcing governments to think the need to advance in the implementation of information and communication technologies (ICTs) for the improvement of the citizenry's participation in decision-making processes, the increasing efficiency of public services, and the improvement of transparency and accountability. In this regard, governments in smart cities are called to play a key role in promoting and managing these cities. Based on prior research, this chapter seeks to analyze the role of governments in smart cities, trying to identify different patterns of management styles in these cities. Also, differences between theoretical and empirical studies about the role of governments in smart cities are identified proposing new research themes for the future.

Keywords Smart cities · Smart governance · Role of governments · Management styles

Abbreviations

ICTs Information and communication technologies
PPP Public–private partnerships

1 Introduction

In the last years, cities are becoming smart not only in terms of the way we can automate routine functions serving individual persons, buildings, and traffic systems but also in ways that enable us to monitor, understand, analyze, and plan the city to improve the efficiency, equity, and quality of life for its citizens in real time (Batty et al. 2012). This has led to an increasing research on this subject that is being recently published in international leading journals or books.

M. P. Rodríguez Bolívar (✉)
Department of Accounting and Finance, Faculty of Business Studies, University of Granada, C/ Campus Universitario de Cartuja, s/n, 18071 Granada, Spain
e-mail: manuelp@ugr.es

J. R. Gil-Garcia et al. (eds.), *Smarter as the New Urban Agenda A Comprehensive View of the 21st Century City,* Public Administration and Information Technology 11,
DOI 10.1007/978-3-319-17620-8_3

Prior research has been focused mainly on business-led urban development, on the social inclusion agenda, on the role of creative industries in urban growth, on the importance of social capital in urban development, and on the urban sustainability. Nonetheless, the growth of smart cities is promoting the increasing use of information and communication technologies (ICTs) by governments to improve political participation, implement public policies, or provide public sector services.

In this regard, the role played by governments in these cities seems to be essential. Hollands (2008) indicates that the need for technologies to be smarter is not just in the way they make it possible for cities to be intelligent (as an institutional agent) in generating capital and creating wealth, but in the ways they operate their governments. In addition, Leydesdorff and Deakin (2011) highlight the role of governments in creating cultural development.

Therefore, governments in smart cities are called to play a key role in promoting and developing smart cities, using ICTs in creating interactive, participatory, and information-based urban environments (Bătăgan 2011; Batty et al. 2012), as well as in improving public services (Deakin 2012) and the functioning of the administration through the increasing collaboration across government departments and with communities (Bătăgan 2011). This role of governments in smart cities is what has been called as "Smart Governance" (Giffinger et al. 2007).

In brief, the new wave of "smart cities" makes to think in the idea of the wider debate about decentralization of governance in the information age (Innes and Booher 2010) and makes governments to think in another way of communication, interaction, and provision of public sector services (Giffinger et al. 2007). Nonetheless, little research has been undertaken in order to know the role of governments to promoting smart cities. Indeed, up to now no research has been undertaken focused on the management styles in public administrations under the smart cities' framework.

Therefore, this chapter seeks to analyze the role of governments in smart cities, trying to identify different patterns of management styles in these cities. Also, differences between theoretical and empirical studies about Smart Governance are identified proposing new research themes for the future.

To achieve the objective of this chapter, we first identify the main attributes that characterize the role of governments in smart cities. These attributes seek to recognize the key dimensions that characterize the role of governments in smart cities. Later, we analyze empirical and theoretical research in order to identify patterns of management styles in smart cities according to the attributes previously defined. Finally, differences between empirical and theoretical research are identified and mapped according to the attributes previously defined. This analysis seeks to identify whether the empirical experiences are according to theoretical models or, else, if empirical experiences are different from theoretical studies and, therefore, a "gap" between theory and practice exists.

Concretely, based on prior research, the main research questions of this chapter are:

RQ1. What are the main attributes that characterize the role of governments in smart cities?

RQ2. Is there a homogenous role of governments in managing smart cities? How can we map them?

RQ3. Are there differences between theoretical and empirical research about the role of governments in smart cities?

The chapter is organized as follows. In the second section, a theoretical analysis of governance models under networking environments is performed. Then the empirical research of this chapter is presented and the results of our analysis are described. Finally, discussions and conclusions, as well as implications for future research are shown.

2 Smart Cities and Smart Governance

Many authors think that ICTs are central to the operation of the future city (Aurigi 2005; Komninos 2002; Odendaal 2003; Walravens 2012), which has provoked the emergence of similar terms such as intelligent cities, virtual cities, knowledge-based cities, digital cities, or information cities, which have caused some problems regarding the identification of the smart city concept[1]. Others highlight the political aspect of this concept in Europe to mobilize all knowledge centers into innovation hubs in order to strengthen the socioeconomic progress in EU member states (Kourtit et al. 2012).

Therefore, although the smart city is in the current arena of research on urban planning and eGovernment, among other fields of knowledge, it is a fuzzy concept that has not been used consistently within the literature (Tranos and Gertner 2012). In fact, one of the main challenges still is to fully understand how to conceptualize smart cities. This has led authors to define this term in a comprehensive way (Caragliu et al. 2011) and through propositions (Nam and Pardo 2011), in which aspects such as investment in social and human capital, the performance of sustainable policies into the environmental area, policies of sustainable economic growth, investment in modern transport technologies, investment in aspects of quality of life (health, safety, housing, etc.), and participatory governance are included (Ferro et al. 2013; Giffinger et al. 2007).

In this regard, in the past few years many groundbreaking promises have been made about the potential of the smart city. Nonetheless, criticism of smart cities is beginning to take place from the analysis of the different aspects mentioned before. For example, the idea of sustainable degrowth, defined as an equitable downscaling of production and consumption that increases human well-being and enhances

[1] In this chapter, smart cities are analyzed but not other terms such as "intelligent city," "virtual city," "innovative city," "knowledge city," or "creative city," since these terms have some overlap and focus on other discourses such as attracting urban professionals and creating an urban cyberspace. It is also argued that the smart city concept is in fact a more user-centered evolution of the other city concepts which seem to be more technological deterministic in nature (Schuurman et al. 2012). Therefore, as this chapter aims to explore the different models of smart city governance specifically, it limited our search to that term.

ecological conditions at the local and global level, in the short- and long term (Schneider et al. 2010), is emerging as a response to the triple environmental, social, and economic crisis (Flipo and Schneider 2008; Martinez-Alier 2009). Although the degrowth theories are not new (Hueting 1980; Fotopoulos 1997; Odum and Odum 2001), academic literature on sustainable degrowth has recently surged in ecological economics and related fields, marking the comeback of the economic growth debate (Berg and Hukkinen 2011).

The proponents of idea of degrowth indicate that technological progress is welcome, but only when it embodies appropriate and chosen limits, rather than continuous innovation to spur consumption (Schneider et al. 2010). In addition, D'Alessandro et al. (2010) present a model exploring the contradiction between increased investments in renewable sources and the acceleration of exhaustibility of resources, highlighting the danger of high growth rates.

Also, political challenges are raised by smart cities, given the need to maintain employment and social stability while decreasing resource use in order to become more sustainable cities. According to Spangenberg (2010), the number of jobs can only increase if the economy grows faster or degrows slower than per capita productivity and resource consumption can only decrease if the economy grows slower or degrows faster than resource productivity. In this regard, a contradiction exists if a city is labeled as "smart" and it cannot support high rates of employment.

Finally, the growth in population in cities is making them to be more compact cities, and although increasing population densities within a city's urban center drastically reduce congestion, emissions, and gasoline consumption (Behan et al. 2008), the compact city fallacy holds that the compact city is neither a necessary or sufficient condition for a city to be sustainable and that the attempt to make cities more sustainable only by using urban form of strategies is counterproductive (Neuman 2005). Therefore, in order to operate in a more environmentally friendly, socially equitable, and economically viable manner, the only issue remaining relates to the question of how to get there (Daneshpour and Shakibamanesh 2011).

In brief, transforming urban processes will only be achieved with better urban governance (Puppim de Oliveira et al. 2013). Cities are therefore increasingly seen as not only the engines of innovation and economic growth but also the level at which solutions to wicked problems need to be produced (Koppenjan and Klijn 2004). The idea of smart city governance fits well within the public management perspective that highlights that solving societal problems is not merely a question of developing good policies but much more a managerial question of organizing strong collaborations between government and other stakeholders (Torfing et al. 2012). Indeed, forms of government are an important direct influence on the approach that communities take to sustainability (Bae and Feiok 2013).

Under this framework, governments must take a leading role in coordinating and managing the city efficiently. It means the need of governments for reconsidering their role in this cities as well as the governance model to be adopted by them with the aim of facilitating and promoting the smart cities' development. This is what has been called as "smart governance" (Giffinger et al. 2007) and it is a much stronger intelligence function for coordinating the many different components that comprise the smart city (Batty et al. 2012).

According to Fazekas and Burns (2012), governance refers to the process of governing societies in a situation where no single actor can claim absolute dominance. Thus, governance is not about what governments do but about the outcomes of interactions between all actors in the public domain. In smart cities, governance encapsulates collaboration, cooperation, partnership, citizen engagement, and participation (Coe et al. 2001). In this regard, it is acknowledged that to improve the quality and performance of cities, it is recommended to involve all interested parties to implement efficiently smart systems, making operations and services truly citizen centric (Bătăgan 2011).

Therefore, smart cities have really become in relational networks of actors in which local governments are key actors to create an interactive, participatory, and information-based urban environment with the ultimate aim of producing increasing wealth and public value, achieving higher quality of life for citizens. As an example, one of the main exponents of this network is the triple-helix model designed to produce knowledge in a city through the collaboration between university, industry, and government (Leydesdorff and Deakin 2011).

Although there are different approaches to the concept of smart city governance in prior research ranking from *institutional conservation* (traditional governance of a smart city) to *institutional transformation* (smart urban governance) (Meijer and Rodríguez 2013), none is said to be the best way of governing smart cities. Therefore, based on the network governance literature (Koiman 2003; Koiman et al. 2008; Provan and Kenis 2008; Fazekas and Burns 2012) and the coproduction literature (Linders 2012; Span et al. 2012; Mergel 2013), we can define the principal dimensions of governance in order to identify them into the smart cities' literature—see Table 1 in appendix.

In brief, under the new wave of smart cities, local governments are forced to take greater advantage of modern ICT infrastructure, eGovernment, and the newly empowered citizenry, seeking to produce higher outcomes regarding wealth and public value. To achieve this aim, local governments must define the role they think that must play and the adoption of different models of governance under these new networking environments.

These networking environments introduce new ways of governance different from traditional bureaucracy, with the use of nonhierarchical, nonmarket forms of organization in the public sector (Considine and Lewis 1999), and are becoming important for public management given that the management of smart cities relies on complex networks of interdependent organizations. In this regard, these models of governance can range from that in which smart cities may be governed completely by the organizations that comprise the network (self-governance model) to that in which local government acts as a highly centralized network broker, or lead organization, and manages the development of the smart city (bureaucratic model).

Table 1 Dimensions of governance. (Source: Author's own elaboration based on Koiman (1993), Peters and Pierre (1998), Denhart and Denhart (2000), Kettl (2006), Span et al. (2012), and Fazekas and Burns (2012))

Dimensions of governance and their definition	
Steering	Steering should be understood about setting priorities and defining goals (Peters and Pierre 1998)
Boundary conditions	Conditions that determine the mission, resources, capacity, responsibility, and accountability of the task to be performed (Kettl 2006)
Alignment dimension	Alignment refers to which party coordinates the smart city development (Span et al. 2012)
Conceptions of public interest	Public administrators must contribute to building a collective, shared notion of the public interest, not to find solutions driven by individual choices (Denhart and Denhart 2000)
Role of local governments	This could be arrayed along a spectrum, as a continuum of top-down to bottom-up processes (Span et al., 2012), ranging from the role of "executor (commissioner)", to the role of "initiator (facilitator)"
Structure of governance	The pattern or structure that emerges in a sociopolitical system as a common result or outcome of the interacting efforts of all involved actors (Koiman 1993)
Governance model	It refers to the model of governing societies in a situation where no singe act can claim absolute dominance (Fazekas and Burns 2012)

Based on a review on the extensive literature of smart cities, this chapter seeks to identify what attributes best characterize the role of governments in smart cities, whether there is a homogenous pattern in managing smart cities, and whether differences between theoretical and empirical research about it exist in the literature about smart cities. The premise of this chapter is that an examination of the dimensions being used by local governments in the smart cities' literature will inform the research questions mentioned above.

3 Data and Method

3.1 Data Collection

The data collection of this chapter was carried out in two stages. The first one was addressed to undertake the literature review needed for achieving the objectives of this chapter.

In this regard, the literature review consisted of a three phases. First, a keyword search was conducted in the three library databases suggested by Webster and Watson (2002)—ABI/Inform (ProQuest), ISI Web of Science and Scopus EBSCO Host. This search, which covered a large number of journals from many disciplines for the obvious reason of not missing any important article, was performed using the following words: "smart cities" and "governance," "smart governance," and "smart administration." The search queries entered in ISI Web of Knowledge, Scopus EBSCO Host, and ABI/INFORM (ProQuest) databases let us to obtain, respectively, a total of 52, 104, and 76 papers about smart cities and governance and/or smart governance.

Second, the full contents of the following academic journals, which were found to be the core outlets for eGovernment publications by Scholl (2009), were examined by means of browsing through titles, author-supplied keywords, and abstracts: (1) *Inderscience's Electronic Government, An International Journal (EGaIJ),* (2) *ACI's Electronic Journal of E-Government (EJEG),* (3) *Elsevier's Government Information Quarterly (GIQ),* (4) IOS Press' *Information Polity (IP),* (5) IGI's *International Journal of Electronic Government Research (IJEGR),* (6) Taylor & Francis' *Journal of Information Technology and Politics* (formerly *Journal of E-Government (JITP/JEG),* and (7) Emerald's *Transforming Government: Process, People, and Policy (TGPPP)*. This second search let us to obtain a total of 76 papers published in these journals regarding the objective of this chapter.

Third, a search was conducted in the proceedings of International IFIP EGOV Conference which, according to Scholl (2009), is one of the three core conferences for eGovernment research. The IFIP EGOV conference was chosen as representing one out of these three quite similar, in terms of the nature of the content, eGovernment conferences. These searches let us obtain seven papers regarding the objective of this chapter.

In sum, our search method combining an open search with a strategic selection of journals and conferences has generated a reasonably representative set of articles (a total of 315 papers).

The second stage consisted of a selection of relevant papers on the basis of abstracts and objectives of the paper. The large volume of collected papers had to be filtered as only articles focused on smart city governance were of interest. The abstract and the introduction section, especially the objective of the paper, were read, and an overview of the structure of the paper was checked. Those papers of specific technical nature without examining any of the domains analyzed in this chapter were eliminated from the sample. In addition, literature reviews were removed from the sample since they do not make a new contribution on the domains. Finally, double counting of papers was avoided by counting only the papers that were different across the databases and the searches performed in the specialized journals and conferences. These processes resulted in a final sample of 40 papers.

To sum up, the methodology for data collection has been designed to conform as closely as possible to the recommendations in the review methods literature (Webster and Watson 2002; Scholl 2009). Therefore, the results of this chapter account for the Smart Governance research as situated within the boundaries of the eGovernment domain where most discussions take place in the form of journal articles.

3.2 Method

The overall analytical approach adopted largely followed the conventions of template analysis, where the researcher produces a list of codes (template) representing themes identified in the textual data (King 2004). These codes can be developed after some initial exploration of the data or they can be defined by the research previously and refined and modified during the analysis process (Crabtreee and Miller 1999). According to King (2004), in our study, the attributes of governance were coded into broad themes (the dimensions of governance are shown in Table 1 in appendix) based on the research aims to create an initial template. Each code was then subjected to a more detailed manual analysis by the researcher, which led to the formation of more specific categories within each theme. In this regard, categories for each one of the dimensions (codes) defined in Table 1 were identified—see Table 2 in appendix

Both codes and categories were identified after reading prior research on network governance and coproduction literature and they were refined during the analysis process. Broad higher-order codes help provide a general overview of the analysis of prior research, while detailed lower-order codes enable fine distinctions to be made, both within and between prior research on Smart Governance (King 2004). This classification system illustrates relationships between higher- and lower-order code names, gives an idea of how various themes are interconnected (King 2014), and allows the exploration of the data and a comparison of the similarities and differences. Dey (1993) explains that codes must be meaningful with regard to the data and in relation to other categories.

To sum up, all selected papers have been analyzed with the template analysis technique in order to answer the research questions previously mentioned and to map the governance models in smart cities, which is presented in the next section of this chapter.

Table 2 Codes and categories. (Source: Author's own elaboration based on Koiman (1993), Peters and Pierre (1998), Denhart and Denhart (2000), Kettl (2006), Span et al. (2012), Linders, (2012), and Mergel (2013))

Dimensions of governance (codes) and categories	
Steering	(1) Local government, (2) joint steering, (3) self-steering
Boundary conditions	(1) Fixed boundaries by local governments, (2) jointly set boundaries, (3) boundaries set by the parties
Alignment dimension	(1) Strategic planning of local governments, (2) joint alignment, (3) alignment by the parties
Dependency dimension	(1) Formal dependency, (2) informal
Conceptions of public interest	(1) Politically defined and expressed in law, (2) result of negotiating values, (3) result of a dialogue about shared values, (4) represents the aggregation of individual self-interests
Role of local governments	(1) Commissioner or executer, (2) coproducer, (3) facilitator
Structure of governance	(1) Corporate governance, (2) lead organization-governed network, (3) interactive governance, (4) governance as the minimal state
Governance model	(1) Bureaucratic model, (2) collaborative model, (3) participatory model, (4) self-governance model

3.2.1 Analysis of Results

Before analyzing the attributes that characterize the role of governments in smart cities, the literature review of this chapter has indicated that most of the papers deal with specific smart objects or areas. In this regard, it is noteworthy that a large volume of papers are dealt with “smart growth” or with smart technologies. The “smart growth” papers deal with the need of a new regionalism, identifying ways of coordination in different areas (legal, economic, etc.) in order to make efficient the policies taken for developing a region (Scott 2007; Mohamed 2008; Hawkins 2011). The papers about smart technologies focus mainly on smart technologies such as smart cards, smartphones, etc., and analyze the risks (for example, a decrease in privacy and security issues) and advantages of their use in smart cities (Margetts 2005).

All this leads to a small number of papers that explicitly mention the need for Smart Governance and the model of governance to be adopted by governments in these smart cities. Some of these papers are general, showing some models regarding two or three characteristics (Walravens and Ballon 2013; Kourtit and Nijkamp 2012). Others are focused on specific areas, no attributes, to be dealt with when governments are prone to implement ICTs under a smart city framework (Sauer 2012; Rohman 2012; Barbry 2012).

In addition, our revision shows that empirical experiences are more prolific than theoretical studies regarding Smart Governance.

3.2.2 RQ1. What Are the Main Attributes that Characterize the Role of Governments in Smart Cities?

Table 3 shows the attributes (dimensions) of governance models that prior research has indicated as essential in both empirical and theoretical studies—see Table 3 in appendix.

According to prior research, steering of smart projects should be managed by the local government or by the local government jointly with stakeholders. This result is also corroborated when these papers analyze the organization that imposes the boundaries of the smart city development. Many of them indicate that the accountability of smart projects should be demanded to public administrations although the steering is joined with stakeholders. This is clearly shown in the alignment dimension because most of sample papers point out that the local governments should design strategic planning in which smart projects should be included and monitored.

Therefore, results seem to indicate that the involvement of stakeholders to develop smart cities is mainly focused on the collection of knowledge and ideas about the smart projects that should be undertaken into the smart city. Nonetheless, the local government is the organization that has the responsibility to undertake the strategic planning of the city and, only in very few cases, they promote the involvement of the stakeholders in this task—see value of category 3 in the alignment dimension in Table 3 in appendix.

In any case, the previously mentioned way of the involvement of stakeholders into the smart projects performed by local governments seems to be the reason why

Table 3 Mapping attributes that characterize the role of governments in smart cities. (Source: Author's own elaboration)

Dimension\categories	1	2	3	4	Total
Steering	13	13	1	–	*27*
Boundaries	14	10	0	–	*24*
Alignment	14	7	2	–	*23*
Dependency	3	29	–	–	*32*
Conception of public interest	6	6	14	2	*28*
Role of local government	7	23	3	–	*33*
Structure of governance	7	15	14	2	*38*
Governance model	5	14	19	2	*40*

The number for each one of the attributes (dimensions) represents the categories shown in Table 2

authors indicate that, up to now, informal dependency is told to be the best way in which the power should be wielded. No rules or procedures are introduced to regulate the relationship between governments and stakeholders.

Nonetheless, increasing involvement of stakeholders is promoted by prior research, because authors think that governments should be open to introduce dialogues with stakeholders regarding the smart projects (dimension of "conception of public interest") and should be prone to the coproduction of public services and public values (dimension of "role of local government").

The involvement of stakeholders into the public sector management regarding smart projects does not mean that they have the strongest power in the relationship with the local governments. Indeed, literature indicates the need of local governments to govern the network (dimension of "structure of governance"), using either a collaborative or a participatory environment (dimension of "governance model").

Therefore, although prior research seems to agree with the need of transforming governments under the era of smart cities to be more collaborative and participative, theoretical and empirical studies seem to advocate that local governments keep playing a key role in the accountability and in the monitoring activities. In fact, local governments govern the network created over the smart projects performed into the smart city. This structure of governance can be exerted directly by the local government (corporate governance), or the government-led organization that governs the network.

3.2.3 RQ2. Is There a Homogenous Role of Governments in Managing Smart Cities? How Can We Map Them?

In order to analyze this research question, as all research questions are based on qualitative attributes, the $\chi 2$ test was chosen to check the association between the dimensions of the role of governments in smart cities—see Table 4 in appendix. This methodology of research has been broadly recognized in many research studies as the best statistical tool to test hypotheses of independence for $r \times c$ contingency tables in which row and column categories are both nominal and mutually exclusive categories (Snedecor and Cochran 1981; Sheskin 2007; Sprent and Smeeton 2007).

In addition, contingency tables were performed to gain an overview of the distribution of the responses—see Table 5 in appendix. Finally, this paper makes a cluster analysis to group homogeneous categories of attributes according to their options collected from the sample papers. Cluster analysis is the most appropriate method because it allows grouping prior research works that have similar categories of dimensions about governance models under smart cities' framework, thus leading to homogeneous empirical types (Rapkin and Luke 1993). Therefore, cluster analyses were performed using the k-means algorithm with the aim to identify homogenous groups of categories of attributes—see Table 6 in appendix.

According to the data in Table 4, all dimensions are associated, except for the association between the alignment and dependency dimensions—see Table 4 in appendix. It means that all governance policies suggested by prior research to transform local governments under smart cities' framework seem to attend a common aim: the collaboration and participation of stakeholders in the design and performance of smart projects, although local governments should govern these projects.

This result is confirmed by the contingency tables between the governance dimensions analyzed in this paper, which are shown in Table 5 in appendix.. As can be seen in Table 5, the steering, boundary, and alignment dimensions are mainly concentrated in the options 1, 1, and 1, respectively. Prior research has indicated that local governments should steer the smart projects, should fix the boundaries of those projects, and should include them into the strategic planning of the city.

Nonetheless, local governments should engage stakeholders into the smart projects, promoting their involvement to share knowledge and ideas for the development of the smart city. It makes that the informal dependency be the main link between stakeholders and local governments. In addition, the role of coproducer has been indicated as the main role of governments in smart cities. This result shows the commitment of the local governments with their stakeholders that are involved in the process of developing smart projects. In this regard, prior research has indicated that local governments do not have enough resources or knowledge to undertake all smart projects in the city. In brief, technical expertise is insufficient; social expertise is indispensable (Stoker et al. 2003). It has made some local governments to use the figure of public–private partnerships (PPP) to undertake these projects (Farris 2001).

This coproduction model proposed by authors and implemented in many smart cities has pressured local governments to create networks that are mainly led by the local governments. Under this framework, according to our results, the collaborative and the participatory models of governance have been implemented in local governments.

Regarding the cluster analysis, we consider two clusters for empirical (cluster 1) and theoretical studies (cluster 2), respectively. The number of clusters is low, thus enhancing model parsimony and achieving within-cluster homogeneity and between-cluster heterogeneity, as the ANOVA test in Table 6 shows. This test is highly significant for all variables in our study. This allows us to draw conclusions about the dimensions of governance of each group, as each cluster is quite dissimilar to another one.

Table 4 Statistical test of association between attributes (dimensions) that characterize the role of governments in smart cities. (Source: Author's own elaboration)

	Steering		Boundaries		Alignment		Dependency		Conception of public interest		Role of Local Government		Structure of Governance	
	χ2	p value	χ2	p value	χ2	p value	χ2	p value	χ2	p value	χ2	p value	χ2	p value
Boundaries	19.221	0.000***												
Alignment	34.821	0.000***	19.000	0.000***										
Dependency	4.052	0.044**	2.895	0.089*	2.353	0.308								
Conception of public interest	28.389	0.000***	6.807	0.033**	22.950	0.001***	16.388	0.001***						
Role of Local Government	18.228	0.001***	8.069	0.018**	11.239	0.024**	14.850	0.001***	35.052	0.000***				
Structure of Governance	36.991	0.000***	9.746	0.008***	35.247	0.000***	16.667	0.001***	44.535	0.000***	26.789	0.000***		
Governance Model	37.234	0.000***	11.017	0.004***	34.295	0.000***	23.172	0.000***	53.148	0.000***	34.177	0.000***	70.473	0.000***

Note: *Significant at 10%; ** Significant at 5%; ***Significant at 1%.

Table 5 Frequency in contingency tables between attributes (dimensions) that characterize the role of governments in smart cities. (Source: Author's own elaboration)

		Steering			Boundaries			Alignment			Dependency		Conception of public interest				Role of Local Government			Structure of Governance			
		1	2	3	1	2	3	1	2	3	1	2	1	2	3	4	1	2	3	1	2	3	4
Boundaries	1	13	1	0																			
	2	0	9	0																			
	3	0	0	0																			
Alignment	1	11	2	0	12	0	0																
	2	0	7	0	0	7	0																
	3	0	0	1	0	0	0																
Dependency	1	3	0	0	3	0	0	3	0	0													
	2	8	13	0	9	10	0	9	7	1													
Conception of public interest	1	6	0	0	6	0	6	6	0	0	3	1											
	2	1	3	0	1	3	4	1	2	0	0	5											
	3	3	6	0	4	5	9	3	3	0	0	13											
	4	0	0	1	0	0	0	0	0	2	0	1											
Role of Local Government	1	7	0	0	7	0	7	7	0	0	3	2	6	0	1	0							
	2	3	10	0	4	8	12	5	5	0	0	20	0	5	12	0							
	3	1	1	1	1	1	2	1	1	1	0	2	0	1	0	1							
Structure of Governance	1	6	0	0	6	0	0	6	0	0	3	2	5	1	0	0	5	1	0				
	2	5	6	0	6	4	0	8	3	0	0	13	1	2	6	0	2	11	1				
	3	1	7	0	1	6	0	0	4	0	0	11	0	3	7	0	0	10	1				
	4	0	0	1	0	0	0	0	0	2	0	1	0	0	0	2	0	0	1				
Governance Model	1	5	0	0	5	0	5	5	0	0	3	1	5	0	0	0	5	0	0	5	0	0	0
	2	6	4	0	7	2	9	7	1	0	0	12	1	1	7	0	2	9	0	2	8	2	0
	3	2	9	0	2	8	10	2	6	0	0	15	0	5	7	0	0	14	2	0	7	12	0
	4	0	0	1	0	0	0	0	0	2	0	1	0	0	0	2	0	0	1	0	0	0	2

Notes: (1) The number for each one of the attributes (dimensions) represents the categories shown in table 2; (2) the table only shows the results that are collected in sample studies when the two dimensions analyzed have an option chosen at the same time.

Table 6 Cluster analysis and ANOVA test for clusters. (Source: Author's own elaboration)

Dimension	Cluster 1	Cluster 2	ANOVA test	
			F	Significance
Steering	1.00	2.00	30.462	0.000***
Boundary dimension	1.00	2.00	12.692	0.004***
Alignment dimension	1.00	2.00	12.692	0.004***
Dependency dimension	1.00	2.00	5.923	0.033**
Conception of public interest	1.00	3.00	10.632	0.008***
Role of government	1.00	2.00	20.034	0.001***
Structure of governance	1.00	3.00	22.214	0.001***
Governance model	1.00	3.00	17.875	0.001***
Empirical/theoretical study	1.00	0.00	3.808	0.077*

The number for each one of the attributes (dimensions) represents the categories shown in Table 2. Regarding the empirical or theoretical nature of the study, the figure 1 means "empirical study" and 0 means "theoretical studies" *Significant at 10 %; **Significant at 5 %; ***Significant at 1 %

In this regard, according to the results in Table 6, when the study comes from an empirical experience, the model used by the smart city has been mainly based on the bureaucratic model of governance, in which noncollaborative and nonparticipatory models of governance have been implemented—see Table 6 in appendix. In this regard, the smart city is governed by the local government and stakeholders are not involved in the design of the smart projects or in their performance.

By contrast, theoretical studies advocate a change of governance models adopted by public administrations, from the Bureaucratic Model of governance to other more collaborative and participatory ones. In fact, the participatory model of governance is that promoted by prior research in which the involvement of stakeholders is high. This model proposed by theoretical studies could lead to an effective collaboration and participation of stakeholders and it could enrich the debate of smart projects to improve smart cities. In addition, the shared knowledge could make smart cities to be developed in an efficient and effective way.

3.2.4 RQ3. Are There Differences Between Theoretical and Empirical Research About the Role of Governments in Smart Cities?

As noted previously in the cluster analysis, some differences between empirical and theoretical studies seem to exist. Whereas empirical studies seem to hold the Bureaucratic model as the most appropriate model of governance for smart cities' management, theoretical studies seem to advocate the involvement of stakeholders through the use of other more collaborative models.

In addition, in order to test statistical differences between empirical and theoretical studies, the Mann–Whitney test has been used. The Mann–Whitney test calculates a statistic that is a very useful measure of effect size, particularly suited to situations in which differences are measured on scales that either are ordinal or use arbitrary scale units (Conroy 2012). Also, nonparametric tests, including the Mann–Whitney test, are used when the measures are not normally distributed and serve to perform group comparisons (Lee and Park 2012). Besides, this test has been demonstrated to be a good test for informetric or scientometric analysis (Huber and Wagner-Döbler 2003). Thus, the Mann–Whitney test can compare two informetric or scientometric samples and calculate the probability from different distributions.

As the categories in each one of the dimensions of governance could be arrayed along a spectrum, as a continuum of top-down to bottom-up processes in accordance with the interaction between citizens and government and with how citizens are involved in the governance of the smart city, the Mann–Whitney test has been performed in order to identify differences regarding each one of the dimensions of governance in empirical versus theoretical studies.

Table 7 shows that there are significant differences between empirical and theoretical studies regarding the dimensions of conception of public interest, role of governments, and structure of governance—see Table 7 in appendix. No significant differences seem to exist in the dimensions of steering, boundary, alignment, and governance model, and the existence of differences in the dimension of dependency is not clear.

In this regard, empirical studies indicate that although stakeholders are involved in developing the smart city (see the governance model dimension), the main role is played by the local government—see Table 8 in appendix. Indeed, empirical experiences have been mainly characterized by the leading role of governments both in the

Table 7 Mann–Whitney test for group comparisons (empirical *versus* theoretical studies). (Source: Author's own elaboration)

Dimension	Mann–Whitney test	
	U Mann–Whitney	Significance
Steering	54.000	0.315
Boundary dimension	46.500	0.334
Alignment dimension	60.000	1.000
Dependency dimension	102.000	0.165
Conception of public interest	42.500	0.022**
Role of Government	86.000	0.098*
Structure of Governance	111.500	0.095*
Governance Model	137.000	0.227

*Significant at 10%; ** Significant at 5%; ***Significant at 1%

Table 8 Mapping attributes that characterize the role of governments in Smart Cities (empirical studies). (Source: Author's own elaboration)

Dimension\categories	1	2	3	4	Total
Steering	11	8	1	–	*20*
Boundaries	11	6	0	–	*17*
Alignment	9	5	1	–	*15*
Dependency	3	17	–	–	*20*
Conception of public interest	6	5	7	1	*19*
Role of local government	7	13	2	–	*22*
Structure of governance	7	9	8	1	*25*
Governance model	5	9	12	1	*27*

The number for each one of the attributes (dimensions) represents the categories shown in Table 2

Table 9 Mapping attributes that characterize the role of governments in Smart Cities (theoretical studies). (Source: Author's own elaboration)

Dimension\categories	1	2	3	4	Total
Steering	2	5	0		*7*
Boundaries	3	4	0		*7*
Alignment	5	2	1		*8*
Dependency	0	12			*12*
Conception of public interest	0	1	7	1	*9*
Role of local government	0	10	1		*11*
Structure of governance	0	6	6	1	*13*
Governance model	0	5	7	1	*13*

The number for each one of the attributes (dimensions) represents the categories shown in Table 2

planning and accountability of smart projects (steering, boundaries, and alignment dimensions) and in the process of management of these projects (role of government and structure of governance mainly). Therefore, these smart experiences could be catalogued as government-centric form of public management in smart cities.

By contrast, theoretical studies advocate more collaborative models of governance in smart cities—see Table 9 in appendix. In this regard, the main differences are focused on the conception of public interest, the role of local governments, and the structure of governance. In fact, theoretical studies indicate the need for engaging stakeholders in the design, performance, and monitoring of smart projects. Thus, "pull" and "networking" strategies are promoted by the literature.

Nonetheless, no clear participation patterns exist. In this milieu, some authors think that stakeholders should collaborate with local governments, but these governments must play a key role in the process. This view is prone to carry out "pull" strategies of governance in which the collaborative model of governance is told as the best for smart city development.

On the other hand, other authors think that stakeholders must be more active in the process of developing smart cities, promoting the effective participation of these stakeholders in the decision-making process and in the performance of smart projects into the city. The coproduction of smart projects is the most characteristic way of producing public value in the city. Under this framework, the local government acts as an actor joined with the rest of stakeholders in the smart city (with the same power as these stakeholders) and performs a "networking" strategy of governance.

4 Conclusions and Discussions

Previous studies of smart cities emphasize smartness of government, administration, and public management as core factors in the creation of a smart city (Nam and Pardo 2012). Indeed, Smart Governance principles could guide the relatively complex administrative enactment of smart and open government more intelligently than traditional static and inflexible governance approaches could do (Scholl and Scholl 2014). Thus, new ways of reengineering cities to make them smart, responsive, competitive, and equitable will require new forms of governance for an online world (Batty et al. 2012).

Despite the previous comments, up to now, little literature on smart cities addresses issues related to governance. It could be a reason why the concepts of Smart Governance and Smart Government have only been rudimentarily developed (Scholl and Scholl 2014). In fact, prior research has been mainly focused on the implementation of ICTs as a key tool to improve smart systems such as in the field of sustainable environment or smart mobility, but little research has been addressed to analyze the field of Smart Governance.

Only recent research has dealt with the issue of Smart Governance, focusing mainly on highlighting some select areas that have been put into focus and are likely candidates for smart governance initiatives (Scholl and Scholl 2014). Therefore, there is a need to redefine how the municipal government managed the smart cities and there is much room for improvement in this subject for future research.

Based on the network governance literature (Koiman 2003; Koiman et al. 2008; Provan and Kenis 2008; Fazekas and Burns 2012) and the coproduction literature (Linders 2012; Span et al. 2012; Mergel 2013), this chapter has set the principal dimensions of governance in order to identify them into the smart cities' literature. Each one of the dimensions has been characterized by categories, which have been defined in accordance with the increasing order of interrelationship between local governments and their stakeholders.

One important finding of this chapter is that empirical experiences are more prolific than theoretical studies regarding Smart Governance. It mainly means that the research on Smart Governance is descriptive in nature. Thus, future research should analyze governance models in smart cities from a theoretical point of view. It could lead to the creation of new theories in public sector reforms for promoting smart cities' development. This research could be based on prior research in the fields of network management, interactive decision-making, and coproduction theories, because all suggest that direct interaction between government agencies, social organizations, and individual citizens has become a necessity (Lowndes et al. 2001; McLaverty 2002).

Another finding is the leading role that local governments are told to play in the smart cities. In fact, despite that the principles of open, transparent, and collaborative government appear to be integral part of the smart city initiatives analyzed by prior research (Scholl and Scholl 2014), most of the sample papers point out that the local governments should design strategic planning in which smart projects should be included and monitored. This result is mainly obtained analyzing the empirical research on Smart Governance.

In this regard, empirical studies could be catalogued as government-centric form of public management in smart cities. This result could be explained because the organizing capacity of a city government is often viewed in terms of internal organization (i.e., commanding the right instruments and achieving adequate cohesion) but, under the smart city framework, it ought to be viewed much more in external or relational terms. Thus, under the smart city framework, public decision-making is not exclusively a task for government, which has led to a shift from government to the concept of governance (Kooiman 2003; Stoker 2003; Rhodes 1996).

This new perspective of governance highlights the importance of new approaches to enhancing governing capacity in which governments reach outwards and downwards to localities, engage with markets, and move out to civil society (Pierre and Peter 2000). This multilevel and multi-actor approach therefore relies more on collaboration, networking, and learning (Goodwin and Painter 1996; Gouldson et al. 2008; Mah and Hills 2012). Although this approach is very broad and encompasses a number of different lines of thought (Torfing et al. 2003), the underlying central assumption is that the classic hierarchical model of public administration does not work, and that a number of forms of "horizontal arrangements" and interrelationship with stakeholders have arisen in its place.

In this regard, our findings indicate that theoretical studies advocate more collaborative models of governance in smart cities. The main differences as for empirical studies are focused on the conception of public interest, the role of local governments, and the structure of governance. Under theoretical studies on Smart Governance, "pull" and "networking" strategies should be adopted by local governments in smart cities. These strategies are included in the collaborative and participatory models of governance, which have been told by theoretical studies as relevant for governing a smart city.

This difference between empirical and theoretical studies could be due to the time sequence of the studies published about smart cities. In this regard, data indicate that empirical studies have been published earlier than theoretical ones. It could mean that local governments have faced the development of the smart city before a deep analysis of this new "wave" of cities. Therefore, local governments have tried to control the development of these cities and they have played the key role in the management of the cities.

Theoretical studies have been published later and these studies have analyzed the phenomenon in order to propose efficient models of managing these cities once experiences have been produced. In the last years, with the advent of the new technologies, open government is proposed, and it is thought that the involvement of citizenry in the public sector management, claimed by the New Public Management philosophy, is relevant and possible. This could explain why more collaborative and participatory models of governance are proposed in theoretical studies than in the empirical ones.

In addition, recent research has indicated that the most committed cities to pursuing sustainability policies do tend to be more participatory (Portney and Berry 2010). In fact, recent empirical results suggest that effective, sustainable smart cities emerge as a result of dynamic processes in which public and private sector actors coordinate their activities and resources on an open innovation platform (Lee et al. 2013). Therefore, future research should analyze whether governance models in the smart cities are being moving over time to more collaborative and participatory models of governance, as theoretical studies seem to advocate.

In any case, the mayor's commitment will be a key aspect for developing smart cities (Ortiz-Fournier et al. 2010). In this sense, future research could analyze the role of the mayor in smart cities' development and the relationship between his/her profile and the governance model adopted in the smart city.

In conclusion, the implementation of new governance models under the smart city frameworks into the empirical area will need the understanding as well as the analysis of their relevance in the current networking environment produced by the development of these cities. First, government needs to use the capability of internet-based technologies to understand "smart society" more fully. Second, governments need to develop a more relational style of contracting and learn from the sharp variations in contracting regimes across the world. Third, leaders of governmental organizations need to accept that digital technologies are now at the core of a wide range of their activities and adapt policymaking processes and innovation accordingly (Margetts 2005). Their knowledge of and commitment to smart cities is recognized as key factor in the successful implementation of smart governance initiatives.

Acknowledgments This research was carried out with financial support from the Regional Government of Andalusia (Spain), Department of Innovation, Science and Enterprise (Research project number P11-SEJ-7700).

References

Aurigi, A. (2005). *Making the digital city: The early shaping of urban internet space*. Farnborough: Ashgate Publishing Company.

Bae, J., & Feiock, R. C. (2013). Forms of government and climate change policies in U.S. cities. *Urban Studies, 50*(4), 776–788.

Barbry, E. (2012). The internet of things, legal aspects. What will change (everything). *Communications and Strategy, 87*(3), 83–100.

Bătăgan, L. (2011). Smart cities and sustainability models. *Informatica Economică, 15*(3), 80–87.

Batty, M., Axhausen, K. W., Giannotti, F., Pozdnoukhov, A., Bazzani, A., Wachowicz, M., Ouzounis, G., Portugali, Y., et al. (2012). Smart cities of the future. *The European Physical Journal Special Topics, 214*(1), 481–518.

Behan, K., Maoh, H., & Kanaroglou, P. (2008). Smart growth strategies, transportation and urban sprawl: Simulated futures for Hamilton, Ontario. *The Canadian Geographer, 52*(3), 291–308.

Berg, A., & Hukkinen, J. (2011). The paradox of growth critique: Narrative analysis of the Finnish sustainable consumption and production debate. *Ecological Economics, 72,* 151–160.

Caragliu, A., Del Bo, C., & Nijkamp, P. (2011). Smart cities in Europe. *Journal of Urban Technology, 18*(2), 65–82.

Coe, A., Paquet, G., & Roy, J. (2001). E-governance and smart communities: A social learning challenge. *Social Science Computer Review, 19*(1), 80–93.

Conroy, R. M. (2012). What hypotheses do "nonparametric" two-group tests actually test? *Stata Journal, 12*(2), 182–190.

Considine, M., & Lewis, J. (1999). Governance at ground level: The front-line bureaucrat in the age of markets and networks. *Public Administration Review, 59*(6), 467–480.

Crabtree, B., & Miller, W. (Eds.). (1999). *Doing qualitative research*. London: Sage.

D'Alessandro, S., Luzzati, T., & Morroni, M. (2010). Energy transition towards economic and environmental sustainability: Feasible paths and policy implications. *Journal of Cleaner Production 18*(4), 291–298.

Daneshpour, A., & Shakibamanesh, A. (2011). Compact city; dose it create an obligatory context for urban sustainability? *International Journal of Architectural Engineering and Urban Planning, 21*(2), 110–118.

Deakin, M. (2012). Intelligent cities as smart providers: CoPs as organizations for developing integrated models of eGovernment services. *Innovation: The European Journal of Social Science Research, 25*(2), 115–135.

Denhardt, R. B., & Denhardt, J. V. (2000). The new public service: Serving rather than steering. *Public Administration Review, 60*(6), 549–559.
Dey, I. (1993). *Qualitative data analysis: A user-friendly guide for social scientists*. London: Routledge.
Farris, J. T. (2001). The barriers to using urban infill development to achieve smart growth. *Housing Policy Debate, 12*(1), 1–30.
Fazekas, M., & Burns, T. (2012). *Exploring the complex interaction between governance and knowledge in education*. OECD Education Working Papers, No. 67. OECD Publishing. http://dx.doi.org/10.1787/5k9flcx2l340-en.
Ferro, E., Caroleo, B., Leo, M., Osella, M., & Pautasso, E. (2013). The role of ICT in smart cities governance. *Proceedings to the International Conference for E-Democracy and Open Government*, Krems, Austria, May 22–25, 2013.
Flipo F., & Schneider, F. (2008). *Proceedings of the first conference for ecological sustainability and social equity*. Paris: Research & Degrowth, Telecom Sud-Paris.
Fotopoulos, T. (1997). *Towards an inclusive democracy: The crisis of the growth economy and the need for a new liberatory project*. London: Cassell.
Giffinger, R., Fertner, C., Kramar, H., Meijers, E., & Pichler-Milanović, N. (2007). Smart cities: Ranking of European medium-sized cities. Vienna. http://www.smart-cities.eu/download/smart_cities_final_report.pdf.
Goodwin, M., & Painter, J. (1996). Local governance, the crises of Fordism and the changing geographies of regulation. *Transactions of the Institute of British Geographers, 21*(4), 635–648.
Gouldson, A., Hills, P., & Welford, R. (2008). Ecological modernisation and policy learning in Hong Kong. *Geoforum, 39*(1), 319–330.
Hawkins, C. V. (2011). Smart growth policy choice: A resource dependency and local governance explanation. *Policy Studies Journal, 39*(4), 679–707.
Hollands, R. G. (2008). Will the real smart city please stand up. *City, 12*(3), 303–320.
Huber, J. C., & Wagner-Döbler, R. (2003). Using the Mann–Whitney test on informetric data. *Journal of the American Society for Information Science and Technology, 54*(8), 798–801.
Hueting, R. (1980). *New scarcity and economic growth: More welfare through less production?* Amsterdam: North-Holland.
Innes, J. E., & Booher, D. E. (2010). *Planning with complexity: An introduction to collaborative rationality for public policy*. London: Routledge.
Kettl, D. F. (2006). Managing boundaries in American administration: The collaboration imperative. *Public Administration Review, 66*(s1), 10–19.
King, N. (2004). Using templates in the thematic analysis of text. In C. Cassell & G. Symon (Eds.), *Essential guide to qualitative methods in organizational research* (pp. 256–270). London: Sage.
King, N. (2014). Template analysis [www page]. School of Human and Health Sciences, University of Huddersfield Web. http://hhs.hud.ac.uk/w2/research/template_analysis/. Accessed 9 March 2014.
Komninos, N. (2002). *Intelligent cities*. London: Spon.
Kooiman, J. (2003). *Governing as governance*. London: Sage.
Kooiman, J., Banvick, M., Chuenpadgee, R., Mahon, R., & Pullin, R. (2008). Interactive governance and governability: An introduction. *The Journal of Transdisciplinary Environmental Studies, 7*(1), 1–11.
Koppenjan, J., & Klijn, E.-H. (2004). *Managing uncertainties in networks*. London: Routledge.
Kourtit, K., & Nijkamp, P. (2012). Smart cities in the innovation age. *Innovation: The European Journal of Social Science Research, 25*(2), 93–95.
Kourtit, K., Nijkamp, P., & Arribas, D. (2012). Smart cities in perspective—A comparative European study by means of self-organizing maps. *Innovation: The European Journal of Social Science Research, 25*(2), 229–246.
Lee, M., & Park, H. W. (2012). Exploring the web visibility of world-class universities. *Scientometrics, 90*(1), 201–218.

Lee, J. H., Hancockb, M. G., & Hu, M.-G. (2013). Towards an effective framework for building smart cities: Lessons from Seoul and San Francisco. *Technological Forecasting and Social Change*. In press, corrected proof. http://www.sciencedirect.com/science/article/pii/S0040162513002187. Accessed Jan 2014.

Leydesdorff, L., & Deakin, M. (2011). The triple-helix model of smart cities: A neo-evolutionary perspective. *Journal of Urban Technology, 18*(2), 53–63.

Linders, D. (2012). From e-government to we-government: Defining a typology for citizen coproduction in the age of social media. *Government Information Quarterly, 29*(4), 446–454.

Lowndes, V., Pratchett, L., & Stoker, G. (2001). Trends in public participation: Part 1—local government perspectives. *Public Administration, 79*(1), 205–222.

Mah, D. N.-Y., & Hills, P. (2012). Collaborative governance for sustainable development: Wind resource assessment in Xinjiang and Guangdong provinces, China. *Sustainable Development, 20*(2), 85–97.

Margetts, H. (2005). Smartening up to risk in electronic government. *Information Polity, 10*(1–2), 81–94.

Martinez-Alier, J. (2009). Socially sustainable economic de-growth 2009. *Development and Change, 40*(6), 1099–1119.

McLaverty, P. (Ed.). (2002). *Public participation and innovations in community governance*. Aldershot: Ashgate.

Meijer, A. J., & Rodríguez Bolívar, M. P. (2013). *Governing the smart city: Scaling-up the search for socio-techno synergy*. Paper presented at EGPA Conference 2013, Edinburgh, Scotland.

Mergel, I. (2013). Social media adoption and resulting tactics in the US federal government. *Government Information Quarterly, 30*(2), 123–130.

Mohamed, R. (2008). Who would pay for rural open space preservation and inner-city redevelopment? Identifying support for policies that can contribute to regional land use governance. *Urban Studies, 45*(13), 2783–2803.

Nam, T., & Pardo, T. A. (2011). Smart city as urban innovation: Focusing on management, policy, and context. *Proceedings of the 5th international conference on theory and practice of electronic governance*, Tallinn, Estonia, 26–28 September 2011. http://dl.acm.org/citation.cfm?id=2072100.

Nam, T., & Pardo, T. A. (2012). Transforming city government: A case study of Philly311. *Proceedings of the 6th international conference on theory and practice of electronic governance, ICEGOV 2012*. Albany, NY, United States, 22–25 October, 2012, pp. 310–319.

Neuman, M. (2005). The compact city fallacy. *Journal of Planning Education and Research, 25*(1), 11–26.

Odendaal, N. (2003). Information and communication technology and local governance: Understanding the difference between cities in developed and emerging economies. *Computers, Environment and Urban Systems, 27*(6), 585–607.

Odum, H. T., & Odum, E. C. (2001). *The prosperous way down*. Boulder: University Press of Colorado.

Ortiz-Fournier, L. V., Márquez, V., Flores, F. R., Rivera-Vázquez, J. C., Colón, P. A., et al. (2010). Integrating educational institutions to produce intellectual capital for sustainability in Caguas, Puerto Rico. *Knowledge Management Research and Practice, 8*, 203–215.

Peters, B. G., & Pierre, J. (1998). Governance without government? Rethinking public administration. *Journal of Public Administration Research and Theory, 8*(2), 223–243.

Pierre, J., & Peter, G. (2000). *Governance, politics and the state*. Macmillan Press: Basingstoke, Hampshire.

Portney, K. E., & Berry, J. M. (2010). Participation and the pursuit of sustainability in US cities. *Urban Affairs Review, 46*(1), 119–139.

Provan, K. G., & Kenis, P. (2008). Modes of network governance: Structure, management, and effectiveness. *Journal of Public Administration and Research Theory, 18*(2), 229–252.

Puppim de Oliveira, J. A., Doll, C. N. H., Balaban, O., Jiang, P., Dreyfus, M., et al. (2013). Green economy and governance in cities: Assessing good governance in key urban economic processes. *Journal of Cleaner Production, 58*, 138–152.

Rapkin, B. D., & Luke, D. A. (1993). Cluster analysis in community research: Epistemology and practice. *American Journal of Community Psychology, 21*(2), 247–277.
Rhodes, R. A. W. (1996). The new governance: Governing without government. *Political Studies, 44*(4), 652–667.
Rohman, I. K. (2012). Will telecommunications development improve the quality of life in African countries? *Info, 14*(4), 36–51.
Sauer, S. C. (2012). Do smart cities produce smart entrepreneurs? *Journal of Theoretical and Applied Electronic Commerce Research, 7*(3), 63–73.
Schneider, F., Kallis, G., & Martinez-Alier, J. (2010). Crisis or opportunity? Economic degrowth for social equity and ecological sustainability. Introduction to this special issue. *Journal of Cleaner Production, 18*(6), 511–518.
Scholl, H. (2009). Profiling the EG research community and its core. In M. A. Wimmer, H. J. Scholl, M. Janssen, & R. Traunmüller (Eds.), *Proceedings of 8th international conference on electronic government, EGOV 2009* (pp. 1–12). Berlin: Springer LNCS 5693.
Scholl, H. J., & Scholl, M. C. (2014). Smart governance: A roadmap for research and practice. *iConference 2014 proceedings*. Berlin, Germany, 4–7 March 2014, pp. 163–176.
Schuurman, D., Baccarne, B., De Marez, L., & Mechant, P. (2012). Smart ideas for smart cities: Investigating crowdsourcing for generating and selecting ideas for ICT innovation in a city context. *Journal of Theoretical and Applied Electronic Commerce Research*, *7*(3), 49–62.
Scott, J. W. (2007). Smart growth as urban reform: A pragmatic 'recoding' of the new regionalism. *Urban Studies, 44*(1), 15–35.
Sheskin, D. J. (2007). *Handbook of parametric and nonparametric statistical procedures*. Boca Raton: Chapman & Hall/CRC, Taylor & Francis Group LLC.
Snedecor, G. W., & Cochran, W. G. (1981). *Statistical methods*. Iowa: The Iowa State University Press.
Span, K. C. L., Luijkx, K. G., Schols, J. M. G. A., & Schalk, R. (2012). The relationship between governance roles and performance in local public interorganizational networks: A conceptual analysis. *American Review of Public Administration, 42*(2), 186–201.
Spangenberg, J. H. (2010). The growth discourse, growth policy and sustainable development: Two thought experiments. *Journal of Cleaner Production, 18*(6), 561–566.
Sprent, P., & Smeeton, N. C. (2007). *Applied nonparametric statistical methods*. Boca Raton: Chapman & Hall/CRC, Taylor & Francis Group LLC.
Stoker, G. (1998). Governance as theory: Five propositions. *International Social Science Journal, 50*(155), 17–28.
Stoker, G., Smith, G., Malony, W., & Young, S. (2003). Voluntary organisations and the generation of social capital. In M. Boddy & M. Parkinson (Eds.), *City matters. Competitiveness, cohesion and urban governance* (pp. 389–407). Bristol: The Policy Press.
Torfing, J., Sorensen, E., & Christensen, L. P. (2003). *Nine competing definitions of governance, governance networks and meta-governance*. Denmark: Roskilde University.
Torfing, J. B., Peters, G., Pierre, J., & Sörensen, E. (2012). *Interactive governance: Advancing the paradigm*. Oxford: Oxford University Press.
Tranos, E., & Gertner, D. (2012). Smart networked cities? *Innovation: The European Journal of Social Science Research, 25*(2), 175–190.
Walravens, N. (2012). Mobile business and the smart city: Developing a business model framework to include public design parameters for mobile city services. *Journal of Theoretical and Applied Electronic Commerce Research, 7*(3), 121–135.
Walravens, N., & Ballon, P. (2013). Platform business models for smart cities: From control and value to governance and public value. *IEEE Communications Magazine, 51*(6), 72–79.
Webster, J., & Watson, R. T. (2002). Analyzing the past to prepare for the future: Writing a literature review (Editorial). *MIS Quarterly, 26*(2), xiii–xxiii.

Manuel Pedro Rodríguez Bolívar is an associated professor in accounting at the University of Granada. He has authored numerous articles in international journals, among them we can highlight *Public Money & Management, Government Information Quarterly, Public Administration and Development, Online Information Review, International Review of Administrative Sciences, American Review of Public Administration, ABACUS, Academia. Revista Latinoamericana de Administración, International Public Management Journal, Environmental Education Research, INNOVAR, Transylvanian Review of Administrative Sciences, Electronic Commerce Research* and *Administration & Society*. He has been also the author of several book chapters published in prestigious international publishers such as Kluwer Academic Publishers, Springer, Taylor and Francis and IGI Global, and is author of full-length books published by the Ministry of Economy and Finance in Spain. In addition he is editor of international journals and books.

Smart City Governance: A Local Emergent Perspective

Albert Meijer

Abstract This chapter presents a local emergent perspective on smart city governance. Smart city governance is about using new technologies to develop innovative governance arrangements. Cities all around the world are struggling to find smart solutions to wicked problems and they hope to learn from successful techno-governance practices in other cities. Learning about successes of smart city governance is important but lessons need to be contextualized: approaches that work in one city may fail in another one. This chapter presents the local cooperative knowledge potential and the nature of the problem domain as key contextual factors and develops a model for studying and assessing smart city governance in context.

Keywords Smart cities · Contextual approach · Multidimensional evaluation framework

1 Introduction

Cities are increasingly seen as not only the engines of innovation and economic growth but also the level at which solutions to wicked problems need to be produced (Inayatullah 2011; Nijkamp and Kourtit 2013). City governments need to produce adequate and innovative approaches for such diverse issues as sustainable growth, social inclusion, and crime control and prevention (Florida 2002; Landry 2006; Barber 2013). In this context, the current wave of attention for "smart cities" is hardly surprising: the expectations of urban systems are exceedingly high and therefore new and innovative forms of governance are needed to meet these challenges (Caragliu and Del Bo 2012). The key question for urban government is: how can we make cities so smart that they can generate economic growth and also produce sustainability, inclusiveness, and safety?

Technological innovations can help city governments to meet these challenges of urban governance and to improve urban environments (Yigitcanlar et al. 2008; Walravens 2012; Hoon Lee et al. 2013; Washburn et al. 2010). New technological

A. Meijer (✉)
School of Governance, Utrecht University, Bijlhouwerstraat 6, 3511 ZC, Utrecht, Netherlands
e-mail: a.j.meijer@uu.nl

J. R. Gil-Garcia et al. (eds.), *Smarter as the New Urban Agenda A Comprehensive View of the 21st Century City,* Public Administration and Information Technology 11,
DOI 10.1007/978-3-319-17620-8_4

developments increase the intelligence of urban systems by supporting innovative solutions to traffic control, energy production, crime monitoring, etc.: city governments can develop solutions to wicked problems by making good use of the available technologies. Smart city governance is defined as using new technologies to improve urban governance through better use of information and better communications. At the moment, we are witnessing two distinctive waves of technological innovation that connect to different forms of smart city governance (see also Nijkamp and Kourtit 2013, p. 308): technologies for concentrated intelligence and technologies for distributed intelligence.

The rapid development of technologies can strengthen the concentrated intelligence of the government by providing better, more comprehensive, and up-to-date information about relevant developments (Kuk and Janssen 2011). Traffic monitoring systems, for example, with cameras and sensors can provide city governments with precise information about traffic streams and congestions, and these governments can use this information for traffic management (Hoh et al. 2008). In addition, technologies for communication in policy networks and communities can bring together a wide variety of urban actors to generate more distributed urban intelligence (Lathrop and Ruma 2010; Linders 2012; Nijkamp and Kourtit 2013). Open data and social media facilitate new forms of collaborative governance by, for example, combining information about crime patterns from various sources and coordinating preventative efforts of citizens, housing corporations, and the police. These two technological waves are combined in various ways to produce hybrid smart city governance in the form of a rich array of technological options for smart city governance. These hybrid uses of technologies are often presented as promising venues for strengthening city governance (Caragliu and Del Bo 2012), but little is known about the effectiveness of these new forms of governance.

These new technologies are impressive and "sexy," but that does not necessarily mean they are effective. Assessing the effectiveness of smart city governance is complicated since there is no simple indicator for success in the public sector such as profitability in the private sector. While evaluation of technologies tends to assess success in terms of diffusion and adoption of innovation (Rogers 1995), urban governance and planning studies highlight the contribution to the quality of the urban environment, both in terms of the outcomes and the process of realizing these outcomes. Smart city governance may not use the most advanced new technologies and still be qualified as successful for providing better outcomes (in terms of economic growth, more sustainability, more safety, etc.) and also for providing a better process (in terms of the speed of decision-making process, a reduction in the number of conflicts, etc.). The overall objective of smart city governance is not to make use of new technologies but to contribute to the objective and subjective quality of the urban environment through new technologies. This means that the contribution of smart city governance to the urban environment needs to be assessed through a combination of community, network, and participant criteria.

This chapter develops a local emergent perspective on smart city governance. The scientific basis for the guidelines consists of theories of technology in governance and theories of urban planning and governance. This chapter puts an emphasis on (1) the contextual nature of smart city governance, (2) the important role of the degree to which local actors are willing and able to cooperate in smart city governance, and (3) the feedback loops that strengthen or undermine smart city efforts. The review of the literature results in a list of specific expectations concerning smart city governance and an agenda for further research.

2 Contextual Model for Smart City Governance

Studying the effects of smart city governance is complicated since the relations between governance arrangements, use of technologies, and effects on the quality of urban life are contextual. An approach that yields fantastic results in one city may fail in another. The situational nature of these relations creates a dilemma: highlighting the unique nature of each situation results in an in-depth understanding of interrelations between various factors but does not produce any general knowledge that would be useful for other situations. To deal with this dilemma, this chapter identifies two key situational variables that help to understand specific contexts while, at the same time, generating more general knowledge about the effectiveness of smart city governance.

The first contextual variable that is included in this contextual perspective on smart city governance is the local cooperative knowledge potential. The literature highlights that the success of smart city governance depends on the match of new technologies to the knowledge and attitude of the relevant actors: a solution that works in one city may not work in another one for a lack of citizens and other stakeholders that are capable and willing to cooperate with the city government. The local cooperative knowledge potential refers to the availability of relevant knowledge among citizens and stakeholders, and the willingness to contribute this knowledge to collective problem-solving. Citizens and shopkeepers may, for example, be prepared to use their smartphones to provide the police with up-to-date information about the local safety situation in one neighborhood, whereas in another neighborhood, they may see the police as an institution that cannot be trusted and should not be collaborated with (Meijer and Thaens 2013).

The second contextual variable is the nature of the problem domain. Situational characteristics, such as democratic institutions and culture, the physical environment, the economic production, etc., matter for the effectiveness of smart city governance since these characteristics are either conducive or limiting to different modes of smart city governance. These situational characteristics interact with a series of political, administrative, and technological choices regarding the use of new technologies for urban governance. This means that an in-depth analysis of the smart solutions in their (political, institutional, societal, economic, and cultural) context is needed to assess the value of certain successful smart city governance approaches for other cities.

Including these two contextual variables, we present the following model for studying smart city governance. We will briefly explain the key features of this

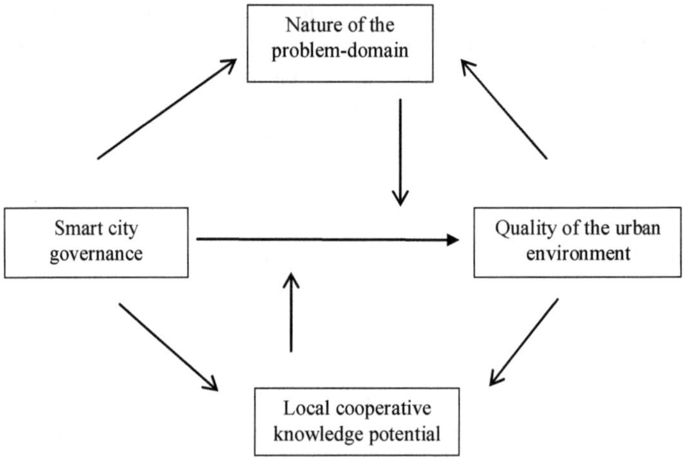

Fig. 1 Contextual model for smart city governance

model and in the remainder of this chapter, we will discuss the various relations more in depth (Fig. 1).

This model highlights the following relations:

- Smart city governance—in different forms—can contribute to the quality of the urban environment by making better use of available resources and producing smarter solutions to problems;
- The effect of different forms of smart city governance on the urban environment depends on the availability of local cooperative knowledge potential and on the question of whether the problem-domain is conducive or limiting to the chosen approach;
- Smart city governance can improve the urban environment not only directly but also indirectly by influencing the nature of the problem-domain and strengthening the local cooperative knowledge potential;
- Improvements in the urban environment can have a feedback effect on the problem-domain and the local cooperative knowledge potential since successes are likely to strengthen the local cooperative knowledge potential and render the problem-domain more conducive.

The model presents an idea of the relations between modes of smart city governance, nature of the problem-domain, local cooperative knowledge potential, and quality of the urban environment. The model identifies various feedback loops and therefore the outcome is nonlinear. The literature highlights mostly positive outcomes, but the moderations and feedback loops can also result in a negative spiral with perverse effects. To position these variables and relations, we will conceptualize them on the basis of the literature from technology studies, governance, and urban planning and we will form expectations regarding the relations between them.

3 Smart City Governance: Concentrated, Distributed, and Hybrid

Urban scholars such as Nijkamp and Kourtit (2013, p. 299) emphasize that "[t]he city is a social fabric based on interaction, participation, and collective responsibility." Our perspective on smart city governance will study the interactions between the variety of actors. The building blocks for a theoretical perspective on smart city governance are theories on technology in (public) organizations (Zuboff 1988; Orlikowski 1992; Fountain 2001; Meijer 2009; Gil-Garcia 2012) and theories on urban governance (Stone 1993; Pierre 1999, 2011; Nijkamp and Kourtit 2013). These theories are used to identify three modes of smart city governance: concentrated, distributed, and hybrid intelligence.

The idea of smart city governance as *concentrated intelligence* stresses that new technologies—big data, data warehousing, monitoring tools—enable central steering actors to strengthen their intelligence, provide more integrated services, develop better policies, and steer other actors in the city more effectively (Leydesdorff and Deakin 2011; Kuk and Janssen 2011). The promise of concentrated intelligence builds upon the foundational work of the Carnegie-Mellon school in organization studies that highlighted that the quality of decision-making in an organization depends on information management (e.g., Galbraith 1973). More recently, Mayer-Schönberger and Cukier (2013) have stressed that the "datafication" and big data will provide invaluable insights that city managers would otherwise not have. These managers can use new technologies such as big data and ubiquitous sensors—referred to as the "Internet of Everything"—to make informed decisions about crime control, traffic management, energy production, etc. Concentrated intelligence may be limited to government but can also take the form of a collaboration between government and a limited number of private partners (i.e., a public–private partnership) or even only a private party when, for example, a company is commissioned to provide urban transport (cf. Driessen et al. 2012).

The idea of *distributed intelligence* highlights that new technologies—social media, Internet, open data—enable the various actors in the city to collaborate more effectively and produce better solutions for the city (Yigitcanlar et al. 2008; Hoon Lee et al. 2013). Theories on collaborative, networked, and coproductive governance highlight that the quality of policies can be strengthened by managing good relations between all stakeholders and tapping into their intelligence. Collaborative learning is at the heart of this approach and (virtual) communities can strengthen the intelligence of the city (Agranoff and McGuire 2003; Koppenjan and Klijn 2004; Torfing et al. 2012). This mode of smart city governance can vary from forms in which city government is still (heavily) involved to self-governance arrangements where private sector, civil society groups, and "social entrepreneurs" are engaged in public problem-solving without any government involvement (Light 2008; Driessen et al. 2012).

The two modes of smart city governance are ideal types and should be seen as extremes on a scale of smart city governance. Intermediate forms are modes of *hybrid smart city governance*. Hybrids may lean towards one of the extremes or form a balanced combination of concentrated and distributed forms of governance. One should note that these configurations are not caused by these technologies but result from (political) choices to focus on certain technological features to attain certain ends: they are emerging modes of governance (Fountain 2001). The mode of smart city governance reflects political choices since they represent different views on the relation between government and society (Koppenjan and Klijn 2004; Pierre 2011). The concentrated intelligence perspective builds upon the idea of delegation of power to the government and accountability through formal mechanisms to the people's representatives or, alternatively, on the idea that government should involve large companies in public–private partnerships. Citizens can choose their democratic representatives but they are subjects of government that can be scrutinized with cameras and other information technology. The distributed intelligence perspective takes direct citizen engagement as its starting point and stresses that citizens—like other stakeholders—are coproducers (Alford 2009). Accountability takes place through more informal and more direct mechanisms in networks of stakeholders (Michels and Meijer 2008). The two modes reflect political choices but, still, they can be assessed in terms of their contributions to the quality of the urban environment.

The modes of smart city governance need to be connected to human capital, (open) innovation, and common pool resources to produce public value (Schaffers et al. 2011). Political and administrative choices and dynamics determine what kind of smart city governance is chosen and developed. The promise of hybrid intelligence holds that cities can find ways to combine concentrated and distributed intelligence but, alternatively, concentrated and distributed intelligence may conflict. City governments can choose to use concentrated and distributed smart city governance in separate domains, they can also choose to use concentrated and distributed smart city governance in the same domain but in different ways, or they can integrate both types of smart city governance in one approach. Little to nothing is known about the effectiveness of these different forms of smart city governance in varying contexts.

4 Quality of the Urban Environment: Community, Network, and Participant Assessments

The public management literature has been struggling to find ways of evaluating collaborative governance. Some authors stress that the performance of collaborative governance should be evaluated in terms of impacts on the economy, mobility, environment, people, and living conditions (Lazariou and Roscia 2012; Winters 2011). Others disagree and emphasize that the basic characteristic of collaborative governance is that participating actors have different objectives. Citizens may regard a project for neighborhood improvement, primarily as a way to improve their natural environment while housing corporations focus on the attractiveness of their property for tenants while the police may highlight the decline of crime in the area. This diversity in objectives means, according to these authors, that the success of collaborative governance can only be assessed in terms of stakeholder satisfaction (Koppenjan and Klijn 2004, p. 124).

The emerging consensus is that a combination of evaluation criteria is needed for a comprehensive assessment (Radaelli and De Francesco 2010; Koppenjan and Klijn 2004; Provan and Milward 2001; Sørensen and Torfing 2009). A sophisticated approach to the evaluation of collaborative governance has been developed by Provan and Milward (2001) that brings together the two other approaches and adds an intermediate level. They highlight that assessing collaborative governance means measuring the effectiveness at the level of the participants (i.e., stakeholder satisfaction), the level of the network, and the level of the community (i.e., the overall impacts on economy, mobility, environment, etc.). This approach highlights that collaborative governance needs to be evaluated both from the perspective of those involved and from the perspective of external stakeholder groups that are confronted with the outcomes of collaborative governance in terms of contributions to the urban environment.

It is important to broaden the assessment to include not only intended effects but also establish side or even perverse effects in terms of, for example, infringements on privacy. The field of surveillance studies highlights that new technologies may turn cities into panopticons where everybody, presumably under the idea that this contributes to urban safety, is always being watched (Lyon 2001). Alternatively, use of new technologies may result in alienation of actors and even new hostilities between certain actors, for example, in relation to traffic management where certain traffic patterns may have an overall positive effect but still a negative effect but specific actors.

Provan and Milward's (2001) approach can be used as a basis for developing a multidimensional approach to assessing the effectiveness of smart city governance. We have used this basis to structure a variety of criteria mentioned either in the literature on smart cities or on collaborative governance. This results in a multidimensional framework for assessing smart city governance (see Table 1).

Applying this framework requires a combination of analyzing data at an urban level about the economic situation, mobility, environment, etc. and collecting perceptual data from network participants through surveys or more qualitative research methods. The results provide rich insights in the objective and subjective effectiveness of smart city governance in different urban contexts.

5 Local Cooperative Knowledge Potential: High or Low

When will smart city governance actually contribute to the quality of the urban environment? Distributed smart city governance taps into local problem-relevant potential, whereas concentrated smart city governance largely ignores or even denies local potential. The presence and availability of local problem-relevant potential re-

Table 1 Multidimensional assessment of smart city governance

Community criteria	Network criteria	Participant criteria
Based on Lazariou and Roscia (2012), Hoon Lee et al. (2013), and Nijkamp and Kourtit (2013)	Based on Provan and Milward (2001)	Based on Koppenjan and Klijn (2004), Sørensen and Torfing (2009), and Meijer et al. (2013)
Economy	*Network constitution*	*Satisfaction with process*
Productivity Gross city product	Strength of relational contacts between actors Diversity of actors involved	Perceived timeliness Perceived openness
Mobility	*Network performance*	*Satisfaction with outcome*
Local accessibility ICT Infrastructure	Range of services provided Integration and coordination of services Absence of service duplication	Perceived desired effects Perceived unintended effects
Environment	*Network cost*	*Learning*
Attractiveness of natural conditions Pollution	Cost of network buildings Cost of network maintenance	Learning about the issue Learning about the actors Learning about institutional context
People	*Unintended or perverse effects*	*Unintended or perverse effects*
Level of education Social and ethnic plurality Participation in public life	Network conflicts Exclusion from dominant networks	Negative image of other actors Unwillingness to collaborate with others
Living		
Cultural facilities Individual safety Education facilities		
Government services		
Speed Accessibility Comprehensiveness		
Unintended or perverse effects		
Privacy Unequal division of benefits		

sults from characteristics of local actors (such as degree of organization, knowledge about the problem, opportunities to collect information, capabilities to contribute to the solution, willingness to engage) and the nature of the policy domain (in terms of capacities and knowledge needed to contribute to a solution). The level of problem-relevant local potential originates from a variety of (political, institutional, societal, economic, and cultural) factors such as legal position of citizens, level of trust in government, level of education, social cohesion, information skills, etc. This means that context does not only differ between cities but also between neighborhoods in cities and, in addition, there are differences between policy areas.

The theoretical expectation is that concentrated intelligence is more effective in a situation of low local problem-relevant potential. An example could be the production of high-tech innovations, such as pharmaceuticals, new materials, and health-care technologies, for urban economic growth. The level of local problem-relevant potential on this issue is generally low since it requires a high level of expertise and formal training and concentrated smart city governance in the form of a so-called triple helix, a collaboration between government, business, and universities, may be expected to be a more effective form of smart city governance than including a broad variety of citizens, NGOs, local groups, etc. (Etzkowitz and Leydesdorff 1997).

Distributed intelligence can theoretically be expected to be more effective in a situation of high local problem-relevant potential. An example could be the maintenance of green facilities. The knowledge of citizens in certain neighborhoods of gardening can be expected to be fairly high and their willingness to allocate time on urban gardening is often also considerable (Schulz et al. 2013). In such a situation, distributed smart city governance using open data and social media can tap into the potential of these actors to produce effective forms of urban green maintenance. Local problem-relevant potential does not always have to align with government objectives: for example, high problem-relevant potential concerning speed controls may be used to undermine the effectiveness of government traffic safety policies.

In theory, combinations of concentrated and distributed intelligence need to fit the level, nature, and form of problem-relevant potential to be effective. An example could be the situation of fighting organized crime in a neighborhood. Local potential in the form of information about illegal prostitution and drug sales can be tapped into through distributed smart city governance, but needs to be supplemented with concentrated intelligence on crime histories of certain persons and criminal networks. The literature highlights that there is a feedback loop between the successes of interactive governance and local problem-relevant potential (Sørensen and Torfing 2009): If citizens see that their efforts result in better solutions to societal problems, they may learn about these issues and be more prepared to contribute to the solution.

These conditions can—in varying degrees—be influenced through various (smart) policy interventions. There is an important difference between the two modes of smart city governance here: whereas distributed smart city governance aims to strengthen the local problem-relevant potential, concentrated smart city governance ignores or even undermines the local potential. An example of these effects on the local cooperative potential could be the different approaches to fighting crime: use of cameras and data surveillance strengthens the police but disempowers citizens while social media networks between community police officers and citizens help to engage citizens in the production of local security.

6 Nature of the Problem-Domain: Conducive or Limiting

The effectiveness of smart city governance for improving the urban environment does not depend only on the local cooperative knowledge potential but also on a variety of external political, institutional, societal, economic, cultural, and physical conditions (cf. Torfing et al. 2012). Altogether, these conditions constitute the nature of the problem-domain in terms of support for or limitations to certain approaches to improving the urban environment. These conditions can either be conducive or limiting to the effect of different modes of smart city governance on the urban environment. Some examples may illustrate the argument:

- Legal frameworks may support open government data practices that can stimulate collective learning around problem-areas, but existing laws may also limit the possibilities (Mayer-Schönberger and Cukier 2013).
- Informal rules for collaboration between a variety of urban actors may help to build learning networks, but they may also result in the exclusion of certain others that could potentially contribute to new solutions (Koppenjan and Klijn 2004).
- The physical conditions of a city in terms of climate and the presence of rivers and hills may be conducive to certain approaches to traffic management or sustainability, but they also limit the effectiveness of these approaches.

These various issues all highlight that the nature of the problem-area can be conducive—or limiting—to the effect of either concentrated or distributed smart city governance on the urban environment. This means that cities cannot simply copy approaches from other cities: they need to explore to what extent and how successes in other cities can be translated to their own city. Differences in legal frameworks, informal rules, and physical conditions matter. This requires an in-depth understanding of the relevance of the various aspects of the problem-domain to compare practices and to learn from successful forms of smart city governance.

While certain aspects of the problem-domain are a given—e.g., the climate—or only change in the long term—e.g., urban culture—other aspects can be influenced by smart city governance. Certain forms of legislations and informal rules can be influenced by smart city governance to render them more conducive to the relation between smart city governance and the quality of the urban environment. Changes in legislation concerning open data, for example, can contribute to the success of distributed smart city governance (Mayer-Schönberger and Cukier 2013).

7 Conclusions and Expectations

This chapter has presented a contextual model of smart city governance and highlights that effects of certain techno-governance arrangements depend on situational factors such as the local cooperative knowledge potential and the nature of the problem-domain. The discussion of the key variables can now be used to formulate more specific expectations concerning smart city governance:

1. Concentrated smart city governance fits a situation where the local cooperative knowledge potential is low. This type of governance can result in improvements of the urban environment at the community level.

2. Distributed smart city governance fits situations where the local potential is high. This type of governance can be expected to contribute to the quality of the urban environment as evaluated from the community, network, and participant level.
3. More distributed smart city governance will strengthen the local cooperative knowledge potential more than concentrated smart city governance. Concentrated smart city governance may even conflict with the local cooperative knowledge potential.
4. The local problem-relevant potential will be strengthened if distributed smart city governance results in improvements of the urban environment as evaluated from the network and participant level.
5. The nature of the problem-area matters. This means that successful forms of smart city governance in one city cannot be copied directly to another city. The role of contextual factors needs to be analyzed and successes need to be translated rather than copied.
6. The problem-area may become more conducive to certain forms of smart city governance if these produce successes at the community level. Successes may help to transform legal frameworks and informal rules for collaboration.

These expectations can form the starting point for local, emergent research and they need to be explored and developed further on the basis of empirical studies.

The contribution of this chapter to the literature on smart city governance is that the effectiveness of techno-governance arrangements depends on situational factors. This does not mean we cannot do comparative research nor does it mean that cities cannot learn from successful practices in other countries. We do, however, need to be careful in our focus on "best practices" of smart city governance. The local emergent perspective highlights that situational factors need to be included in comparative analyses and "best practices" cannot simply be copied from one city to another. There is no one-size-fits-all approach to smart city governance: city governments need to develop techno-governance arrangements that work in their specific urban context.

References

Agranoff, R., & McGuire, M. (2003). *Collaborative public management: New strategies for local governments*. Washington, DC: Georgetown University Press.

Alford, J. (2009). *Engaging public sector clients. From service-delivery to co-production*. Basingstoke: Palgrave McMillan.

Barber, B. (2013). *If mayors ruled the world: Dysfunctional nations, rising cities*. New York: Yale University Press.

Caragliu, A., & Del Bo, C. (2012). Smartness and European urban performance: Assessing the local impacts of smart urban attributes. *Innovation: The European Journal of Social Science Research, 25*(2), 97–113.

Driessen, P. P. J., Dieperink, C., Laerhoven, F., Runhaar, H. A., & Vermeulen, W. J. (2012). Towards a conceptual framework for the study of shifts in modes of environmental governance—experiences from the Netherlands. *Environmental Policy and Governance, 22*(3), 143–160.

Etzkowitz, H., & Leydesdorff, L. (Eds.). (1997). *Universities in the global economy: A Triple Helix of university-industry-government relations*. London: Cassell Academic.

Florida, R. L. (2002). *The rise of the creative class and how it's transforming work, leisure, community and everyday life*. New York: Basic books.

Fountain, J. E. (2001). *Building the virtual state. Information technology and institutional change*. Washington, DC: The Brookings Institute.

Galbraith, J. R. (1973). *Designing complex organizations*. Boston: Addison-Wesley Longman Publishing.

Gil-Garcia, J. R. (2012). *Enacting electronic government success: An integrative study of government-wide websites, organizational capabilities, and institutions*. New York: Springer.

Hoh, B., Gruteser, M., Herring, R., Ban, J., Work, D., Herrera, J. C., Bayen, M., Annavaram, M., & Jacobson, Q. (2008). Virtual trip lines for distributed privacy-preserving traffic monitoring. *Proceedings of the 6th international conference on mobile systems, applications, and services* (pp. 15–28). ACM.

Hoon Lee, J. H., Phaal, R., & Lee, S.-H. (2013). An integrated service-device-technology roadmap for smart city development. *Technological Forecasting and Social Change, 80*(2), 286–306.

Inayatullah, S. (2011). City futures in transformation: Emerging issues and case studies. *Futures, 43*(7), 654–661.

Koppenjan, J., & Klijn, E. H. (2004). *Managing uncertainties in networks: A network approach to problem solving and decision-making.* London: Routledge.

Kuk, G., & Janssen, M. (2011). The business models and information architectures of smart cities. *Journal of Urban Technology, 18*(2), 39–52.

Landry, C. (2006). *The art of city making*. New York: Routledge.

Lathrop, D., & Ruma, L. (2010). *Open government: Collaboration, transparency, and participation in practice*. Beijing: O'Reilly.

Lazariou, G., & Roscia, M. (2012). Definition methodology of the smart cities model. *Energy, 47*(1), 326–332.

Leydesdorff, L., & Deakin, M. (2011). The triple-helix model of smart cities: A neo-evolutionary perspective. *Journal of Urban Technology, 18*(2), 53–63.

Light, P. C. (2008). *The search for social entrepreneurship*. Washington, DC: The Brookings Institute.

Linders, D. (2012). From e-government to we-government: Defining a typology for citizen coproduction in the age of social media. *Government Information Quarterly, 29*(4), 446–454.

Lyon, D. (2001). *Surveillance society. Monitoring everyday life*. Buckingham: Open University Press.

Mayer-Schönberger, V., & Cukier, K. (2013). *Big data. A revolution that will transform how we live, work and think*. London: John Murray.

Meijer, A. J. (2009). Complex responsibilities. An empirical analysis of responsibilities and technological complexity in Dutch immigration policies. *Public Management Review, 11*(6), 771–790.

Meijer, A. J., & Thaens, M. (2013). Social media strategies: Understanding the differences between North American police departments. *Government Information Quarterly, 30*(4), 343–350.

Meijer, A. J., Bannister, F., & Thaens, M. (Eds.). (2013). *ICT, public administration and democracy in the coming decade*. Amsterdam: IOS Press.

Michels, A., & Meijer, A. (2008). Safeguarding public accountability in horizontal government. *Public Management Review, 10*(2), 165–173.

Nijkamp, P., & Kourtit, K. (2013). The "New Urban Europe": Global challenges and local responses in the urban century. *European Planning Studies, 21*(3), 291–315.

Orlikowski, W. J. (1992). The duality of technology: Rethinking the concept of technology in organizations. *Organization Science, 3,* 398–427.

Pierre, J. (1999). Models of urban governance the institutional dimension of urban politics. *Urban Affairs Review*, *34*(3), 372–396.

Pierre, J. (2011). *The politics of urban governance*. London: Palgrave Macmillan.

Provan, K. G., & Milward, H. B. (2001). Do networks really work? A framework for evaluating public-sector organizational networks. *Public Administration Review, 61,* 414–23.
Radaelli, C., & De Francesco, F. (2010). Regulatory impact assessment. In M. Cave, R. Baldwin, & M. Lodge (Eds.), *Oxford handbook of regulation* (pp. 279–301). Oxford: Oxford University Press.
Rogers, E. M. (1995). *Diffusion of innovations* (4th ed.). New York: Free Press.
Schaffers, H., Komninos, N., Pallot, M., Trousse, B., Nilsson, M., & Oliveira, A. (2011). Smart cities and the future internet: Towards cooperation frameworks for open innovation. In J. Domingue, A. Galis, A. Gavras, T. Zahariadis, & D. Lambert (Eds.), *The future internet* (pp. 431–446). Heidelberg: Springer.
Schulz, M., van der Steen, M., & van Twist, M. (2013). *De koopman als dominee. Sociaal ondernemerschap in het publieke domein.* Den Haag: Boom Lemma.
Sørensen, E., & Torfing, J. (2009). Making governance networks effective and democratic through metagovernance. *Public Administration, 87*(2), 234–258.
Stone, C. N. (1993). Urban regimes and the capacity to govern: A political economy approach. *Journal of Urban Affairs, 15*(1), 1–28.
Torfing, J., Peters, B. G., Pierre, J., & Sørensen, E. (2012). *Interactive governance: Advancing the paradigm*. Oxford: Oxford University Press.
Walravens, N. (2012). Mobile business and the smart city: Developing a business model framework to include public design parameters for mobile city services. *Journal of Theoretical and Applied Electronic Commerce Research, 7*(3), 121–135.
Washburn, D., Sindhu, U., Balaouras, S., Dines, R. A., Hayes, N. M., & Nelson, L. E. (2010). *Helping CIOs understand "Smart City" initiatives: Defining the smart city, its drivers, and the role of the CIO*. Cambridge: Forrester Research, Inc.
Winters, J. V. (2011). Why are smart cities growing? Who moves and who stays. *Journal of Regional Science, 51*(2), 253–270.
Yigitcanlar, T., Velibeyoglu, K., & Martinez-Fernandez, C. (2008). Rising knowledge cities: The role of urban knowledge precincts. *Journal of knowledge management, 12*(5), 8–20.
Zuboff, S. (1988). *In the age of the smart machine: The future of work and power*. New York: Basic Books.

Albert Meijer is a professor of public innovation at the Utrecht School of Governance. His research focuses on technologies and innovation in the public sector. He has frequently published on transparency, e-government, coproduction, and innovation in journals such as *Public Administration Review, Government Information Quarterly*, and the *International Review of the Administrative Sciences.*

Rethinking Learning in the Smart City: Innovating Through Involvement, Inclusivity, and Interactivities with Emerging Technologies

H. Patricia McKenna

Abstract This chapter explores the learning dimension of the smart city and the potential for innovation through use of an early-stage social radio tool. Based on use experience with the tool, this study aims to provide an understanding of: (a) how participatory and collaborative engagement can be fostered and (b) the awareness aspect of emerging technologies. Emergence theory (e.g., emergent behaviors) and the key concepts of awareness, creativity, and innovation are used to provide a context and framework for investigation of use experience with a tool designed for learning through involvement, inclusivity, and interactivities. Nam and Pardo's technology–people–institution framework for smart cities provides the basis for expanding upon and rethinking learning in the smart city—specifically, rethinking learning flows and relationships to enable interactivities and mutual learning between local government and educators/learners. A case study approach incorporating under design (e.g., a minimally viable tool) is used and multiple methods of data collection and analysis are employed in generating quantitative and qualitative findings. This work makes several contributions to the eGovernment literature by providing: (a) insight into the value of under-design approaches in understanding and assessing tools at the early development stages for eGovernment, transformational government, and lean government; (b) a framework for rethinking and innovating the learning city; and (c) an expanded way of looking at and working with learning and innovation in the smart city that may have implications for other types of eGovernment relationships (e.g., G2C, G2B, and G2G).

Keywords Adaptability · Awareness · Diversity · Emerging technologies · Learning cities · Smart cities · Under design

Abbreviations

4P	Public–private–people–partnership
G2B	Government and businesses
G2C	Government and citizens
G2G	Interagency or government to government

H. P. McKenna (✉)
AmbientEase, #541, 185-911 Yates Street, Victoria, BC V8V 4Y9, Canada
e-mail: mckennaph@gmail.com

J. R. Gil-Garcia et al. (eds.), *Smarter as the New Urban Agenda A Comprehensive View of the 21st Century City*, Public Administration and Information Technology 11,
DOI 10.1007/978-3-319-17620-8_5

l-Government	Lean government
OSS	Open source software
R5 Framework	Reveal, refine, reuse, release, and run
UNESCO	United Nations Educational, Scientific and Cultural Organization

1 Introduction

This chapter provides insight into use experience with an aware-enabled technology as a way of engaging people locally in the early-stage development of a learning tool having the potential to generate smart city solutions. The technology under study represents a social media tool designed to foster an environment supportive of the social, technical, and learning dimensions of smart cities (Nam and Pardo 2011a). Key elements of this environment under study include creativity, innovation, and context awareness in relation to involvement, inclusivity, and interactivities. In keeping with "strategies to innovate while avoiding risks" as articulated by Nam and Pardo (2011b), this study employs the under-design concept (Fischer 2013) using a minimally viable tool. Concerned also with the "contexts underlying innovation and risks" (Nam and Pardo 2011b), this study involves diverse demographics (e.g., age, interests, and skills).

The significance of this chapter is that it uses a simple social radio tool to explore aware-enabled technology in order to expand upon and rethink the learning component of smart cities identified by Nam and Pardo (2011a) and Chourabi et al. (2012). This chapter also highlights the increased blurring of boundaries across formal and informal places of study, work, and everyday life, characteristic of a twenty-first century city.

The main aim of this work is to contribute to the learning dimension of smart cities by advancing the importance of enabling people to: (a) experience new, early-stage technologies; (b) use a minimally viable tool as a way to think about and participate in the design and generation of new ideas; and (c) gain new understandings of human-centered smart environments in order to meaningfully collaborate in the development of insights, recommendations, and directions, critical to innovation for a twenty-first century city. This multipart aim gives rise to the key research questions guiding this study, identified in Sect. 3.7.

Using a case study approach, the experimental setup for this research involved the download, installation, and use experience of a social radio tool in a virtual, distributed environment. The tool was made available for study and exploration through the School of Information Studies, Syracuse University. Described in more detail under methodology in Sect. 4, this study used multiple methods of data collection and analysis, identified by Snead and Wright (2014) as important for eGovernment research.

A background and context is provided for this chapter in terms of definitions for key terms used in this work. A selective review of the literature is provided for cities as innovative, transformative, open, interactive, learning, and is followed by a context for learning in the twenty-first century. The methodology, research questions, and study findings are presented. The need for rethinking the learning city is advanced and a conceptual framework for learning and innovation in the smart city is proposed. Challenges and mitigations are addressed and implications for research and practice are identified.

2 Background

Chourabi et al. (2012) argue that "making a city 'smart'" is "emerging as a strategy to mitigate" the complex, wicked, and tangled challenges generated by rapid urbanization. Eight elements constituting an integrative framework are identified by Chourabi et al. as critical to the success of smart city initiatives—management and organization; technology; governance; policy context; people and communities; economy, built infrastructure, and natural environment. Within this framework, education forms a component of the people and communities dimension while technology refers to "the intensive use of information and communication technologies (ICTs) to better serve citizens."

Nam and Pardo (2011a) conceptualize the smart city in relation to technology, people, and institutions where *technology factors* encompass the city as digital, intelligent, ubiquitous, wired, hybrid, and as an information city. For Nam and Pardo (2011a), *human factors* focus on the city as creative, learning, humane, and as a knowledge city.

This chapter explores learning and innovation for the smart city through use experience with an aware-enabled social media technology.

In order to understand the aware-enabled technologies discussed in this chapter within the context of the smart city, key definitions are presented in Sect. 2.1.

2.1 Definitions

Definitions for key terms used in this chapter are provided, based on the research literature for ambient awareness, the learning city, the smart city, and social media.

2.1.1 Ambient Aware

McCullough (2013) claims that "ambient awareness can reflect a more general mindfulness" where "almost any use of the word *ambient* suggests some aspect of sensibility" and "sensibility to surroundings has become important again." In twenty-first century environments, McCullough (2013) describes the shifting "role of technology" as a movement "away from a means to overcome the world toward a means to understand it."

Ambient refers to “a complex set of phenomena” (McCullough 2013) with “a new attitude about attention.” Ambient is understood as “that which surrounds but does not distract” and “a continuum of awareness and an awareness of continuum.”

2.1.2 Learning City

Nam and Pardo (2011a) claim that “a smart city is also a learning city, which improves the competitiveness of urban contexts in the global knowledge economy.” United Nations Educational, Scientific and Cultural Organization (UNESCO) and the International Conference on Learning Cities (2013) define a learning city as one which “mobilizes human and other resources to promote inclusive learning from basic-to-higher education; revitalizes learning in families and communities; facilitates learning for and in the workplace; extends the use of modern learning technologies; enhances quality in learning; and nurtures a culture of learning throughout life.”

2.1.3 Smart City

Nam and Pardo (2011b) identify a smart city as “a city’s effort to make itself smart” through “a comprehensive commitment to innovation in technology, management and policy.” Drawing on the public sector research literature, Nam and Pardo (2011b) indicate that, “innovation simply denotes ‘novelty in action’ and ‘new ideas that work’.”

2.1.4 Social Media

Criado et al. (2013) define social media “as a group of technologies that allow public agencies to foster engagement with citizens and other organizations” and “can be understood as platforms to interact with citizens and organizations with innovative potentialities.”

3 The Smart and Learning City: A Selective Literature Review

The smart city research literature is selectively reviewed in terms of innovation, eGovernment transformation, open government, interactive practices (Nam and Pardo 2011a), (UCLG 2012), and learning.

3.1 *Innovative City*

Resistance to innovation is said to be "rampant, devious, and well-organized" (Eesley 2013), and yet, a recent international study of cities describes "the rapid development of new technologies and of innovation processes" contributing to "a new city model" referred to as the "smart city" (UCLG 2012). Nam and Pardo (2011b) argue that "making a city smart is a new approach to urban development" in order to "solve the tangled and wicked problems" inherent in rapid urbanization and the associated social, political, and organizational issues, where making a city smart constitutes the innovation (Nam and Pardo 2011b). Chan (2012) contrasts "traditional closed innovation approaches where the locus of innovative endeavors lies within the boundary of an organisation" to open innovation where "an open boundary between an organization and its surrounding environment" exists "such that innovative endeavors can interweave across the boundary." Schaffers et al. (2011) argue for open, user-driven innovation ecosystems based on sustainable partnerships and cooperation strategies where the sharing of common resources can be leveraged for real-world experimentation and development as living labs to achieve socioeconomic development and the quality of life.

3.2 *Transformative City*

Nam and Pardo (2011a) note that the term "smart" "captures innovative and transformative changes driven by new technologies" but that a more comprehensive socio-technical and socioeconomic (Nam and Pardo 2011b) understanding is required. Hunnius and Schuppan (2013) explore hindrances, uncovering the importance of competencies to transformational eGovernment. Katz and Bradley (2014) claim that "cities are networks rather than governments" and that the recession is contributing to opportunities resulting in "new institutional arrangements," placing "disparate players (public, private, university, and civic) around one table." Daniel and Doran (2013) argue that change is occurring rapidly, going "much further than interconnections and interactions enabled and facilitated by the use of ICTs and geomatics." Neuroni et al. (2012) point to the importance of institutional transformation enabled by openness.

3.3 *Open City*

Dimensions of the open city include open data, databases, and resources; open government; and open infrastructure.

Open Data Sayogo and Pardo (2013) note that open data now encompasses "the wider environment of government, citizens, business and civil society organizations." Sayogo and Pardo (2013) identify Blue Button initiatives as government led while industry-led initiatives are referred to as Green Button, highlighting the collaboration of many, resulting in "expanding the capability of open data." Kaschesky and Selmi (2013) address the issue of how to make open data actionable, based on the R5 framework encompassing—reveal, refine, reuse, release, and run. Fostering data action-ability points to the importance of more autonomy with and across systems, community partners, and data.

Open Resources The open resources concept is evident in the idea of a "beta city" (Violino 2014) designed "to provide data sets and insights that any government worker, citizen or organization could potentially benefit from having." Nam and Pardo (2011a) argue that a smart city can be built successfully using either a top-down or bottom-up approach but it is essential to have "active involvement from every sector of the community." Khan et al. (2014a) describe an application designed for "the real world environment of four European cities" enabling participatory bottom-up urban planning and policymaking. Building upon this collaborative approach, this chapter encompasses multiple communities, neither top-down nor bottom-up, and focuses upon interactive participation where learning opportunities may emerge for everyone involved.

Open Databases The open database concept emerges when Schaffers et al. (2011) refer to "the city as database" in the governance component of the smart city.

Open Government Neuroni et al. (2012) describe open government as an approach to people-centric governance based on transparency, participation, and collaboration.

Open Infrastructure Open infrastructure can be understood in relation to interoperability, where Hellberg and Grönlund (2013) note that although "interoperability is a top priority today as governments try to integrate services across departments so as to improve effectiveness as well as efficiency," most efforts result in failure due to "complex relationships among government, society, and technology" which are "not fully understood."

Schaffers et al. (2011) maintain that open, user-driven innovation ecosystems "may evolve to constitute the core of '4P' (Public–Private–People–Partnership)" with opportunities for "citizens and businesses to cocreate, explore, experiment and validate innovation scenarios." Yet, Schaffers et al. (2011) make the claim that "it becomes increasingly challenging to design open infrastructures that efficiently support emerging events and citizens' changing needs."

It is precisely this type of challenge that awareness technologies under exploration in this chapter aim to address in coming to understand more about the dynamic, emergent, adaptive, and uncertain side of real-world environments.

3.4 The Interactive and Aware City

The interactive city is collaborative and participatory and people are understood to be contributors, not just consumers (Nam and Pardo 2011a). Gil-Garcia and Aldama-Nalda (2013) state that, "the essence of becoming smarter seems to be related to connectedness, responsiveness, efficiency, and sustainability." Nam and Pardo (2011a) envision smarter governments to be "interconnecting dynamically with citizens, communities, and businesses in real time to spark growth, innovation, and progress," becoming more *citizen centric* in the design, planning, and development of operations and services (Nam and Pardo 2011a).

Kelly and Hamm (2013) point to "the importance of developing technologies in close collaboration with the people who will eventually use it." This means that, "one of the key elements of innovation will be the willingness to use the real world, in all of its messiness and complexity, as a living laboratory for developing new technologies." As an alternative to smart cities, de Lange and de Waal (2013) propose ownership as a lens where ownership is intended to foster engagement, responsibility, and stewardship (de Lange and de Waal 2013), expanding the discussion "from an infrastructural to a social point of view, or from 'city management' to 'city making'." Nam and Pardo (2011a) point to expansive smart city definitions that include other types of infrastructure such as human infrastructure, along with other terms such as humanware, and soft infrastructure (e.g., knowledge networks).

AlAwadhi and Scholl (2012) describe their research in Seattle where practitioners understand smart city as smart government at the local level, in contrast to the research literature definition that is more expansive, including "the complex and multidimensional urban space of a city" (AlAwadhi and Scholl 2012). However, Mellouli et al. (2014) use the term *smart community* which "refers to the use of information and communication technologies [ICTs] by local governments and cities to better interact with their citizens, taking advantage of all available data to solve important problems." Mellouli et al. (2014) add that *smart government* refers to "extensive use of technology by governments" and *citizens' engagement* refers to "extensive use of technology by citizens to interact with government." For citizen engagement to be meaningful, Mellouli et al. (2014) add that "useful, relevant, and complete information from the government" is needed, providing an opportunity to think about the learning city.

3.5 Learning City

Nam and Pardo (2011a) identify learning (including smart education) as the human component for the smart city. People and how they interact are considered to be critical to cities. Elements such as awareness, social learning, access to learning, training, and skills development contribute to smart education. Creativity is a key driver of the smart city, encompassing people, education, learning, and knowledge (Nam and Pardo 2011a). *Learning cities* are characterized as "actively involved in building a skilled information economy workforce" while the *knowledge city* places an emphasis on innovation and the networked, connected, and competitive economy (Nam and Pardo 2011a).

Fuentes-Bautista (2014) encourages a rethinking of localism, pointing to the importance of "digital community media projects" for "promoting advanced digital skills, economic opportunities, civic engagement, and social cohesion." The Beijing declaration emerging from the International Conference on Learning Cities (UNESCO 2013) stated that, "All sectors of society have a key role to play in learning and education and, therefore should participate in building learning cities."

Neuroni et al. (2012) claim that *collaboration* "implies using innovative tools and systems to cooperate among public agencies" and is a relevant parameter to "engage citizens in the work of their government." An *information city* is a digital space representing all information interactions among all participants (e.g., people, businesses, government, and institutions) (Nam and Pardo 2011a).

Where Lakka et al. (2013) focus on the specific innovative technologies of secure servers and open source software (OSS) that affect eGovernment, this chapter focuses on an exploration of awareness technologies as a means to explore learning and innovation in the smart city. While many smart city initiatives seem to assume the presence of viable technologies and infrastructure in support of eGovernment and citizen engagement, this chapter takes a step back to explore early-stage use of the underlying technologies and the potential for innovating relationships for learning.

3.6 Summary of the Smart City and the Context for Learning

Johnson et al. (2014) identify "wicked challenges" for education and learning going forward, as in, "too complex to even define, much less address," adding to the wicked and tangled challenges facing the city (Chourabi et al. 2012). The response is to innovate through a movement toward creating the smart city and the learning city. Characteristics and requirements for learning in twenty-first century environments in the 2014 Horizon's report (Johnson et al. 2014) have a lot in common with evolving future work skills (IFTF 2011) and transformational e-government competencies (Hunnius and Schuppan 2013). In summary, learning in the twenty-first century smart city is characterized by the following:

- Fluidity of environments (work, study, and everyday life)
- Active, interactive, social, collaborative, and sharing
- Real-world-based learning for meaning making and solution generation
- Relevance of education to work and everyday environments
- Adaptation of work and learning environments and practices to twenty-first century needs (dynamic, emergent, innovative, creative, and uncertain)

Signet (2014) defines emergent learning as pragmatic, presenting the need for "overcoming challenges, especially those that have no simple solutions, but require discipline, ongoing attention, learning through experience, and adaptation." Developed with workplaces in mind, the increased blurring of work, learning, and everyday life means that this definition by Signet has relevance for formal and informal learning environments in the smart city.

The smart city is already showing evidence of public spaces as innovation labs as noted by Schaffers in relation to living labs (Schaffers et al. 2011), seamless learning, and maker spaces (Sharples et al. 2013), seeded with problems that the city and organizations are addressing. A study by Caldwell et al. (2013) aims "to redefine spaces of learning to places of learning through the direct engagement of local communities as a way to examine and learn from real-world issues in the city." The aim is to connect students, educators, businesses, and community members to their city, enabling a mutual understanding of the issues and possibilities that emerge when people live and work together in city spaces. This type of direct connection may provide opportunities for contributing to evolve initiatives to measure the value of eGovernment (Savoldelli et al. 2013). Opportunities for all members of a community to communicate and collaborate in smart city spaces is critical and requires attention to a range of factors including the digital divide (Chourabi et al. 2012) and associated policy issues (Ferro et al. 2011).

Janssen and Estevez (2013) identify lean government (l-Government) as a third wave of eGovernment which "aims at reducing the complexity of the public sector by simplifying and streamlining organizational structures and processes" and "stimulating innovation by mobilizing stakeholders." In brief (Janssen and Estevez 2013), "public organizations introduce platforms facilitating innovation and interactions with other public organizations, business and citizens, and focus on their orchestration role" such that, "Experimentation, assessment, and gradual improvement based on user requirements are key factors for realizing l-Government."

In summary, key elements of the smart city and the learning city are identified in Table 1 as: innovative, transformative, open, interactivity, and awareness. In the right column, why learning matters is described for each element and identified as: creative, meaning/satisfaction, collaborative, social, and adaptability.

Mellouli et al. (2014) describe eGovernment as "the use of information technologies in government" generating three types of interactions: government and citizens (G2C), government and businesses (G2B), and inter-agency or government to government (G2G). While the intent for these interactions may be two-way, this chapter points to a possible gap requiring attention. Specifically, the tendency to emphasize a one-way focus from government to business and from government to citizens may contribute to the loss of innovation potential. Using an under-design approach aligned with agility and leanness, this chapter responds to the need for more fully interactive and adaptive social media tools (Criado et al. 2013) and the potential for innovation in smart city learning, guided by a set of research questions.

Table 1 Summary of smart city and learning city elements and why learning matters

Elements	Smart city	Learning city	Why learning matters
Innovative	✓	✓	Creative
Transformative	✓	✓	Meaning/satisfaction
Open	✓	✓	Collaborative
Interactivity	✓	✓	Social
Awareness	✓	✓	Adaptability

3.7 Research Questions

The potential for innovating the learning dimension of the smart city, supported by the literature review, gives rise to four research questions that will be addressed during exploration of use experience with a social radio tool.

- Q1: Does use experience with an aware-enabled social radio tool contribute to a greater understanding of the nature and potential of learning in smart spaces?
- Q2: Is the social radio tool a social space (e.g., supports interaction, inclusivity, and involvement in terms of collaboration and sharing)?
- Q3: Does the social radio tool foster innovation and creativity?
- Q4: What is the value of an under-design approach for exploring emerging technology environments in community spaces?

4 Methodology

This chapter is based on research with an early-stage aware-enabled social radio tool, over a 4-month period, in early 2012. The research investigation conducted through the School of Information Studies (iSchool), Syracuse University is described in three parts in terms of process, sources of evidence, and data analysis techniques.

4.1 Process

Using a case study approach, the emerging and contemporary issue of smart learning environments was investigated based on use experience with a minimally viable social radio tool, containing awareness features such as presence awareness. The tool also supported elements of autonomy and control, theorized to be features that would foster creativity and innovation. In keeping with emergence theory, a largely unstructured approach was used for the experimental setup, to maximize real-world experience and the potential for emergent and adaptive behaviors to occur. Minimal guidance and supports were provided in the form of two brief instructional videos, which also provided exposure to the tool, if barriers to use occurred.

Participants were recruited through the iSchool and included faculty and students (both distance and on campus) and a new form of student, representative of professionals in business, government, and other community sectors. This new form of student is concerned with lifelong learning and is seeking advanced degrees in the face of rapidly changing work environments and requirements.

Participants were instructed to: download and install the social radio application; create a radio station; develop a radio show with their choice of content; host or cohost the show with another individual; and livestream the show for shared listening within the tool environment, with Facebook friends, and with others who wished to listen to the Web broadcast over the Internet.

4.2 Sources of Evidence

Participant tool use ($n = 34$) was tracked in the form of quantitative activity data. Pretested interview and focus group protocols were used to gather qualitative use experience data. The focus group protocol and the interview protocol each contained 21 questions designed to probe use experience and four questions pertaining to recommendations for improvements. Based on focus group ($n = 6$) and interview responses ($n = 22$), a survey instrument was developed with closed- and open-ended questions; pretested; and administered online, generating quantitative and qualitative data ($n = 20$). Multiple methods of data collection enabled the triangulation of data in support of validity and rigor.

4.3 Data Analysis Techniques

In keeping with the small sample size ($n = 34$) for activity data and survey responses ($n = 20$), descriptive statistics were used in the analysis of quantitative data. Content analysis was used for the qualitative data, employing both deductive and inductive techniques and the development of a coding glossary. One-thousand text segments were extracted from interview, focus group, and open-ended survey data and coded with the assistance of a second coder, achieving an inter-coder reliability rate of 91–94 %.

5 Analysis and Findings

Research findings based on use experience with an aware-enabled social radio tool contribute to a range of insights relevant to the smart city, the learning city, and innovation. Analysis and findings are presented through a response to the research questions identified in Sect. 3.7.

5.1 Response to the Research Questions

The research questions are categorized into four response areas as: learning, social, creativity and innovation, and under design.

Table 2 Recommendations for tool features to enhance awareness and smartness

(n = 20)	Strongly disagree (%)	Somewhat disagree (%)	Neutral (%)	Somewhat agree (%)	Strongly agree (%)
User profile details (recommending users to each other)	0	5	20	55	20
Radio show details (recommending users to each other)	0	0	20	60	20
Ad feature (recommend content to listener to purchase)	5	5	20	60	10
Ability to "like" a broadcast	0	0	10	50	40
Number of listeners display	0	0	5	70	25

5.1.1 Q1: Learning

In response to Q1, *does use experience with an aware-enabled social radio tool contribute to a greater understanding of the nature and potential of learning in smart spaces*, 75 % of survey respondents (n = 20) indicated that a greater understanding of awareness systems and environments occurred. Table 2 identifies what would contribute to greater awareness and smartness of the social radio tool.

Survey response indicates the importance of feedback features in the form of "like," where 40 % strongly agree and 50 % somewhat agree, for a total agreement of 90 %. Popularity of radio broadcast, where the system displays the number of listeners, was rated at 25 % for strongly agree and 70 % for somewhat agree, for a total of 95 %.

Participant interviews revealed a lack of understanding about aware or smart systems and environments. Engagement in discussion to learn more about awareness technologies emerged through the approximately 20 % of text segments coded for some aspect of smartness (e.g., location, presence, recommending, resources, and situation). Participants identified awareness features (e.g., presence awareness) as a key part of social media spaces to the point where this functionality is "taken for granted." Awareness and other smart features, in support of social networking and information interactions, are now expected and enhanced forms of smartness for increasingly aware environments is highly desirable.

5.1.2 Q2: Social

In response to Q2, *Is the social radio tool a social space (e.g., supports interaction, inclusivity, and involvement in terms of collaboration and sharing)*, 70 % of survey responses (n = 20) indicated yes, 5 % indicated no, and 25 % were unsure.

Interview responses revealed that use experience with the aware product motivated some participants to begin thinking about the next frontier of social media. Evolving understandings of sharing, social, trust, and privacy emerged. Adaptability was evident as participants exhibited emergent learning and sharing behaviors. Participants associated with government workspaces exhibited a tendency to avoid social media, influencing innovation potentials (Criado et al. 2013; Khan et al. 2014b; Bekmamedova and Shanks 2014), but did participate in social media related to family connectivities.

Respondents spanning age ranges from the 20 to 60s, contributed a wide variety of interpretations for use, potential uses, and imagined use, illustrative of the involvement, inclusivity and interactivities enabled and fostered by social media technologies and practices.

5.1.3 Q3: Creativity and Innovation

In response to Q3, *Does the social radio tool foster innovation and creativity*, survey responses in Table 3 show how participants ($n = 20$) felt during their use experience in terms of creative, autonomy, in control and innovative. Participants felt more creative (40 % strongly agree; 50 % somewhat agree for a total of 90 %) than innovative (35 % strongly agree; 35 % somewhat agree for a total of 70 %).

Autonomy and control are associated with creativity and innovation. During use experience with the minimally viable tool, measures of autonomy (25 % strongly agree; 45 % somewhat agree for a total of 70 %) and control (25 % strongly agree; 35 % somewhat agree for a total of 60 %) are present.

Survey responses ($n = 20$) for self-assess of creativity, in terms of idea generation, are illustrated in Table 4.

Table 3 Creativity, autonomy, control, and innovativeness

(n=20)	Somewhat disagree (%)	Neutral (%)	Somewhat agree (%)	Strongly agree (%)
Creative	10	0	50	40
Autonomy	5	35	45	25
In control	5	35	35	25
Innovative	5	25	35	35

Table 4 Self-assessments for idea generation during tool use experiences

(n=20)	No (%)	Yes (%)
I created one or more new ideas	40	60
I thought about creating one or more new ideas	30	70
I noticed that other users created new ideas	50	50
Other people I talked to about the tool came up with new ideas	50	50

Sixty percent of respondents reported to have created one or more ideas during their use experience; 70 % thought about creating one or more ideas; 50 % noticed ideas created by others; and 50 % indicated that ideas were generated during conversations. Approximately 19 % of text segments were coded for creativity, capturing discussion data in this category in support of survey self-assessments. Twenty-eight percent of text segments were coded for innovation, and of these, 27 % pertained to possibilities and 38 % pertained to interpretations.

When asked whether the aware tool fosters innovation, respondents indicated some degree of innovativeness or innovation potential (30 % indicated very innovative; and 60 % indicated somewhat innovative for a total of 90 %.)

Probing further for innovativeness, participants were asked to assess the disruptiveness of aware environments in terms of the potential for new and transformative outcomes for people, information, and technology. On a 5-point scale (1 = not really and 5 = absolutely); 5 % responded at the fifth level; followed by 40 %; and then 25 % at the mid-level, suggestive of an overall cautiously optimistic perception for disruptive and transformative outcomes.

Self-assessment survey data on idea generation in Table 4 is supported when compared with actual use of the tool (activity data) and data from interviews and focus groups.

5.1.4 Q4: Under Design

Using the under-design concept advanced by Fischer (2013), where partially formed products contribute to learning opportunities for creativity, the social radio tool represents an early-stage technology, in a state of sufficient readiness to elicit valued feedback and interpretations from participants. In response to Q4, *What is the value of an under-design approach for exploring emerging technology environments in community spaces*, survey participants ($n = 20$) were asked to rate their level of satisfaction with the early-stage tool. Participants responded positively to use experience satisfaction as shown in Table 5, (15 % very satisfied; 65 % satisfied for a total of 80 % satisfaction).

Approximately 90 % of interview ($n = 22$) and focus group ($n = 6$) participants indicated a willingness to continue using the tool when iterated to incorporate suggested improvements, suggestive of the perceived value and potential.

Table 5 Satisfaction

(n=20)	Neutral (%)	Satisfied (%)	Very satisfied (%)
Use experience satisfaction level	20	65	15

5.2 *Summary*

Use experience with an under-designed tool proved to be an effective approach for supporting learning about aware-enabled technologies (e.g., smartness) and social media potentials for creativity (idea generation), innovation and transformativeness, enabled by environments that foster feelings of autonomy and control.

6 Discussion: Rethinking and Innovating the Learning City

Exploring use experience with an early-stage social radio tool in a virtual, distributed real-world setup, this study demonstrated the fluid, emergent, dynamic, creative, and adaptive nature of twenty-first century smart communities. The university environment is the real-world workspace for faculty and students, blurring the boundaries of study, work, and everyday life. Faculty and students were both on campus and at-a-distance, rendering local to be fluid, as in, the community where one is located, while social radio interactivity and cohosting often involved more than one local community. Participants encompassed educational environments and extended to government, business, and other organizational settings. The shifting landscape of formal and informal learning, occurring anywhere anytime gives rise to the need for rethinking the learning city.

Based on understandings and insights emerging from this study, rethinking the learning city is centered on awareness; adaptability; and creativity, incorporating the critical components of autonomy and control, fostering greater involvement, inclusivity, and interactivities, as depicted in Fig. 1.

In keeping with the idea of lean government (l-Government), the under-design approach with a minimally viable social radio tool assists in understanding the underlay of an aware technology embedded in a smart city or a smart government service. Potential for the smartness in this information landscape to be influenced interactively among government and citizens emerges, in terms of: students, faculty,

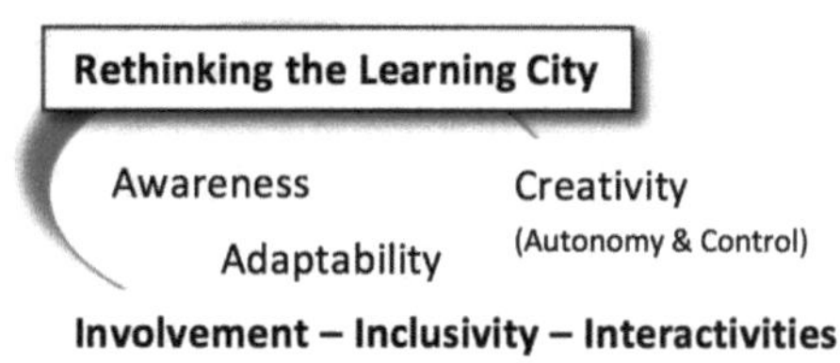

Fig. 1 Rethinking the learning city

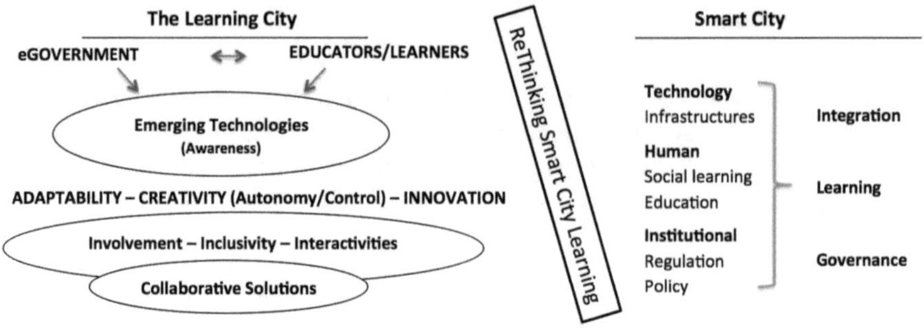

Fig. 2 Rethinking smart city learning

and the increasing numbers of business, government, and other community individuals engaged in learning.

Depicted in the left portion of Fig. 2, interactivities between eGovernment and educators/learners are enabled through emerging technologies of the awareness type, in the form of an aware-enabled social radio tool. The tool is designed to minimally foster adaptability, creativity (including autonomy and control), and innovation. In this under-design tool environment, use experience is explored, encompassing involvement, inclusivity, and interactivities, contributing to the potential for emergent collaborative solutions.

The under-design approach to the learning city enables a rethinking of smart city learning in terms of Nam and Pardo's (2011a) technology–people–institutions conceptualization of smart city, outlined in the right portion of Fig. 2. Learning is integrated with technology across all community sectors, illustrative of what is possible in a real-world, local community, suggestive of the role and potential of the aware learning city.

6.1 Role and Potential of the Aware Learning City

Fischer (2013) points to "the unique synergy between learning, creativity, and cultures of participation." In the context of rethinking smart city learning, this chapter proposes that people–technology–information interactions can be innovated through smarter relationships (Chourabi et al. 2012), (Criado et al. 2013). To visualize this thinking, Fig. 3 depicts a proposed conceptual framework for learning and innovation in the smart city.

Aware technologies support learning, through fostering adaptability and creativity. A sense of autonomy and control is an important feature of this environment, contributing to greater involvement, inclusivity, and interactivities in support of innovation and the potential for transformativeness.

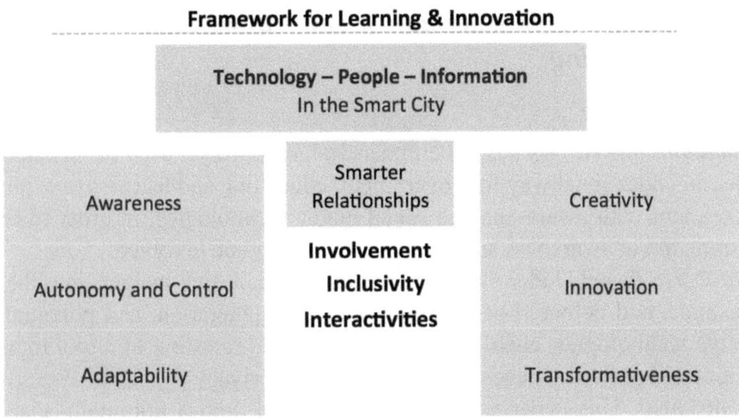

Fig. 3 Conceptual framework for learning and innovation in the smart city

6.2 Summary

A rethinking of the learning city (Fig. 1) was enabled through focusing on key concepts explored during use experience with an early-stage social media tool; supporting a rethinking of smart city learning (Fig. 2); and development of a proposed conceptual framework for learning and innovation in the smart city (Fig. 3).

7 Challenges and Mitigations

The state of readiness of the early-stage, social radio tool was a challenge for participants, which was mitigated by enabling the generation of imaginative uses, illustrating the strength of a minimally viable tool. The small sample size for activity data ($n = 34$) and survey response data ($n = 20$) does not support the development of statistical significance or generalizability. This limitation was mitigated by the use of multiple methods of data collection and analysis. Qualitative data from interviews and focus groups generated rich insights, pointing to the value of the under-design approach for robust idea generation, persisting across diverse demographics (e.g., age, interests, skills, and gender).

8 Implications for Research and Practice

This research work on use experience with an aware-enabled social media tool has implications for innovation in the smart city in terms of practical recommendations and future research.

8.1 Practical Recommendations: Innovating Smart City Learning

1. *Education-infused projects:* The under-design approach used in this study provides an educational way for government, educators, and learners to explore and experiment with aware-enabled social media technologies, in order to enhance learning about awareness and smartness for everyone involved.
2. *Real-world focus:* Under design can be used to gain rich insights into the value, relevance, and potential of existing, new, next generation, and potentially disruptive technologies, enabling rapid iteration and retesting of a tool to adapt to current and projected needs and concerns (e.g., privacy).
3. *Engagement:* Use experience and imagined use with a minimally viable tool contributes to engagement when focused on personally meaningful interpretations for use, in relation to real-world, smart city initiatives.
4. *Change it up:* Risks associated with innovation and change can be minimized in smart city initiatives by using the under-design approach employed in this research and modified for the particular practice setting.
5. *Smart relationships:* Innovate relationships for learning between government, educators, and learners in order to gear learning to real world city needs and to gear government initiatives to the needs of learners for education in the smart community.
6. *Learning and work:* Connect learning with emerging and future work skills (Johnson et al. 2014) (e.g., collaboration, creativity, innovation, novel thinking, adaptability, and capacity to respond to uncertainty).

8.2 Future Research: Innovating Smart City Learning

1. *Smart city learning:* As an early-stage exploratory study, this research contributes to the awareness, smart city, and learning city research literature; serves as a bridge study to further work on innovating through learning in the smart city; and initiates a rethinking of smart city learning.
2. *Learning and innovation:* A conceptual framework for learning and innovation in the smart city is proposed, calling for further investigation and development.
3. *Smart relationships:* Building upon earlier work (Chourabi et al. 2012), (Criado et al. 2013), this chapter develops a space for smarter relationships in the proposed framework for learning and innovation in the smart city, calling for further research and development.
4. *Awareness:* Aware environments invite a rethinking of notions of involvement, inclusivity, and interactivities for learning cities. Drawing on understandings of the human side of infrastructure (Dourish and Bell 2011), use experience with aware-enabled tools provides an opportunity for further work on Dourish and Bell's *experience of infrastructure* and *infrastructure experience* (Dourish and Bell 2011)

5. *Smart modeling:* This chapter emphasizes the value of under design in the exploration of emerging technologies contributing to the potential for smarter modeling of innovating learning in the smart city.

9 Conclusions

This chapter explored the learning dimension of the smart city and the potential for innovation using an emerging aware-enabled technology. Focusing on emergent behavior in relation to awareness, autonomy, creativity, and innovation, this study contributes insight into mechanisms for participatory and collaborative engagement using a social media tool designed for learning, involvement, inclusivity, and interactivities. A rethinking of the learning city is presented, featuring interactivities between government, learners, and educators in collaborative idea generation. A case study approach, incorporating under-design principles and multiple methods of data collection and analysis was employed in generating triangulated quantitative and qualitative findings. This work has implications for smart city research and practice, making recommendations for practice related to: education-infused projects; real-world focus; engagement; change; smart relationships; and learning connected to work. Contributions to the research literature across multiple domains include a rethinking of the learning city; under design for learning and innovation in the smart city; and smarter relationships for learning. Recommendations are made for further research in these areas.

The intended audience for this chapter encompasses research and practice; including educators, learning researchers, students, policy makers, designers, developers, and anyone concerned with livable, learning, and innovative cities.

A key take-away from this chapter is that for many participants, this study provided a possibly rare opportunity to think about aware and smart environments; to discuss issues and challenges posed by smart environments (e.g., privacy); and participate in the actual interactive and iterative design of an aware-enabled tool through interpretations for use and recommendations for improvement. Smarter relationships through smart connectivities (e.g., involvement, inclusivity, and interactivities) make way for innovating education to be geared to a smart city and innovating a smart city to be geared to learning and education.

References

AlAwadhi, S., & Scholl, H. J. (2012). *Aspirations and realizations: The smart city of Seattle*. Proceedings of the 46th HICSS, 1695–1703.

Bekmamedova, N., & Shanks, G. (2014). *Social media analytics and business value: A theoretical framework and case study*. Proceedings of the 47th Annual HICSS, 3728–3737.

Caldwell, G. A., Foth, M., & Guaralda, M. (2013). An urban informatics approach to smart city learning in architecture and urban design education. *Interactive Design & Architectures, 17,* 7–28.

Chan, C. M. L. (2012). *From open data to open innovation strategies: Creating e-services using open government data*. Proceedings of the 26th HICSS, 1890–1899.
Chourabi, H., Nam, T., Walker, S., Gil-Garcia, J. R., Mellouli, S., Nahon, K., Pardo, T. A., & Scholl, H. J. (2012). *Understanding smart cities: An integrative framework*. Proceedings of the 45th HICSS, IEEE, 2289–2297.
Criado, J. I., Sandoval-Almazan, R., & Gil-Garcia, J. R. (2013). Government innovation through social media. *Government Information Quarterly, 30*(4), 319–326.
Daniel, S., & Doran, M. A. (2013). *GeoSmartCity: Geomatics contribution to the smart city*. Proceedings of the 14th Annual International Conference on Digital Government Research, 65–71.
Dourish, P., & Bell, G. (2011). *Divining a digital future: Mess and mythology in ubiquitous computing*. Cambridge: MIT Press.
de Lange, M., & de Waal, M. (2013). Owning the city: New media and citizen engagement in urban design. *First Monday, 18*(11), online.
Eesley, C. (2013). *Technology Entrepreneurship MOOC* (Massive Open Online Course). Stanford University.
Ferro, E., Helbig, N. C., & Gil-Garcia, J. R. (2011). The role of IT literacy in defining digital divide policy needs. *Government Information Quarterly, 28*(1), 3–10.
Fischer, G. (2013). Learning, social creativity, and cultures of participation. In A. Sannino & V. Ellis (Eds.), *Learning and collective creativity: Activity-theoretical and sociocultural studies*. New York: Taylor & Francis/Routledge.
Fuentes-Bautista, M. (2014). Rethinking localism in the broadband era: A participatory community development approach. *Government Information Quarterly, 31*(1), 65–77.
Gil-Garcia, J. R., & Aldama-Nalda, A. (2013). *Making a city smarter through information integration: Angel network and the role of political leadership*. Proceedings of the 46th HICSS, 1724–1733.
Hellberg, A. S., & Grönlund, Å. (2013). Conflicts in implementing interoperability: Re-operationalizing basic values. *Government Information Quarterly, 30*(2), 154–162.
Hunnius, S., & Schuppan, T. (2013). *Competency requirements for transformational e-government*. Proceedings of the 46th HICSS, 1664–1673.
IFTF. (2011). *Future work skills 2020*. Palo Alto: Institute for the future for the University of Phoenix Research Institute.
Janssen, M., & Estevaz, E. (2013). Lean government and platform-based governance: Doing more with less. *Government Information Quarterly, 30*(Suppl. 1), S1–S8.
Johnson, L., Adams Becker, S., Estrada, V., & Freeman, A. (2014). *NMC Horizon report: 2014 higher education edition*. Austin: The New Media Consortium.
Kaschesky, M., & Selmi, L. (2013). *Fusepool R5 linked data framework: Concepts, methodologies, and tools for linked data*. Proceedings of the International Conference on Digital Government Research, 156–165.
Katz, B., & Bradley, J. (2014). *Lessons from the road: The metropolitan revolution 2014*. Washington, DC: Brookings Institution.
Kelly, J. E., & Hamm, S. (2013). *Smart machines: IBM's Watson and the era of cognitive computing*. New York: Columbia University Press.
Khan, Z., Ludlow, D., Loibl, W., & Soomro, K. (2014a). ICT enabled participatory urban planning and policy development: The UrbanAPI project. *Transforming Government: People and Policy (TGPPP), 8*(2), 205–229.
Khan, G. F., Swar, B., & Lee, K. S. (2014b). Social media risks and benefits: A public sector perspective. *Social Science Computer Review, 32*(5), 606–627.
Lakka, S., Stamati, T., Michalakelis, C., & Martakos, D. (2013). What drives eGovernment growth? An econometric analysis on the impacting factors. *International Journal of Electronic Governance, 6*(1), 20–36.
McCullough, M. (2013). *Ambient commons: Attention in the age of embodied information*. Cambridge: MIT Press.
Mellouli, S., Luna-Reyes, L. F., & Zhang, J. (2014). Smart government, citizen participation and open data. *Information Polity, 19*(1–2), 1–4.

Nam, T., & Pardo, T. A. (2011a). *Conceptualizing smart city with dimensions of technology, people, and institutions*. Proceedings of the 12th Annual Conference on Digital Government Research, 282–291.

Nam, T., & Pardo, T. A. (2011b). Smart city as urban innovation: Focusing on management, policy, and context. Proceedings of the 5th International Conference on Theory and Practice of Electronic Governance, 185–194.

Neuroni, A. C., Riedl, R., & Brugger, J. (2012). *Swiss executive authorities on open government data—policy making beyond transparency and participation*. Proceedings of the 46th HICSS, 1911–1920.

Savoldelli, A., Misuraca, G., & Codagnone, C. (2013). Measuring the public value of e-Government: The eGEP2.0 model. *Electronic Journal of e-Government, 11*(1), 373–388.

Sayogo, D. S., & Pardo, T. A. (2013). *Understanding smart data disclosure policy success: The case of green button*. Proceedings of the 14th Annual International Conference on Digital Government Research, June 17–20, Quebec City, Canada, 72–81.

Schaffers, H., Komninos, N., Pallot, M., Trousse, B., Nilsson, M., & Oliveira, A. (2011). Smart cities and the future internet: Towards cooperation frameworks for open innovation. In J. Domingue et al. (Eds.), *Future Internet assembly, LNCSS 6656* (pp. 431–446). Heidelberg: Springer.

Sharples, M., McAndrew, P., Weller, M., Ferguson, R., FitzGerald, E., Hirst, T., & Gaved, M. (2013). Innovating pedagogy 2013: Exploring new forms of teaching, learning and assessment, to guide educators and policy makers. Open University Innovation Report 2. UK: Institute of Educational Technology, The Open University.

Signet. (2014). What is emergent learning? Signet Research and Consulting. http://www.signet-consulting.com/concepts/emergent_learning.php. Accessed 10 June 2015.

Snead, J. T., & Wright, E. (2014). E-government research in the United States. *Government Information Quarterly, 31*(1), 129–136.

UCLG. (2012). *Smart cities study: International study on the situation of ICT, innovation and knowledge in cities*. Bilbao: The Committee of Digital and Knowledge-based Cities of UCLG (United Cities and Local Government).

UNESCO. (2013). Beijing declaration on building learning cities—Lifelong learning for all: Promoting inclusion, prosperity and sustainability in cities. Working document 1. Prepared for adoption at the International Conference on Learning Cities, Beijing, China, Oct 21–23.

Violino, B. (2014). Smart cities are here today—and getting smarter: Big data, mobile, sensors, social media are already in use, but security and privacy are issues. *Computerworld, 12*. http://www.computerworld.com/article/2487526/emerging-technology-smart-cities-are-here-today-and-getting-smarter.html. Accessed 10 June 2015.

H. Patricia McKenna is the founder and president of AmbientEase, a Canadian company focused on the conceptualization, development, and testing of new and emerging environments crossing the boundaries of work, study, and everyday life. Patricia has extensive experience in the start up of information systems and services for government, non-profit, and academic organizations. Patricia holds a BA degree from the University of New Brunswick, a MLS degree from McGill University, and a doctorate degree in information management from Syracuse University. Patricia is a practitioner and researcher in the area of human-centered computing (HCC), placing an emphasis on creativity, autonomy, innovation, and awareness as people interact in technology-pervasive environments. Patricia is a member of the technology & society global research space.

Ad Hoc BYOD Information Services in Public Places of Smart Cities

Jarogniew Rykowski and Wojciech Cellary

Abstract In this chapter, we focus on public information broadcasting as an important component of a smart city. To broadcast information, we propose to install miniature and very cheap access points in public places in a smart city, capable to interact with mobile personal devices according to the bring your own device (BYOD) idea, in ad hoc and anonymous mode, by means of Bluetooth connection. The broadcasted information is automatically adjusted to the specificity of a given time/place, limiting the need for further processing at the user side. The proposal is not a competitor to current solutions, such as generic mobile applications providing city services based on geo-location. It is rather a natural supplement and extension to the already existing intelligent transportation systems. According to the idea "right information at the right place and the right time," the proposed system provides the information well suited for the place of its access, as well as minimizes the efforts related to parameterized access and personalized filtering.

Keywords Smart city · Public information · BYOD · Intelligent transportation systems

1 Introduction

People always need to be well informed to better understand their situation. The more complex the environment, the more information is needed to deal with this complexity. This is particularly visible in crowded and continuously evolving places, such as modern cities.

Throughout the history, cities have always tried to inform their citizens and visitors about the most important things, both permanent and temporary. To that end, several methods, techniques, and technologies have been used, starting from heralds

J. Rykowski (✉) · W. Cellary
Department of Information Technology, Poznan University of Economics,
Niepodległości 10, 61-875 Poznań, Poland
e-mail: rykowski@kti.ue.poznan.pl

W. Cellary
e-mail: cellary@kti.ue.poznan.pl

J. R. Gil-Garcia et al. (eds.), *Smarter as the New Urban Agenda A Comprehensive View of the 21st Century City,* Public Administration and Information Technology 11,
DOI 10.1007/978-3-319-17620-8_6

in the Middle Ages, through wall hanging posters to contemporary light-emitting diode (LED) displays. For both historical and organizational reasons, city information has been divided into independent sectors, each sector using a different way of information dissemination—from paper to mobile-phone applications (Hadidi and Rezgui 2010). Even if the information provided by these methods is useful, still the whole city information system is not consistent, holistic, and comprehensive, not to mention data quality and accuracy, as well as cultural and ecological factors (Eom and Kim 2014). As a result, to be well informed in a particular instance, a person has to collect partial information from different sectorial systems. For example, if someone who does not know the city well would like to get to a certain institution, first he/she has to consult the city Web site to get its location, and to identify the bus/tram line to get there. Then, at the bus stop, he/she can use graphical displays to learn about bus schedule. At the destination, he/she has to consult a map, either a paper one or an electronic one supported by a GPS-operated (GPS 2014) personal device to find the way to the building. At the entrance to the building, he/she has to read a table to know the location of a given office. Sometimes it may create a problem if the person cannot clearly state the purpose of the visit and does not understand well the internal organization of the institution. Finally, if he/she is lucky and passes all the previous steps, he/she presents him/herself to the right clerk.

From the above example, even if theoretically a person has full access to information, he/she has to perform a number of actions and use several techniques, devices, and senses to be successful. If we take into account other sectors of public life, such as health care, police, or sport and cultural facilities (city stadium, concert halls), we can see that city information tailored to various sectors and provided via different communications channels is inefficient and inconvenient. Thus, we should enumerate several challenges for the modern city administration: making communication attractive to citizens, stimulating the information exchange without forcing people to limit their privacy, and looking beyond the technology and understanding its potential (Meijer et al. 2012), while moving toward smart cities (Nam and Pardo 2011).

People, if lost, always search for help. In the above example, many people instead of using a GPS map or a timetable of a bus line would simply ask somebody in the street for advice. Such an interaction is anonymous and depends on the actual context—place and time. For example, a person at a bus stop asks a question "when does the next bus arrive?" seeing a street closed for renovation asks: "what is the best detour?" or seeing a tourist attraction—"how to get to the entrance?"

Nowadays, citizens are supported by electronic displays, located in the streets. Such displays are usually big enough to be visible from a distance (tens of meters for pedestrians, hundreds of meters for vehicle drivers), showing such information as the number of non-occupied parking places nearby, a schedule of the next trams/buses at a stop, weather conditions at the location, etc. Such graphical displays are expensive and hard to maintain—these are complicated and fragile electronic devices hung outdoors. Also, direct sun or snow may strongly limit their visibility; so may the crowd or a bigger vehicle on the road.

To improve the situation in a smart city, in this chapter we recommend using a network of specialized electronic devices to propagate city-related information in a consistent way, regardless of the abovementioned sectors, information type, and purpose. Therefore, we propose to install small, cheap, distributed access points to city information which may be accessed anonymously in an ad hoc manner by any person equipped with a personal mobile device, such as a mobile phone, smartphone, or tablet, according to the bring your own device (BYOD) idea (BYOD 2014). From the technical point of view, an access point is a miniaturized, Linux-based computer equipped with wide area networking (Ethernet or GSM-based) and short distance networking (e.g., Bluetooth 2014), such as Raspberry (Raspberry 2014). An access point located at a given place is continuously provided via the wide area network with time-dependent information that is specific for that place. This information may be read on a personal mobile device by any person that is close enough to the access point via a short distance network, which preserves anonymity and privacy. If, for example, an access point is located at a bus stop, then it stores, among others, the timetable for this bus stop. Moreover, if the information stored in access points is updated in real time according to the situation (e.g., current delay of a bus), the answers to local users' requests reflect actual rather than planned situations. Some access points may also be located on board of public vehicles—buses and trams—to distribute information among their passengers. The proposed solution not only provides a reasonable trade-off between "being well informed" and "remaining anonymous," but it also limits the need for self-learning, thus reducing social divide and making the city information available to anybody. However, as it is restricted to the access to public information, the proposal is not a cure for every smart city problem—it should be treated as an extension to nowadays intelligent transportation systems (ITS) services.

The remainder of the chapter is organized as follows. First, current solutions in the area of city information systems including intelligent transportation systems already installed in some cities are presented followed by a discussion on their disadvantages and restrictions. Then, basic technologies for ad hoc networking, as well as architectures for existing systems are described. Next, main objectives of ad hoc networking in public places within a city are discussed. Then basic architecture and technologies used to implement our proposal are presented, followed by descriptions of several usage scenarios and economic analysis. Finally, some conclusions are drawn and directions for possible extensions are pointed out.

2 Current Public Information Systems and Services

In modern society, efficient access to information becomes crucial (Pardo et al. 2012). Due to the progress in information and communication technology, especially mobile communications, up-to-date, real-time information services became a must for many people (Ojo et al. 2013; Scholl 2012). The need for online information concerns also domains traditionally using paper-based technologies, such as central and local governments. They are evolving toward eGovernments, eGovernance, and recently smart governments (Scholl et al. 2010, 2012), and smart cities (Chourabi et al. 2012). Further progress is, however, envisaged.

Public information systems have been used for many years, starting from medieval heralds, via plates on doors and posters on walls, to modern solutions, such as electronic displays of various kinds. Although the basic ways of receiving the information—involving voice and sight—have not changed, recently we have observed an evolution in strategic goals of such systems addressed to urban areas. The highest attention is focused on transportation. In recent years, due to the growing car population, the main goal of such systems has been to better inform drivers about the current situation on roads. Nowadays, a common trend is to encourage drivers of private cars to switch to public transportation within the city area. Thus, the main goal is shifted toward better quality information for pedestrians and passengers of trams and buses. Certainly, public transportation is less convenient than a private car, therefore, to counterbalance the advantages of private transportation, potential passengers have to be provided with instantaneously updated information enabling them to use city infrastructure efficiently. The question arises what to use as the base for such information broadcasting—voice messages or optical (graphical) signs.

Voice-based public information systems have been quite popular in developed countries for the past 20 years. For example, traffic message channel (TMC) service (TMC 2014) is used in many European countries to inform drivers about potential threats and troubleshooting on the road. TMC service notifications are broadcasted as voice messages via standard ultra high frequency (UHF) channels, and as such received by most of on-board radio receivers. The system is more efficient than its main counterpart—CB radio—as it is accessible over a wider area and provides more trusted information. The main disadvantage of the system is the fact that the same information (1) reaches all drivers, not necessarily interested in particular messages due to, for example, different location, and (2) provokes the so-called stampede effect when several drivers take the same decision at the same time, resulting in moving the traffic jam to an alternate street. TMC is also useless for pedestrians and cyclists, not to mention the passengers of public transportation.

Optical-sign-based information systems, usually identified with ITS term (ITS 2014), use dedicated graphical signs for dynamically displaying and updating important information. Basically, these signs show the way to the nearest parking place and inform about traffic jams (interesting to drivers), and provide timetables for buses and trams (interesting to passengers). The main goal of such systems is to encourage car drivers to use public transportation, by leaving the car and immediately switching to a well-scheduled tram/bus. As neither private cars nor public vehicles are able to fully satisfy alone the transportation needs of the citizens, intelligent interconnection of them may result in (1) reduction of car traffic in the city center, and (2) increased satisfaction of the passengers efficiently guided to their destinations. Synchronization of different transportation means is followed by improved throughput of the city transportation system, smaller air pollution, time saving, etc.

Munich COMFORT with its Elektronische Fahrplanauskunft (EFA) large-scale displays and park-and-ride utilities is one of the first successful ITS installations

(COMFORT2014). RegLog in Regensburg (Regensburg 2014) is another example of successful implementation of the ITS idea. This German city is famous for its charming, narrow streets in the historic center. Thus, an idea to limit the traffic there for both private cars and business-transport vehicles was put into practice. It was detected that majority of supplies could be delivered with smaller cars, moreover, shared by several suppliers and navigated by the system from place to place. As a result, car traffic in the old town area was greatly reduced.

3 Rationale Behind Improved City Information Systems and Services

Past experience in the usage of ITS and related systems (Anthopoulos and Fitsilis 2013; SmartCities 2014) in many cities across Europe, the USA (IBM 2014), Korea, and Japan showed many difficulties and obstacles. Below, we provide their analysis from two points of view: The city (administrator) and the end-user.

A typical ITS only partially fulfills information needs of the citizens. Therefore, its large graphical displays are still accompanied by door plates and posters, planned schedules hanging on walls of bus/tram stops, entrance/door labels with office locations or office clerk names, making the whole city information system tailored to individual sectors.

As mentioned above, nowadays this information is fed to the Internet, as a part of city Web pages. However, this information must be searched for prior to use, thus forcing the user to provide several parameters limiting his/her anonymity and privacy. Even if some of these parameters may be generated automatically by a mobile device (e.g., a geo-location), such automation is of limited use due to both organizational and physical restrictions (e.g., the lack of availability of GPS data inside buildings and the underground railroads).

The city information system is also hardly ever adjustable to sudden changes and incidental cases, such as changes in communication lines due to an accident or festivals (such as Pulaski Day or Saint Patrick's Day parades in the US cities, completely stopping the traffic in the city center). In such cases, graphical displays are usually switched off not to mislead people, and some paper-based posters are hung on the walls. Such a situation is paradoxical—people are deprived of actual information when they really need it.

From the city point of view, the main problem of broad deployment of an ITS is related to its economy. The system is useful only if it covers a substantial part of the city. However, covering a large area requires spending a lot of money on hundreds of expensive, large-scale displays at every crucial point (e.g., all the tram/bus stops), thousands of sensors and detectors, continuous monitoring of the traffic, etc. The installed hardware must be maintained and repaired if needed, which in turn involves street cranes, highly-qualified repair teams, and more. Also the software and back-office servers and services must be continuously administrated and maintained. Dedicated to and written especially for a given location/city, the software needs a lot of maintenance effort to correct detected errors, improvements at users' requests, etc. And, as in the case of every distributed system, maintaining changes for the whole city requires efficient remote-communication channels or frequent on-site visits.

As mentioned above, most of the city ITSs provide access to information via dedicated applications installed on mobile devices in parallel to large-scale information displays. However, this information channel poses some specific problems. First, as the mobile application is hardly personalized for the reasons described above, even if it needs no user manual and intensive learning, it is sometimes very tedious in use and requires many manual operations from the user. Second, usually "free" Wi-Fi access is not possible due to the limited number of public access points across the city, thus people are forced to communicate via paid channels such as GPRS/EDGE, HSPA (HSPA 2014), or recently LTE (LTE 2011). If the system is frequently used and raw data are sent and processed by mobile applications, the personal traffic is quite big and costly. Third, the application is hardly accepted by handicapped people as it does not assume nonstandard usage of the information. For example, sand-blind people cannot use voice messages instead of screen-displayed timetable of a bus, wheelchair-bound person cannot get additional information about the height of curbs and stairs, etc. Even if the abovementioned extensions are feasible by means of current technologies, they are not observed in current implementations for economic and organizational reasons.

ITS infrastructure is not only related to the front-office hardware and software but one has to also mention core servers and databases required to deal with user-defined queries, both long-lasting (such as information about planned schedule of buses at a particular bus stop) and incidental ("when will the next bus arrive at a given bus stop?"). Back-office infrastructure must deal with uneven load distribution due to periodic rush hours, when the system is overloaded, and calm hours (e.g., deep in the night), when almost nobody accesses the information. For economic reasons, not to over invest in unnecessary system elements, ITS services usually concentrate on core city-related services, such as public transportation and tourist information at certain places, but nothing more.

For some cities, a generic application is available, automatically parameterized by the GPS location known by the mobile. However, in such a case users must pay for their mobile access to the Internet and accept the fact of limited privacy. Moreover, if the GPS signal is not accessible, the location must be provided manually, which may be difficult or at least annoying with the small-size screen/keyboard of a mobile.

In this chapter, we propose providing city information in a different way. First of all, we assume one single broadcasting channel for any information, thus abandoning traditional split into individual sectors, such as public transportation, city offices, tourist information, etc. The purpose of the information does not matter—we assume one method for presenting all information. Second, we propose using personal mobile devices to access the city information system, with access mode and information content depending on place and time of access. In doing so, we propose applying BYOD idea and short-distance radio communication in ad hoc (incidental), anonymous mode (Wei et al. 2013).

We consider our proposal not as a competitor to current solutions (such as generic mobile application with city services based on geo-location), but rather as a natural supplement and extension to already existing ITS applications. We tend to provide the same information as it is now available within the scope of any ITS and city Web pages, however, in a ready-to-use manner. According to the idea "right information at the right place and time," the proposed system provides information well suited for the place of its access, minimizing efforts related with parameterized access and personalized filtering.

BYOD is not only applied to personalize visualization of the information, but also to enable ad hoc, incidental interaction with the system. The next section is devoted to such ad hoc interaction, presenting a discussion on basic requirements and technologies to provide such access to ITS information.

4 Ad Hoc Networking in Public Places—Requirements and Technologies

In our approach we propose using a simplified ad hoc architecture aimed at a set of fixed, distributed access points, and mobile stations capable of connecting to such points. Such an assumption greatly reduces overall complexity of the system. The fixed part of the system is to be administrated by the city, whereas mobile stations act according to the BYOD idea. Access points are distributed over the city area, at certain fixed places-of-interest, such as bus/tram/metro stops, inside public-transportation vehicles, at public buildings (offices), stadiums, and interesting places (monuments, viewpoints, museums, etc.).

We assume in this chapter that ad hoc access is restricted to local (closed-neighborhood) information only. In other words, ad hoc access points provide information specific to the location of a given point and time, in a raw format. All mobile devices connected to the access point get the same, restricted, locality, and time-dependent information that may be further personalized inside the device.

Ad hoc communication naturally deals with anonymity and there is no need for prior registration. We assume that anybody can get access to any local information provided by any city-administered access point. How this information is further processed and used is a personal decision of every individual. As the access is anonymous, the system never registers connections and information exchanged, and stores no history of previous interactions with users.

With respect to technology choice, we are restricted by the devices owned by people currently, on one hand, and the physical possibilities of radio-based communication at a short distance on the other hand. As for the devices, we assume general accordance with the BYOD idea: everybody carries a mobile device, such as a smartphone, pad or tablet, capable of processing information in a personalized manner. As for the communication scheme at a short distance, we propose to apply Bluetooth with its broadcasting extension. Bluetooth capabilities are sufficient to deal with transmission of the information available in city access points, such as traffic info or bus schedule. The useful signal range (usually a few meters) is also reasonable, as well as low energy consumption.

5 Objectives of Ad Hoc Networking in Public Places Within City Area

Ad hoc anonymous local networking for smart city information contributes to the solution of two basic problems namely: efficiency of user-raised queries, and reasonable cost of system deployment and maintenance. First, let us discuss the way of improving of user queries. In the classical approach (Fig. 1), a centralized three-stage architecture is used: (1) client-side query formulation based on local data (filtering), (2) query interpretation at the server-side, and (3) sending results for client-side presentation.

The first stage is related to preparation of some information to be used for filtering out the information to be fetched, based on user-defined set of keywords (or other kinds of patterns), or pulled-down menu choice. User-defined data are completed by some automatically added information, such as the geo-location of a mobile device fetched from its GPS unit, some predefined values stored as user preferences, access rights, session processing (user login/password), etc. The completed query is sent to the centralized service where (Stage 2) it is interpreted, and the results are collected. These results are in turn sent back as server response (Stage 3), to be finally formatted there (i.e., adjusting to the screen size, coloring, etc.) and presented to the user.

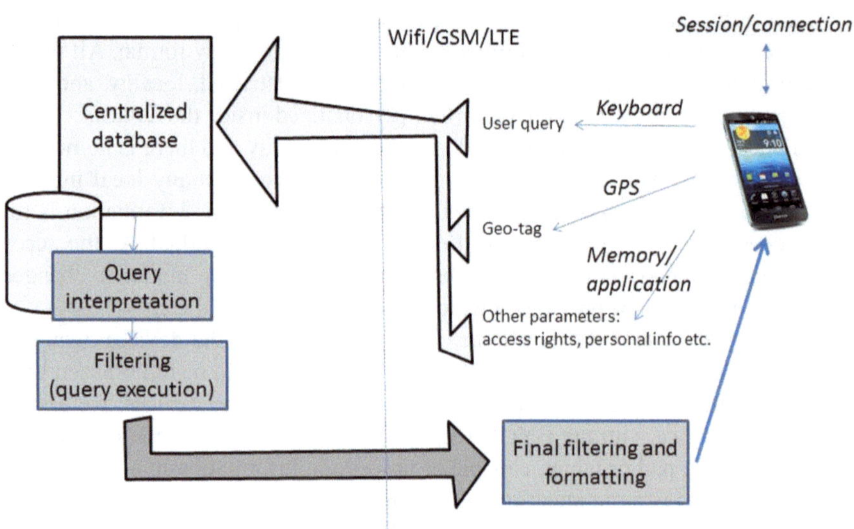

Fig. 1 Classical queries in a typical centralized service

For example, a user is interested when the next bus No. 5 comes to the stop he/she is close to. The user choses the bus number from the pull-down menu of the list and presses “Schedule” button. This request is supplemented with the GPS location of the mobile and sent to the service. The server retrieves a full schedule of a given bus from the centralized database and compares the geo-locations of the stops with the received parameter. Once the “closest” stop is identified, partial information from the schedule is sent back. Once this information is returned, it is displayed in user-preferable format.

Note that the above process is sensitive to namely: (1) rush hours (for a few hours a day, the queries may be substantially delayed due to a huge number of requests, mostly very similar or even identical, but issued by different users), and (2) user education level—one must know the application very well to compose right queries in given circumstances. Moreover, usually several interactions are needed to get access to useful information, including the login and session maintenance. In the approach proposed in this chapter, information stored in the central server is preprocessed according to locality and sent to access points (Fig. 2). For example, if we take into account a planned and real-time schedule of a bus, this information is divided into parts and each part is sent to a corresponding point, according to the idea “right information at the right place.” Such parted information needs (almost) no further processing, neither at the server nor at the client side as it is already adjusted to the specificity of the location where it is offered to the end users.

Information fetched from a local access point may be directly presented to the user in the raw format. Such information is equivalent to the information presented on large-scale information displays. However, if needed, this information may be further personalized. For example, instead of displaying the information about all the “next” buses, the device may filter out only the line related with the bus the user

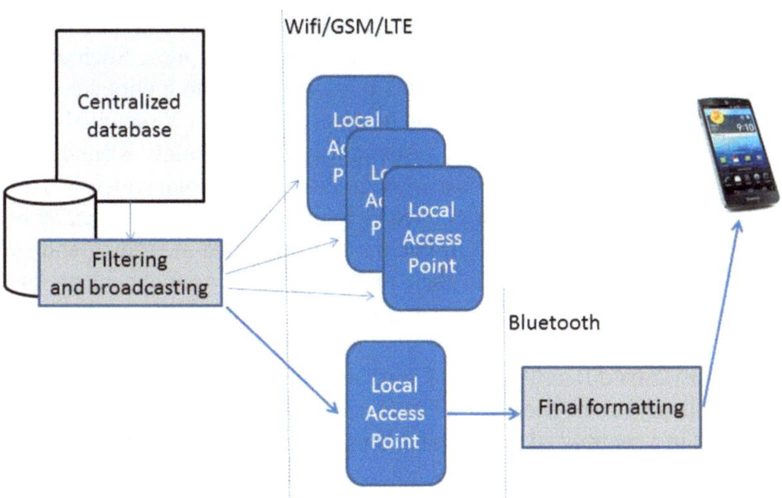

Fig. 2 At-the-place information broadcasting

is waiting for, assuming that earlier he/she provided the bus line number. Moreover, if the information is to be presented to a sand-blind user, such a limited message may be read by means of voice synthesizer instead of displaying it on the screen, for instance, 1 min before the expected arrival of the bus.

Note that the notion of "the next" bus or tram is different for different users, and even for different places and moments in time for the same user. Thus, it cannot be fixed and should be automatically set up based on local data and user requirements and habits. Also note that filtering out the right "next" bus from local information concerning only a few buses and one single stop is trivial as compared to centralized filtering of any bus at any place and at any moment.

To recapitulate, by restricting local access to local network traffic, the need for network throughput is reduced, so is the need for big computational resources at the server side and expensive information displays at the client side. Due to the nature of the proposed information access points, there is limited need for exhaustive communication among key servers and the access points. Instead, local information is broadcasted to access points at every change, not necessary frequently (e.g., every few minutes or even hours). There is no rush hour problem as the local queries are restricted to the local access point, not influencing broadcasting the information by key city services. Finally, even if the bandwidth of proposed local connection (Bluetooth) is not big, the interaction is substantially faster as it is restricted to a single, few-meter-long radio connection.

6 Local Announcements

City information access points may also serve as places of local announcements. Traditionally, to publish an announcement locally, one uses small sheets of paper stuck in places frequently attended by local community, usually on special advertising columns. Passersby may stop for a moment and take a look. Such announcement may concern very different things, from a small sale to a public concert. An announcement is valid for certain period of time, afterward it is removed or simply covered by other announcements. We propose using access points as announcement repositories. Anyone may post such a message in an access point which will be visible for the next few days to people connected to that access point. Especially people who are bored waiting for a tram or bus may be interested in reading and storing such electronic announcements in their mobile devices. Announcements may be posted in a traditional form, not only as a standard unformatted text, but also as a multimedia message—voice or even a short movie clip.

We propose two different kinds of announcements namely: local, to be stored and accessible only at the access point that was used for their upload, and general, to be processed in a centralized manner after summarizing the announcements coming from different access points. Local announcements (Fig. 3, bottom-right side) are used mainly to inform other people about different things, as presented above. These announcements are never sent outside the access point, and they are automatically

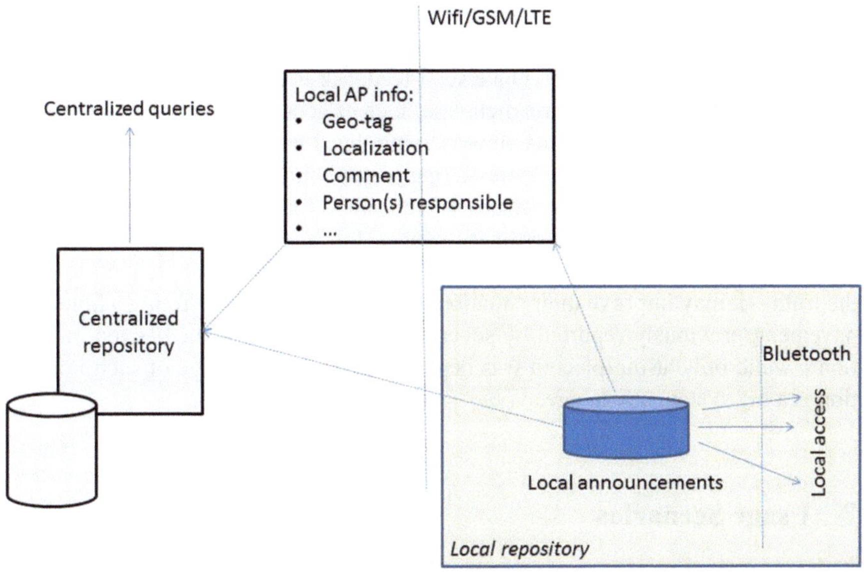

Fig. 3 Local-access announcements

removed after some time (reasonably after a few days). Generic announcements are used to report some problems to the city administration. For example, a ticket machine that is out of order should be reported. Nowadays, a traditional phone call is used for that, which is not very efficient (Fig. 4).

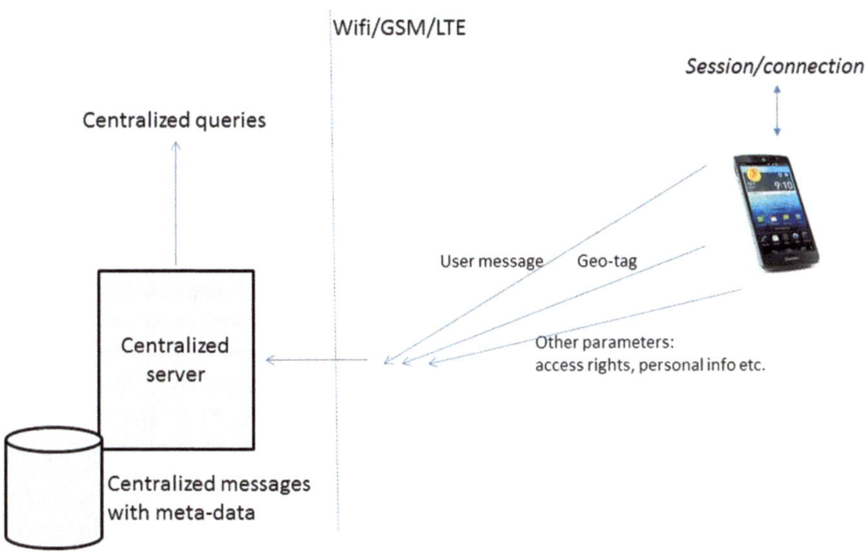

Fig. 4 Traditional distributed commenting scheme

We propose to use city information access points to collect such feedback sent as a text, voice, or video comment from a personal mobile device, similar to local announcements described above. Once such feedback is received, it is propagated to the central database with certain metadata, such as geo-location of the access point, date and time, etc. (Fig. 3, data flow is symbolized by dotted lines—right/central part of the figure). Note that the messaging is anonymous.

The proposed commenting scheme is well suited for the "smart" society (SmartSociety 2014) idea: closer to ordinary people, fast and efficient contact and feedback, high-quality information, even if anonymous (as most people tend to report the truth—it may improve their situation, for example, by faster fixing a hole in the pavement, previously reported by several people). Most of the collected information is valid only at-the-place it was deposited, thus local storage of such information is a big system advantage.

7 Usage Scenarios

To present the potential of the proposed idea, in this section we discuss a few basic usage scenarios, addressed to both ordinary and handicapped users.

7.1 Passenger at a Bus Stop

Let us first discuss a trivial but probably the most frequent usage scenario: a potential passenger of a city bus waiting at a certain bus stop. Traditionally, this passenger gets the bus schedule from a timetable hanging nearby, from a large-scale graphical display, or from his/her mobile device by means of GSM networking. In any case, this information is tabularized, that is, either the whole schedule is presented as a big table divided into hours and stops (paper-based info, GSM phone), or as prospect schedule of a few "next" vehicles (large-scale display). In the first case, one has to manually filter out the exact hour and stop to get useful information, whereas in the latter case he/she has to wait for the "right" bus, continuously observing the display. Both these operations are quite annoying.

In the proposed approach, a passenger simply opens a small application on his/her mobile. This application connects to the nearest access point remaining within radio signal range and fetches the broadcasted information. In a situation when user preferences are not declared, the whole fetched information is displayed on the mobile's screen. This is similar to the usage of the abovementioned large-scale display, except that one does not have to pay attention to the surroundings, may hide under sun or rain shield, etc. However, if the user is going to set up some preferred values, such as the number of the bus he/she is waiting for or the destination point, then the information may be filtered out to display only the exact values of any interest to the user. Moreover, an alarm may be set, for example, 1 min before prospective arrival of the "right" (from the user's point of view) bus.

One may claim that the same functionality can be achieved with a generic application parameterized with geo-location. This is true in general, however: (1) such applications are not very common, even in the most modern solutions, (2) one has to pay for GSM traffic, and (3) GPS location cannot be determined in certain places, such as inside buildings, under the ground (metro stations), or even among high buildings in the street. Last but not the least, ad hoc interaction needs only one effort—just being close to the information point, whereas GSM-based interaction involves several manual actions, including starting the application, manually choosing the needed function, sending a query, and waiting for the results.

The presented application is very simple in both implementation and usage. It simply collects all the broadcasted information and displays it according to user's preferences. This interaction generates very low cost—large-scale displays are not needed. Instead, only a cheap and almost maintenance-free access point is to be accessed by the public, with minimized network traffic, even at rush hours.

The proposed approach may be useful to passengers requiring some assistance of the bus driver. For example, a woman with a small child in a baby stroller may ask the bus driver for help in advance, just by sending a message from her mobile phone via an access point to "the next" vehicle. Similarly, a wheelchair-bound person may ask to extend a special platform to get on the bus. Nowadays, such activities are very limited—a woman with a baby stroller must first present herself to the driver (usually by approaching the front door and waving her hand) and then move to the dedicated door (usually—in the middle of the vehicle). She is noticed by the driver only after the bus stops at the platform. Unfortunately, a typical platform to facilitate boarding with a baby stroller may be extended only when the doors are closed. As the driver usually first opens the doors, and then detects the problem, the woman must wait until all the new passengers are inside and the doors are closed. Then, the platform is extended and the doors are opened once again, enabling the stroller to enter the vehicle. However, as a result, instead of being the very first, a person needing assistance is usually the last one to board the vehicle.

7.2 Commenting on Troubleshooting and Problems

Almost everybody was in a situation when he/she wanted to "ask a civil servant to do something" with problems such as: a hole in the sidewalk, an overflowing trash can, a broken ticket machine, graffiti on a wall, etc. Some people report such incidents by phone, but it is quite a time consuming and difficult task for both sides—the caller and the civil servants. The caller has to describe in detail the reported case and provide some additional information, such as geo-location, level of public risk (e.g., for small children and vehicles). The civil servant has to understand and verify the information, and then refer the case to appropriate city services to fix the problem. These activities may seriously increase reaction time and even discourage such citizens' behavior.

Within the proposed approach, one may find a simpler solution to the problem of efficient reporting of small city-related problems. Once a person has noticed a problem and is near the access point, he/she may use this access point as a broker to the proper city service. All the person has to do is to describe the problem in a few words (or even get a predefined description from a pull-down list) and send it. The messages are stored in the access point, and may be easily collected from this place by city administration staff, together with its geo-location, thus creating a kind of a distributed database of comments on reported problems.

The reporting is not restricted to some cases lasting for hours or even days, such as a damaged ticket machine or a full trash can. It may also apply to more dynamic situations, such as a robbery or a car accident. In such cases, however, the comment should be immediately forwarded to the centralized safety system, to initiate instant reaction.

7.3 Unified Guide to Complex City Services

In the previous scenarios, we have shown how to solve single problems, one at a time. However, in real life we deal with more complex cases. Imagine, for example, a person living in the suburbs who must visit a given city office to deal with his/her specific case. The person knows the approximate location of the office building and basic directions of main bus/tram lines, but has very limited knowledge of the city and even smaller of the organization of the offices inside the building. Once this person gets off his/her car at the park-and-ride parking place, he/she connects to the access point, obtaining the location of the nearest bus stop. There, his/her device automatically connects to another information point, providing the schedule of "next" vehicles. On the bus, the device provides information about next stops, possible changes (with real-time schedule), "important" places nearby (such as the office the person is going to) and so on. Finally, at the entrance to the office, he/she connects to the building information point, obtaining the information about the internal organization of the office (such as "Personal taxes—first floor," "VAT—second floor"). The person chooses a location and goes there, switching to the next information point, with detailed data about the public servant names and their responsibility area.

As can be seen, the interaction involves almost no manual operations of the user (except for some scrolling of small-screen contents at the mobile device), and the information presented is continuously updated as the person changes his/her location.

Again, one may claim that the abovementioned information can be provided in a traditional way, for instance, by using generic Web applications and location-based services (LBSs), such as Google Maps. This is true, but such activities require extensive navigation on Web pages and search engines and manual filtering of the information coming from different places and presented in different formats. Moreover, the exact GPS location is hardly achieved inside the building, thus the guidance stops at the entrance to the building, leaving the person alone with paper-based information or forcing human-based help.

8 Economic and Social Aspects of the Proposed City Information System

As mentioned in the introduction, the economic aspects of the proposed city information system are very important because the very low cost of the installation and maintenance of access points is one of its main advantages. Below we discuss these aspects in detail.

8.1 Cost of Networking Infrastructure

To analyze the overall cost of the networking infrastructure, we have to distinguish two cases: the city already operates an ITS system, or there is no such installation yet. In the first case, we can easily extend the existing infrastructure by adding small, cheap, and maintenance-free access points at every place where any ITS device is distributed, such as a graphical display or an interactive kiosk. The new devices, consuming relatively low energy, could be powered by the existing infrastructure.

In the latter case, the system should be implemented in two steps. First, the access points should be connected via a traditional GSM networking scheme (or WiMAX, if available) to the central system. Power supply of these access points can be achieved locally. So the cost of installation is very low, whereas the cost of maintenance is mainly limited by the costs of GSM networking. As the amount of data transmitted is not large (just periodically broadcasted information), one may expect that the transmission cost will be low as well.

In the second step, one may wire all the access points, thus achieving zero-cost transmission and power supply. These activities can start from the city center (shortest connections) and spread toward the suburbs. As this would be a long-lasting or even never-ending process, a mixed solution will be used in practice—the more points are wired, the lower the overall cost of everyday maintenance.

8.2 Cost of Hardware and Software

Ad hoc access points may be achieved at minimum cost of installation: a very small computer running Linux or a similar system with wide area networking—Ethernet or GSM-based, and short distance networking—Bluetooth, no screen/mouse/keyboard, no case (totally hidden and publicly inaccessible). Consequently, there is almost no need for maintenance. Recently introduced microboards, such as Raspberry

(Raspberry 2014) or Atmel-based solutions are the best candidates, showing required functionality and accessible in retail at the cost less than $30. When we compare this cost to expensive large-scale graphical displays, we can see that the proposed solution is characterized by minimum investment and low-maintenance cost.

Centralized (back-end) hardware is also implemented at minimum cost as there is no problem of rush hours and the system does not have to be oversized. There is no need for an expensive server farm and powerful system to interpret user queries; moreover, if implemented as an extension to an existing ITS, both the information and propagation methods are already there. The same comes for software for data distribution—standard solutions and technologies, such as SQL, Java, Tomcat, or Apache are sufficient to achieve this goal.

8.3 Cost of the Preparation of the Information to be Broadcasted

Most of the broadcasted information concerns diverse public services, including real-time schedule of city transportation system. All the data about the schedules are already stored in a centralized database owned by the city (or city-related company maintaining public transportation), thus, they may be retrieved at no cost. Moreover, this information is already parted to geo-locations (mainly bus/tram stops), so there is no need for expensive processing, structuring, or formatting.

8.4 Risk and Security Analysis

As for any other anonymous or semi-anonymous systems, there is a risk that some people will try to cheat or abuse the service. There is no such problem for broadcasted data as they are created by the system and cannot be changed by end users. We conclude that the only case when an end user is able to abuse the service is citizens' comments. We have described three main points of usage of such comments namely: announcements about local problems, local announcements, and notifications to drivers of the "next" vehicles. For the first case, it is sufficient not to react to a single comment but wait for at least a few comments on the same topic to start an action. For the second case, the expected level of fraud and abuse is probably similar to the one of nowadays paper-based columns—rather low, and it is up to the reader to accept or react to the information or not. And as for the last case, even if the information addressed to a driver is a joke or fraud, it poses no danger—the driver always takes a look to see if the case needs a special activity, thus the message is treated only as a warning rather an order.

To sum up, as can be seen from the above short discussion, even if no security measures are applied, the whole system is not vulnerable to fraud.

9 Conclusions and Directions for Future Work

As nowadays almost everybody appearing in a smart city area is equipped with a mobile device, a city information system may be organized following the BYOD idea instead of large, public displays. As city information access points proposed in this chapter are very cheap and need no maintenance, for the same amount of city money their network may be much denser than the network of public displays. They may be accessed anonymously preserving people's privacy, and then may provide personalized information and communication services that are more useful and convenient. As the proposed system uses standard communication technologies, hardware and software, it may be immediately implemented and deployed at minimum cost in almost any city. If a city is already equipped with a public ITS, the proposal may be a cheap and efficient extension to the current installation, attracting more users and providing more services.

What we propose in this chapter is a generic solution. Like every generic solution, also this one may be extended toward other application areas, also business-oriented. We have described a clear case when the city is responsible for the installation and maintenance of the access points, as well as for most information content, paid from public money. However, one may propose a mixed solution in which this infrastructure is also used for some business purposes, also provided in the scope of public–private partnership and citizen initiatives. For example, certain access points, partially financed by shopping centers nearby, may distribute some information related to the shops and their offer. One may propose a city game, addressed to those who are bored waiting at a bus/tram stop. Fun group of city history and tradition may prepare extended information about interesting and historical places, such as monuments, environment protection agency—information about pollution level at a given place/time important to allergy sufferers, etc. The extensions are accessible on condition that the local-networking channel is not overloaded (i.e., not during rush hours and at crowded places), and after installing separate, dedicated software in the personal mobile devices of potential users. As this functionality is not mandatory for average users only looking for city information, the main advantages of the proposed ad hoc interaction are still preserved.

References

Al-Hadidi, A., & Rezgui, Y. (2010). Adoption and diffusion of m-government: Challenges and future directions for research. *IFIP Advances in Information and Communication Technology, 336,* 88–94.

Anthopoulos, L., & Fitsilis, P. (2013). Using classification and roadmapping techniques for smart city viability's realization. *Electronic Journal of e-Government, 11*(1), 326–336.

Bring Your Own Device BYOD idea. (2014). Gartner, Inc. http://www.gartner.com/technology/topics/byod.jsp. Accessed March 2014.

Chourabi, H., Nam, T., Walker, S., Gil-Garcia, J. R., Mellouli, S., Nahon, K., Pardo, T. A., & Scholl, H. J. (2012). Understanding smart cities: An integrative framework. Proceeding of 45th Hawaii International Conference on System Sciences, 2012. doi:10.1109/HICSS.2012.615.
Eom, S.-J., & Kim, J. H. (2014). The adoption of public smartphone applications in Korea: Empirical analysis on maturity level and influential factors. *Government Information Quarterly*. doi: 10.1016/j.giq.2014.01.005.
Feasibility study for Further Advancements for E-UTRA (LTE-Advanced). (2011). 3GPP TR 36.912. Accessed March 2014.
High Speed Uplink/Downlink Packet Access. (2014). 3GPP Global Initiative portal. http://www.3gpp.org/HSPA. Accessed March 2014.
Intelligent Transport Systems. (2014). http://www.etsi.org/technologies-clusters/technologies/intelligent-transport. Accessed March 2014.
Meijer, A., Koops, B., Pieterson, W., Overman, S., & ten Tije, S. (2012). Government 2.0: Key challenges to its realization. *Electronic Journal of e-Government, 10*(1), 59–69.
Munich COMFORT—cooperative transport management for Munich and its surrounding area. (2014). http://www.mobinet.de/Fachinformation/pdf/mc_e.pdf. Accessed March 2014.
Nam, T., & Pardo, T. A. (2011). Conceptualizing smart city with dimensions of technology, people, and institutions. The Proceedings of the 12th Annual International Conference on Digital Government Research, pp. 282–291.
Ojo, A., Janowski, T., & Awotwi, J. (2013). Enabling development through governance and mobile technology. *Government Information Quarterly, 30,* 32–45.
Pardo, T. A., Nam, T., & Burke, G. B. (2012). E-government interoperability: Interaction of policy. *Management, and Technology Dimensions, Social Science Computer Review, 30*. doi: 10.1177/0894439310392184.
Raspberry Pi—Quick Start Guide. (2014). Raspberry Pi Foundation. http://www.raspberrypi.org/quick-start-guide. Accessed March 2014.
Regensburg „RegLog®—City-Logistik". (2014). http://www.werkstatt-stadt.de/en/projects/8/. Accessed March 2014.
Scholl, J. (2012). Five trends that matter: Challenges to 21st century electronic government. *Information Policy Volume, 17*(3), 317–327.
Scholl, H., Janssen, M., Traunmüller, R., & Wimmer, M. (2010). Electronic government. *Advances of Management Information Systems, 17,* 11.
Scholl, J., Janssen, M., Wimmer, M. A., Moe, C. E., & Flak, L. S. (2012). Electronic government: 11th IFIP WG 8.5 International Conference. EGOV 2012.
Smarter cities: Infrastructure. Operations. People. (2014). IBM project "Smart Cities". http://www.ibm.com/smarterplanet/us/en/smarter_cities/overview/. Accessed March 2014.
The Traffic Message Channel (TMC). (2014). http://www.tmcforum.com/technologies/tmc/. Accessed March 2014.
Wei, H. Y., Rykowski, J., & Dixit, S. (2013). *WiFi, WiMAX and LTE multi-hop mesh networks: Basic communication protocols and application areas*. New York: Wiley & Sons, ISBN: 978-0-470-48167-7 May 2013.
What is Bluetooth technology? (2014). http://www.bluetooth.com/Pages/what-is-bluetooth-technology.aspx. Accessed March 2014.
What is GPS? System Overview. (2014). http://www.gps.gov/systems/gps/. Accessed March 2014.
When People Meet Machines to Build a Smarter Society. (2014). Smart society project. http://smart.inf.ed.ac.uk/. Accessed March 2014.
Why Smart cities? Comprehensive study on intelligent solutions provided in small- and medium-size cities across Europe. (2014). Smart Cities Programme—Vienna University of Technology, OTB Research Institute for Housing, Delft University of Technology, University of Ljubljana. http://www.smart-cities.eu/why-smart-cities.html. Accessed March 2014.

Jarogniew Rykowski received the M.Sc. degree in computer science from the Technical University of Poznan, Poland in 1986 and the Ph.D. degree in computer science from the Technical University of Gdansk, Poland in 1995. In 2008 he received habilitation degree from the Institute of Computer Science, Polish Academy of Science (Warsaw, Poland). From 1986 to 1992 he was with the Institute of Computing Science at the Technical University of Poznan. From 1992 to 1995 he worked in the Franco-Polish School of New Information and Communication Technologies in Poznan. Since 1996 he has been with the Poznan University of Economics. His recent in-terests have gone toward applications of Internet of Things and calm-computing devices, including "intelligent buildings and workplaces," semantic support for IoT systems, telematics, ad hoc and multi-hop networking. He is the author or coauthor of 3 books, over 45 papers in journals and conference proceedings, and 2 patents.

Wojciech Cellary is a computer scientist, head of the Department of Information Technology at the Poznan University of Economics. In his professional career he worked at nine universities in Poland, France, and Italy. His research interests are currently focused on electronic business and economy, electronic government, information society, smart cities, and multimedia. He is an author of 11 books and over 150 scientific papers. He was a leader of many national and international scientific and industrial projects supported by Polish, French, and American industry, Polish ministries, and EU Framework Programs. He served as a consultant to many Polish ministries, Parliament and Senate, as well as European Commission. Recently, he served as a chairman of the Council of Digitization that is advisory body to Polish government concerning eGovernment solutions. In 2011 he chaired the 6th European Ministerial eGovernment Conference which was a major European event in the area of eGovernment.

Toward a Methodological Approach to Assess Public Value in Smart Cities

Michele Osella, Enrico Ferro and Elisa Pautasso

Abstract This chapter proposes a novel framework aimed at measuring performances of smart cities. The methodological approach underlying the framework has its roots in an in-depth analysis of the smart city paradigm conducted from the perspective of urban governance. In this context, the notion of public value is seen as a backdrop for exploring the various ways in which a value for society can be created in a smart city. With this respect, a multidisciplinary synthesis of various strands of literature related to smart cities paves the way to the conceptualization of a framework meant to evaluate the "smartness" of a city through the lenses of economic, social, and environmental performances, in line with the "triple sustainability" principle. This vision is subsequently operationalized by means of a harmonized set of key performance indicators (KPIs) that can be grouped into two categories (called "core" and "ancillary"): whilst "core" indicators are identified with the intent to allow international comparability and to help policy makers in benchmarking their city on a global scale, and "ancillary" indicators are crafted considering the peculiarities of the city local context. Finally, the Italian city of Turin is used as a case study for testing the proposed assessment tool.

Keywords Smart city · KPI · Public value · Measurement · Sustainability

1 Introduction

An overwhelming body of scientific evidence now clearly indicates that the world is facing a serious sustainability challenge (Stern 2007). The unprecedented growth in the world population occurred over the last centuries coupled with the gradual

M. Osella (✉) · E. Ferro · E. Pautasso
Istituto Superiore Mario Boella, Via Pier Carlo Boggio, 61, 10138 Torino, Italy
e-mail: osella@ismb.it

E. Ferro
e-mail: ferro@ismb.it

E. Pautasso
e-mail: pautasso@ismb.it

J. R. Gil-Garcia et al. (eds.), *Smarter as the New Urban Agenda A Comprehensive View of the 21st Century City,* Public Administration and Information Technology 11,
DOI 10.1007/978-3-319-17620-8_7

increase in developing countries' spending power has contributed to exacerbate the unsustainability of existing consumption patterns. The drawing of world's natural resources at a faster pace than they can be restored has been proven over the decades to be one of the main pitfalls of modern socioeconomic systems (Meadows et al. 2004). The combined effect of the above phenomena is progressively exacerbating global environmental, economic, and social issues. Mankind is called upon to act united and focused to face a number of major societal challenges. Contemporary governments, businesses, and individuals are faced with an unprecedented responsibility toward future generations. This situation calls for a quick and significant reconceptualization of current economic and societal models, and furthermore, the governance of the required change poses complex policy challenges.

In such a scenario, cities have been identified by many commentators as the battleground, in which global issues may be addressed locally. As a matter of fact, cities are responsible for over 70% of the world's greenhouse emissions (UN-HABITAT 2011) and for 60% of planetary energy demand (Van der Hoeven 2012), but at the same time, they are places where the greatest efficiencies may be obtained. Moreover, half of the world's population already lives and works in cities, (OECD 2006) generating more than 80% of global gross domestic product (GDP) (Dobbs et al. 2011); this approximate power-law distribution is testified by the World Health Organization and UN-HABITAT (2010), according to whom 150 metropolitan urban regions across the world generated almost 50% of the global gross domestic product (GDP). In other words, cities are the locus where a process of deep societal and economic reform should start from. In the "city planet" (Brand 2006) that we observe nowadays, urban centers have a sufficient critical mass in both demographic and economic terms to ignite a planetary revolution. At the same time, cities are also characterized by relevant contextual aspects that need to be duly taken into consideration when translating global recipes into local practices.

In spite of the role played by cities in the global quest for sustainable development, the paucity of systematic frameworks—crafted neither at national level nor at city level—aimed at operationalizing the outcome-based evaluation of public undertakings undermines the assessment of the achievement of medium-term or long-term goals. It seems to be of paramount importance to establish a sound basis to shift policy makers' emphasis from activities to results, from outputs to outcomes, and from how a program operates to the good it accomplishes.

Taking stock of such a gap, research findings illustrated in this chapter provide a contribution to the strand of literature devoted to smart cities by suggesting a methodological approach for the measurement of their performances which comprises economic, social, and environmental aspects. This approach is inspired by the public value theory and incorporates the analysis of the needs expressed by the constellation of actors that interact in a city.

The main research questions the chapter aims to answer are the following ones:

- Which dimensions should be considered for measuring the smartness of a city?
- What are the main criteria that must be adopted in the selection process of performance indicators?

- How do context-specificities influence the definition of such measurement tool?
- How do policy makers benefit from the methodological approach proposed?

Research findings expounded in this chapter have their root in the strategic program of research on smart cities that the Istituto Superiore Mario Boella (ISMB)[1] launched in 2011 with the intent to provide a significant contribution to the creation of a new integrated vision of cities as well as to support urban decision makers to harness ICT as a key catalyst for social, political, and organizational change.

2 Theoretical Background

2.1 Smart City Paradigm

As many commentators highlight, the term "Smart City" is not new. It probably finds its roots in the late 1990s with the smart growth movement (Bollier, 1998). The "new guard" of urban planners embracing the movement were calling for a new vision and new policies aimed at ameliorating phenomena, such as traffic congestion, disconnected neighborhoods, and urban decay having their roots in haphazard urban sprawl (Ewing et al. 2002), which was seen as a financial and social drain. In presence of increasing urban concentration, smartness—achieved through the coordination of housing, transportation, and other infrastructure investments, accompanied by the involvement of local residents in development decisions—was a *conditio sine qua non* for creating livable communities where people effectively benefit from greater built densities (Danielsen et al. 1999).

Nevertheless, it was not until 2005 that some of the top-tier ICT global players—CISCO[2], SIEMENS[3], and IBM[4]—started referring to smart cities as the integration of information systems with urban processes (Harrison and Donnelly 2011). Since then, the term has evolved to capture a more complex concept that many scholars have ventured in trying to craft a comprehensive definition for.

Chourabi et al. (2012) assert that the idea of a smart city itself is still emerging, and the work of defining and conceptualizing it is in progress. In authors' opinion, the concept is used all over the world with different nomenclatures, contexts, and meanings often replacing the word "smart" with adjectives, such as "digital" (Ishida and Isbister 2000) or "intelligent" (Komninos 2009; Lee and Lee 2014).

The first wave of contributions was heavily dependent on city's endowment of hard infrastructure ("physical capital"). In this school of thought, smart city has

[1] www.ismb.it. Accessed 11 June 2014.

[2] http://www.cisco.com/web/learning/le21/le34/downloads/689/nobel/2005/docs/Abdulhakim_Malik.pdf. Accessed 11 June 2014.

[3] http://www.siemens.com/innovation/en/publikationen/publications_pof/pof_spring_2004.htm. Accessed 11 June 2014.

[4] http://ibm.co/Ja1v83. Accessed 11 June 2014.

been conceptualized by Hall et al. (2000) as the urban center of the future, made safe, secure, environmentally green, and efficient because all critical infrastructures are designed, constructed, and maintained making use of advanced, integrated materials, sensors, electronics, and networks that are interfaced with computerized systems comprised of databases, tracking, and decision-making algorithms. The focus on ICTs was emphasized also by Dutton et al. (1987) and Donath (2011).

Another prominent strand of contributions drifts away from technological determinism to underline the importance of human capital and social capital as game-changers in the quest for sustainability and high quality of life (Nam and Pardo 2011). As put by Caragliu et al. (2011), the availability and quality of ICT infrastructures are not the sole ingredients of smart cities. Berry and Glaeser (2005)—for instance—show that the most rapid urban growth rates have been achieved in cities where a high share of educated labor force is available.

In addition, the networked nature of next-generation cities is well emphasized by Hollands (2008) and Harrison (2010).

The paucity of definitional precision, not to mention an underlying self-congratulatory tendency (Hollands 2008), goes hand in hand with the coexistence of alternative perspectives having to do with the direction envisaged for roadmaps of urban development.

Some lines of thought are inclined to place as "north star" the development of a competitive environment thriving economic growth. In the urban milieu, the blossoming of user-driven open innovation ecosystems leveraging open business models of collaboration between citizens, enterprises, and local governments (Schaffers et al. 2011) has the potential to enable local competitiveness and prosperity (Paskaleva 2009). Moreover, cities—becoming business-friendly environments capable of attracting new businesses (Hollands 2008)—may trigger spillover effects on the surroundings (Komninos 2009).

Slightly different is the school of thought that places sustainability at the heart of urban development. The well-known Brundtland Report states that sustainable development is the development that meets the needs of the present without compromising the ability of future generations to meet their own needs (World Commission on Environment and Development 1987). In a world where resources are scarce, a profound attention is worth to be devoted to social and environmental sustainability as a major strategic component of "Smart City" label (Caragliu et al. 2011). Along these lines, according to the Natural Resources Defense Council, a city striving to make itself "smarter" should be committed to become more efficient, sustainable, equitable, livable, and prudent in the management of natural resources[5]. Making reference to companies, Elkington (1998) coined the "triple bottom line," which focuses not just on the economic value of a company or project add, but also on the environmental and social value they add or destroy. By the same token, this approach has been transferred to the urban realm (Inayatullah 2011).

The approach placing sustainability at the heart of urban development has often attempted to balance the manifold nuances of sustainability by finding a common

[5] http://smartercities.nrdc.org/about. Accessed 11 June 2014.

ground among the stakeholders present in the vibrant city life. This tendency could be framed taking inspiration from the "triple helix." This model has emerged as a reference framework for the analysis of knowledge-based innovation systems, and relates the multiple and reciprocal relationships between the three main agencies in the process of knowledge creation and capitalization namely: university, industry, and government (Etzkowitz and Leydesdorff 2000). In the development of smart cities, the triple helix model makes its appearance to baseline urban ecosystems in terms of their traditional and contemporary roles as follows: first as generators of intellectual capital, creators of wealth, and regulators of standards (university, industry, and government, respectively), then as cities that use such attributes to be smart in supporting the social learning, market-based entrepreneurial capacities and knowledge-transfer abilities which are needed to meet the requirements of their regional innovation systems (Lombardi et al. 2012).

To summarize, a bird's eye view on the numerous facets of smart cities comes from the benchmarking exercise about the "smartness" of 70 medium-sized European cities (Giffinger et al. 2007) conducted by the Centre of Regional Science at the Vienna University. In the context of this research undertaking, the comparison was conducted along six different dimensions they are as follows: smart economy, smart mobility, smart environment, smart people, smart living, and smart governance; each dimension was subsequently operationalized in a number of quantitative indicators.

In addition to this experience, many other analysts have developed activities aimed at defining indicators (key performance indicators (KPIs) for a smart city. The main interesting initiatives in this field will be analyzed in the Sect. 4 of this chapter.

In the *mare magnum* succinctly depicted above, for the purpose of the present chapter, we do not intend to coin a brand-new definition of smart city to be added to the multitude already present in the literature. Therefore, for the purpose of the discussion to be conducted in the next sections, the definition proposed in the above-quoted article written by Caragliu et al. (2011) will be employed. Such definition has been chosen on the basis of its ability to reasonably capture all the relevant aspects previously highlighted: "a city is smart when investments in human and social capital and traditional (transport) and modern (ICT) communication infrastructure fuel sustainable economic growth and a high quality of life, with a wise management of natural resources, through participatory governance."

2.2 *Smart Urban Governance*

The concept of smart city includes the notion of governance. As expounded by the Forrester Research, smart governance is the core of smart city initiatives owing to the relentless need of urban systems to coalesce into a "holistic"-integrated approach (Bélissent 2011). Thus, it becomes a paramount to better understand such concept, to draw its contours, and to single out its components (Misuraca et al. 2011).

In the late 1990s, governance was viewed by international organizations such as UNDP, World Bank, UNESCO, and OECD primarily as a form of political regime (Kraay et al. 1999), as the process of exercising economic, political, and administrative authority in the management of a country's affairs at all levels (UNDP 1997), as the use of political authority and exercise of control in a society in relation to the management of its resources for social and economic development (OECD 1997). Such broad definitions mainly focused on the role of public authorities in establishing the environment in which economic operators function and in determining the distribution of benefits as well as the nature of the relationship between the ruler and the ruled. More recently, the European Union published a white paper on European governance (European Commission 2001). The document proposed to revisit governance practices by introducing the concept of "good governance" based on five pillars, namely, openness, participation, accountability, effectiveness, and coherence. These five principles of good governance reinforced those of subsidiarity and proportionality already part of the EU governance framework. The OECD as well provided a definition of good governance that unfolds along a number of dimensions. According to such organization, good governance is participatory, consensus oriented, accountable, transparent, responsive, effective and efficient, equitable and inclusive, and follows the rule of law. It assures that corruption is minimized, the views of minorities are taken into account, and that the voices of the most vulnerable in society are heard in decision making (OECD 2001).

Little literature on smart cities addresses issues related to governance (Alawadhi et al. 2012; Chourabi et al. 2012; Nam and Pardo 2012; Nam and Pardo 2014). According to Mooij (2003), the presence of leadership is important for good governance. In the same way, Lam (2005) emphasizes the presence of a "champion" that collaborates with all stakeholders as an essential factor for good governance; in addition, while implementing good governance, a smart city should hinge on citizen participation (Giffinger et al. 2007) and private/public partnerships (Odendaal 2003). According to Johnston and Hansen (2011), smart governance depends on the implementation of a smart governance infrastructure that should be accountable, responsive, and transparent (Mooij 2003). This infrastructure helps to allow collaboration, data exchange, service integration, and communication (Odendaal 2003).

Altogether, by looking at the evolution undergone by the concept of governance over the last 15 years, it is possible to notice a gradual shift in focus from a mere application of administrative and political authority toward a bidirectional discourse with a diversified constituency who is more and more recognized as an authoritative interlocutor in the process of value creation for society. In this respect, good smart city governance should attempt to achieve two important operational objectives namely: produce effective decisions—i.e., make the best use of information to optimize decision making—and establish adequate incentives—i.e., given that all individuals act in their own self-interest, provide the incentives that produce the best/desired outcome (Ferro et al. 2013). But, in order to achieve these results, it is vital to have developed a clear and strategic vision detailing what value needs to be generated.

2.3 Public Value

It has been two decades since the public value framework emerged, articulated by Moore (1995). By and large, public managers who have been exposed to the idea have embraced it enthusiastically (Alford and O'Flynn 2009), whereas academics have been divided as follows: whilst some are intrigued by it (Stoker 2003; Talbot 2011), others are quite hostile to it (Rhodes and Wanna 2007).

Put it simply, the value delivered to shareholders is the private sector's ultimate measure of a company's success. However, in the public sector, where stakeholders replace shareholders, there is no single or simple "bottom line" for gauging success. In a broad sense, the focus on public value is the analogue of the desire to maximize shareholder value in the private sector; in fact, according to Kelly et al. (2002), all governments should maximize "public value added," that is, the benefits of government action when weighed against the costs (including the opportunity costs of the resources involved).

It goes without saying that this notion of public value does not connote a monadic structure, but rather a collection of ingredients giving life to a multifaceted "value mix" going far beyond traditional public financials. Such "values" in public value vary considerably according to different authors.

A 2002 report by the UK Cabinet Office's Strategy Unit suggests that public value can be understood in three dimensions, namely, services, outcomes, and trust (Kelly et al. 2002).

Cresswell et al. (2006), for their part, assert that public value may be subdivided in two components, that is, the value to the public that results from improving *intra moenia* the government itself, and the "broader" value that results from delivering specific benefits directly to persons, groups, or to the public at large.

Moreover, taking the stance of Benington and Moore (2011), public value consists of three distinct but interrelated processes, which are as follows: clarifying and specifying strategic goals and public value outcomes; creating the environment necessary to achieve these outcomes; and utilizing the required operational resources such as staff, skills, and technology.

Less schematic is the formulation of Hills and Sullivan (2006), which is founded on clusters of core values, some of these overlap or blend into one another. On the one hand, there are clusters relating to the process of public service delivery: these include new public management (NPM) values (Hood 1991) of efficiency, effectiveness, and cost effectiveness as well as broader values, such as involvement of the public, transparency, equity, authorization, and trust. On the other hand, there are clusters of values that relate to the outcome of public services, namely, examples in this vein are quality of life, wellbeing, and happiness; social capital; social cohesion and social inclusion; safety and security; equality, tackling deprivation, and social exclusion; promoting democracy and civic engagement.

A shift from NPM to public value has been sustained in the literature by the mid-2000s. Wallace (2013) considers the public value approach as an "emerging thinking on an enabling state". The argument in favor of a shift from NPM to outcomes-based performance management is based on the notion that is managing performance on outcomes, rather than inputs and processes, will avoid gaming behavior and more closely align assessments on performance to the experiences of the public.

An additional contribution in this field is provided by Cordella and Bonina (2012), who use the public value paradigm as an alternative approach to NPM to study ICT-enabled public sector reforms, and conclude that new e-government indicators are needed to account for public value creation.

Furthermore, many authors analyze e-government—and in general the process of ICT diffusion and usage in public sector—using the public value perspective (Karunasena and Deng 2012; Savoldelli et al. 2013; Bannister and Connolly 2014).

The documented complexity in following this approach derives from the extremely vast scope of inquiry needed to identify and document public value creation: "how can we observe, measure, and document the creation of value for the public?". Some attempts have tried hitherto to find an answer to this nagging question. In fact, the notion of public value spawned the development of performance measurement/management frameworks, attracting the attention of practitioners and management enthusiasts (Kelly et al. 2002). Cole and Parston (2006) crafted the Accenture Public Service Value Model's methodology for measuring how well an organization achieve outcomes and cost-effectiveness over a period of years and, adopting a sectorial perspective, Cresswell et al. (2006) outlined a public value framework for the ROI analysis of government IT estate. Despite some difficulties in operationalizing the concept through wide-ranging measurement systems, the notion of public value may offer a promising way of measuring government performance and guiding policy decisions.

3 Methodological Approach

The methodological approach—stemming from the in-depth analysis of literature regarding the smart city paradigm, the urban governance and the concept of public value—is presented in this section.

Firstly, ISMB strategic program on smart cities adopts a vision in which a city may be considered on the right track toward smartness only if its policies tend to balance economic, environmental, and sociocultural progress in view of the layered structure of societal needs (Maslow 1943). With this respect, it is necessary to stress that a misalignment of priorities may only be considered a temporary solution, thus rendering the resulting disequilibrium (and, consequently, the city) unsustainable (Ferro et al. 2013).

The advancement brought by this chapter in this sphere is the integration between sustainability and the concept of public value, seen as a backdrop for exploring the various ways in which value may be created in a smart city. Given the inherently intangible nature of public value, indicators about sustainability can be used to appraise benefits and repercussions of policies, following the belief that the noble intent of governments to maximize "public value added" (Kelly et al.2002) results in the measure in which the needs of the present stakeholders are satisfied without compromising the ability of future generations to meet their own needs, in line with the seminal Brundtland Report (World Commission on Environment and Development 1987).

Consequently, a set of indicators—or key performance indicators (KPIs)—can be operationalized as a "workbench" for gauging urban sustainability.

The array of KPIs operates in compliance with "golden rules" required for endeavors in this vein. This list of indicators follows the World Bank's criteria outlined in terms of measuring and monitoring city performances namely (World Bank 2006): sustainability, credibility, understandability, timeliness, importance and relevance, accessibility, limit costs. As this list reveals, a number of trade-offs are likely to occur while delving into KPI selection. For instance, at a first glance analysts are willing to measure—following the principle of "completeness"—all possible phenomena but this occurs only before they realize the labor intensity and the costs which are required; a principle of "sustainability" tends inevitably to counterbalance the inherent tendency to be exhaustive. As a supplement, the local relevance seems physiological to be pursued in order to put under the spotlight phenomena that are very significant for the observer. Howbeit, once ascertained that an analysis does not take place in a vacuum, international comparability becomes pivotal to guarantee that units of analysis may be put in relation with similar cases. Taking stock of such trade-offs, which are accompanied by several others, value for policy makers is sought by striking the right balance between extremes of the continuums representing trade-offs (Fig. 1).

In order to mitigate trade-offs in KPI selection, a two-tiered approach that results in two types of different—yet complementary—indicators is introduced (Fig. 2). "Core" indicators represent a restricted array of indicators conceived to emphasize comparability with international experiences and to create time series, given the

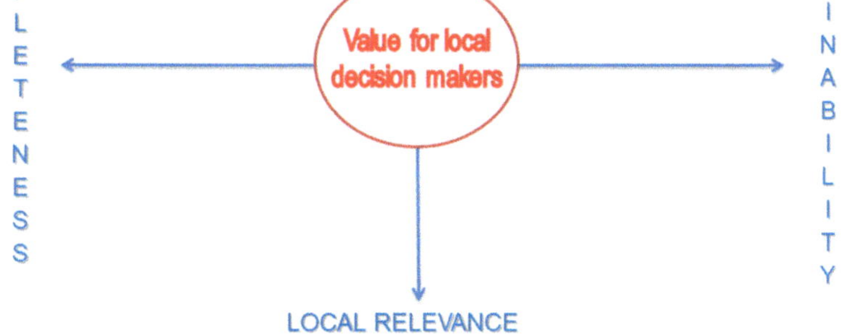

Fig. 1 Trade-offs in key performance indicators (KPIs) selection

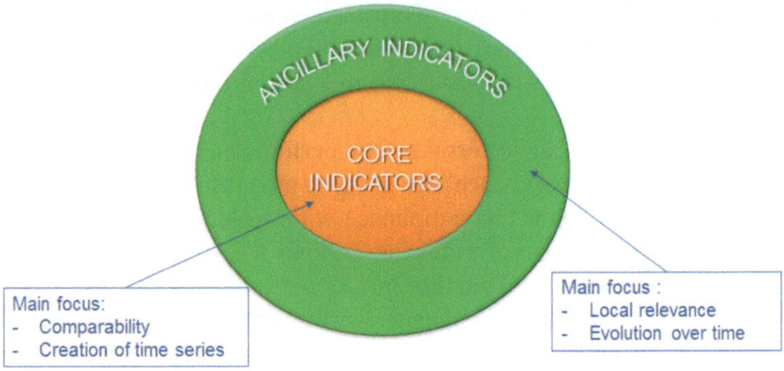

Fig. 2 Two-tiered approach for smart city indicators

recourse to existing archives already established. Conversely, "ancillary" indicators are a wealth of indicators specific to the unit of analysis. This second body of indicators is geared toward local relevance and creation of longitudinal analysis, regardless the comparability at international level.

This dual approach may be better clarified looking at the dichotomy between processes used to generate the two families of indicators (Fig. 3). "Core" indicators stem from a top-down approach; they have their roots in international databases managed by renowned institution and are selectively filtered through the lenses of general relevance for smart cities and, subsequently, data availability. On the contrary, "ancillary" indicators are the fruits of a bottom-up approach sifting local databases—and sometimes coining new indicators from scratch, when needed—to find out measures pertinent in terms of local relevance and (as optional criterion) comparability with other cases of interest.

The dual approach proposed for the smart city dashboard allows tackling major roadblocks on the way to smart city measurements, which are exacerbated by the

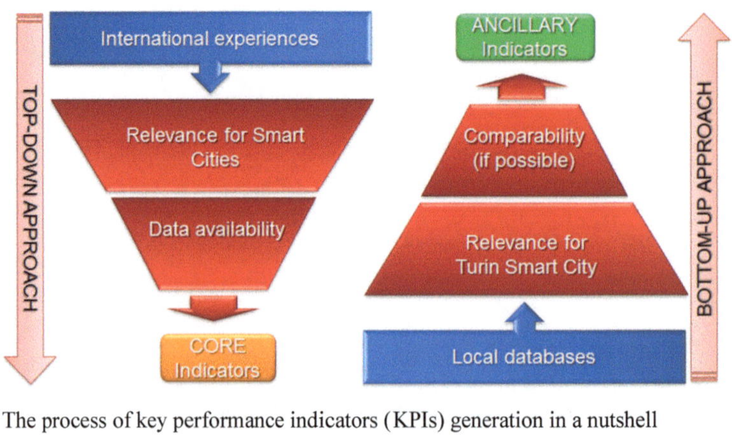

Fig. 3 The process of key performance indicators (KPIs) generation in a nutshell

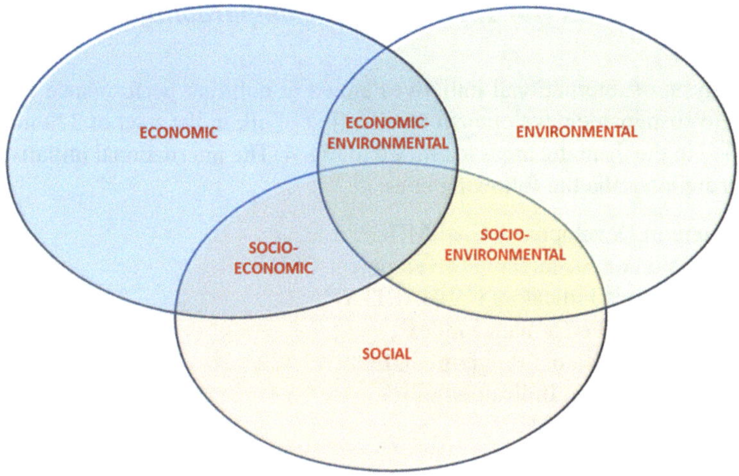

Fig. 4 Spheres of sustainability for key performance indicators (KPIs) selection

paucity of standardized methodologies for KPIs selection in general (Parmenter 2010) and for smart cities in particular (Smart Cities Stakeholder Platform 2013).

Numerous international experiments showed that critical elements are covered only partially and that frequently the focus lies on well-known aspects. As a supplement, in some of aforementioned initiatives, several indicators are improperly used at city level although their granularity is regional or national. Finally, infrequent data updates represent an Achilles' heel hampering the creation of time series. With this respect, the methodological advancement lies in the equilibrium among relevance, comparability, and data availability that makes the indicator system more flexible.

Besides the significant methodological advancement, a novel perspective about the integration between sustainability and the concept of public value is provided in this chapter; for appraising the capacity of policy actions at stake to generate positive net changes compared to the "no change" scenario, indicators of sustainability are elaborated covering the three "traditional" areas of sustainability (i.e., economic, social, and environmental) and their reciprocal overlaps (Fig. 4).

4 Key Performance Indicators for Smart Cities

According to previous considerations, this section illustrates the process of KPIs generation, following the scheme proposed in Fig. 3 and choosing a hybrid "glocal" approach that mingles the research of well-recognized international practices with the peculiar traits of the local terrain. The methodological assessment at a local level, in particular, has been contextualized in the Turin milieu where the Institute, besides being headquartered, supports local policy makers on a regular basis.

4.1 "Core" KPIs (for International Comparability)

The analysis of international initiatives aimed at defining performance indicators related to sustainable development results, first of all, in list a set of 373 indicators classified in the thematic areas identified in Fig. 4. The international initiatives considered are inter alia the following ones:

- Millennium Development Goal-MDG[6];
- Commission on Sustainable Development-CSD[7];
- Habitat Agenda Indicators UN-HABITAT[8];
- Environmental Performance Index[9];
- Global City Indicators Program-GCIP[10];
- European Common Indicators-ECI[11];
- Urban Ecosystem-UE[12].

Starting from the above mentioned long list of indicators, a filtering activity has been conducted in order to pursue international comparability. In particular, the selection process has been carried out bearing in mind the following rules: the relevance of the indicator with respect to the smart city paradigm and the data availability for the territory considered.

The resulting set of 12 "core" indicators is portrayed in Fig. 5.

As schematized in Fig. 5, the final list of "core" indicators is characterized by equilibrium between economic, social, and environmental aspects, as in that three indicators for each domain have been identified. Moreover, three additional indicators have been picked out at the intersection between the main areas. Finally, it is of paramount importance to note that, in addition to aforesaid general selection criteria (i.e., relevance to the smart city paradigm and data availability), the choice of these 12 "core" indicators has been influenced by the following two aspects:

- Only indicators related to city as territorial unit have been considered (often, in fact, indicators used for a urban context make reference to regional or national geographical areas);
- Indicators that are rarely updated have not been considered.

The choice of "core" indicators has been profoundly shaped by the adoption of a responsive approach meant to promptly capture emerging aspects pertaining to the urban context in order to gather a good understanding of any change in acceptable

---

[6] http://www.un.org/millenniumgoals/. Accessed 11 June 2014.

[7] http://sustainabledevelopment.un.org/csd.html. Accessed 11 June 2014.

[8] http://unhabitat.org/. Accessed 11 June 2014.

[9] http://epi.yale.edu/. Accessed 11 June 2014.

[10] http://www.cityindicators.org/. Accessed 11 June 2014.

[11] http://www.cityindicators.org/Deliverables/eci_final_report_12-4-2007-1024955.pdf. Accessed 11 June 2014.

[12] http://www.legambiente.it/sites/default/files/docs/rapporto_eu2013_web.pdf. Accessed 11 June 2014.

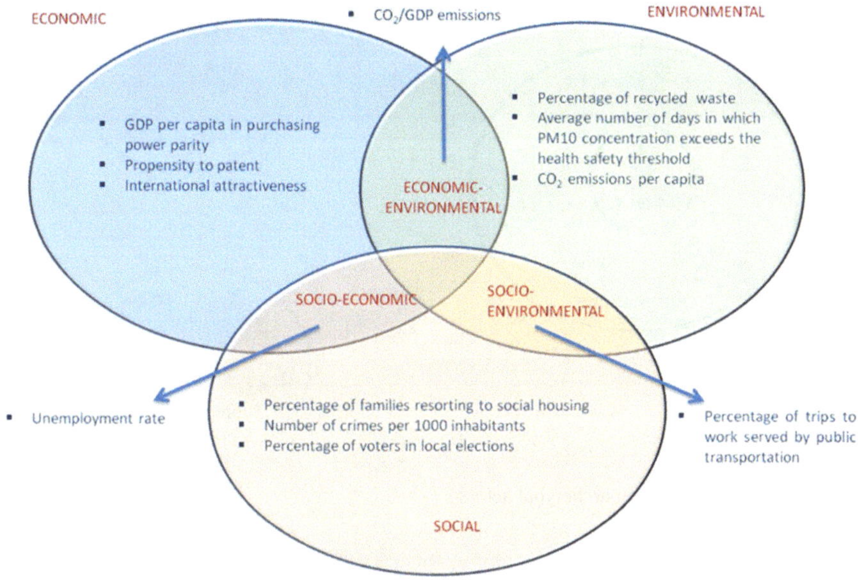

Fig. 5 Classification of "core" key performance indicators (KPIs)

time frames. This modus operandi let us to realize a concise, useful and sustainable over time dashboard for monitoring smart cities. Its strong points consist in the identification of a cohort of indicators related to smart cities that are strictly linked to data available and internationally recognized.

4.2 "Ancillary" KPIs (Local), the Case of Turin City

The methodology that we proposed for the identification of smart cities KPIs, as shown in Fig. 3, has consisted in a dual approach mingling two processes. The former that follows a top-down logic, has led us to identify 12 "core" indicators whereas the latter, driven by a bottom-up logic, has stimulated the generation of a wider set of indicators called "ancillary." This latter panoply of local KPIs has the capability to adhere to the context that goes from time to time under the spotlight.

Thanks to the consultation of local[13], national[14], and international[15] data sources, a set of 29 indicators has been identified. The choice of these indicators—spanning the gamut from the economic to the social sphere, and passing through environ-

[13] E.g., Turin city www.comune,torino.it; local transport company in Turin (GTT) http://www.comune.torino.it/gtt/. Accessed 11 June 2014.

[14] E.g., Italian National Institute of Statistics ISTAT www.istat.it. Accessed 11 June 2014.

[15] E.g., EUROSTAT http://epp.eurostat.ec.europa.eu/portal/page/portal/eurostat/home/. Accessed 11 June 2014.

Fig. 6 Interactions between archetypal actors

mental aspects—heavily depends on data availability and on the seven strategic challenges (called "grand challenges") that have been distilled after a thorough analysis of emerging needs in Turin context.

These strategic challenges have been singled out through a preliminary work on emerging needs in urban contexts. Firstly—drawing on a thorough literature review putting together needs (Maslow 1943; Habermas and Mccarthy 1973; Max-Neef 1989) and, specifically, urban needs (Hallsmith 2003)—emerging urban needs have been identified and categorized in nine "need ideal-types."

Secondly, the kaleidoscope of urban needs has been associated to four different archetypal actors (i.e., citizens, organizations, local authorities, and public service providers). Fig. 6 illustrates the interdependencies between such actors.

In that way, it has become possible to create the "local needs matrix," in which urban needs are mapped along the two above-mentioned dimensions (Fig. 7).

This matrix has a general validity[16] and can be adapted to every urban context[17]. Focusing on Turin city, the table has been filled with specific needs derived from the consultation of local policy documents and from a data-driven desk research analysis. Finally, looking at the resulting matrix, the seven strategic challenges have been distilled in order to propose a synthetic vision of the necessities lying in the city considered. Table 1 summarizes the grand challenges and provides a short description for each one.

As previously said, this second assortment of indicators aims to measure more precisely the ability of a city to foster sustainable development in order to meet

[16] The matrix encompasses a wide-ranging gamut of urban needs by drawing on relevant literature and on the analysis of a wealth of international policy documents.

[17] Policy makers have the chance to appropriately fill-in the matrix in light of the contextual background of each city under the lens.

NEED IDEAL-TYPES

ACTORS	HOUSING	HEALTH	SAFETY	PUBLIC TRANSPORTATION	PRIVATE MOBILITY	INFRASTRUCTURE	EDUCATION	LEISURE	GOVERNANCE
CITIZENS									
ORGANIZATIONS									
LOCAL AUTHORITIES									
PUBLIC SERVICES PROVIDERS									

Fig. 7 "Local needs matrix" scheme

Table 1 Turin and its seven grand challenges

Grand challenge	Description
Turin as city in transition	Reduction of the ecological footprint by achieving greater energy efficiency, widespread use of renewable energy sources, sustainable management of urban solid waste
Turin as eco-mobile city	Creation of a full-fledged logistics platform for sustainable mobility leading to widespread use of public transport, car and bike sharing, as well as electric vehicles
Turin as triple helix city	Implementation of a socioeconomic development model based on the interaction among a profitable entrepreneurial ecosystem, an international research environment, and an agile and financially robust public administration which has to be capable of supporting local actors
Turin as healthy and safe city	Remarkable improvement of the quality of urban environment achievable through a strong reduction of pollutants coupled with adequate measures of safety implemented both at professional level and in private life
Turin as inclusive city	Achievement of harmony between different ethnic groups and generations meant to ensure societal cohesion and fair access to opportunities
Turin as family-friendly city	Creation of a societal environment favorable for families thanks to ad hoc supporting services and forms of work flexibility (e.g., part-time, telecommuting) which are particularly compelling for women willing to balance family life and professional life
Turin as cultural city	Promotion and enhancement of historical and cultural heritage in order to boost tourism attractiveness of the city and its surrounding area

specific needs of the actors interacting in the territorial context under consideration. For this aim, in Fig. 8 "ancillary" KPIs are mapped against the strategic challenges.

In conclusion, "ancillary" indicators can be considered as a bespoke tool for analyzing in a very detailed way the specificities of a milieu under examination.

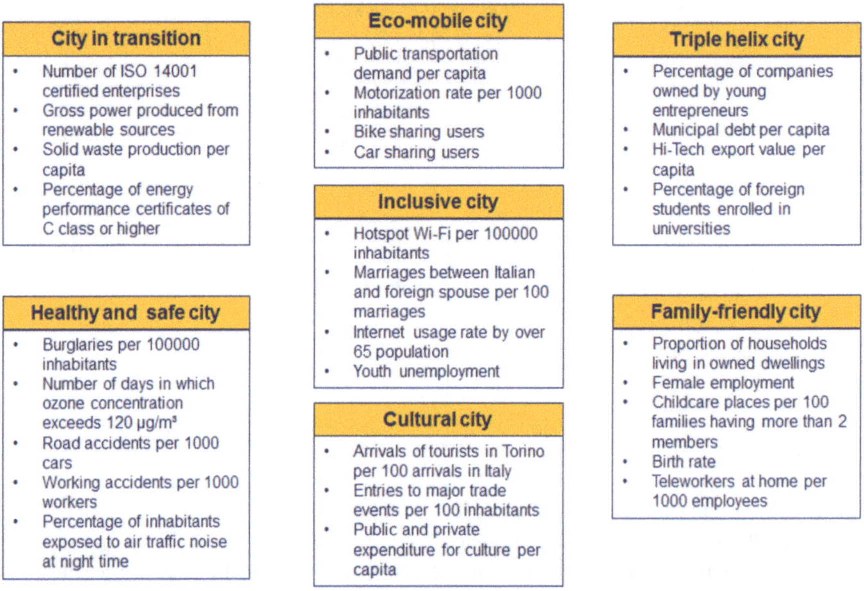

Fig. 8 "Ancillary" KPIs for Turin and their relationship with seven grand challenges

Moreover, such indicators have been studied in order to achieve an adherence to transformations over time of phenomena that are analyzed. Owing to this reason, the list of "ancillary" indicators needs to be regularly refreshed in order to capture the emergence of new challenges and/or the evolutionary patterns of actual ones.

5 Conclusions

The measurement approach proposed in this chapter represents a useful tool for policy makers and practitioners willing to evaluate urban performances against the litmus test of economic, social, and environmental sustainability.

Among the numerous significant elements of novelty brought by this brand-new approach, three of them clearly stand out.

Firstly, the dual approach proposed for the smart city dashboard allows tackling major roadblocks on the way to smart city measurement, which are exacerbated by the paucity of standardized methodologies for KPIs selection in general and for smart cities in particular. Numerous international experiments showed that critical elements are covered only partially and that frequently the focus lies on well-known aspects. As a supplement, in some of the initiatives in the limelight, several indicators are improperly used at city level although their granularity is regional or national. Finally, infrequent data updates represent an Achilles' heel hampering the creation of time series. With this respect, the methodological advancement lies in the equilibrium among relevance, comparability, and data availability that makes the indicator system more flexible.

Secondly, besides the significant methodological advancement, this chapter brings a novel perspective about the integration between sustainability and the concept of public value; for appraising the capacity of policies to generate positive changes, indicators of sustainability are elaborated covering the three "traditional" areas of sustainability (i.e., economic, social, and environmental) and their reciprocal overlaps. Concretely speaking, this orientation constitutes a founding stone in the rationalization of urban policies, prioritizing the ones that privilege sustainability, thus creating value for the society at large.

Thirdly, this chapter chooses a hybrid "glocal" approach that mingles the research of well-recognized international practices with the peculiar traits of the local terrain. The methodological assessment at local level, in particular, has been contextualized in the Turin milieu where the Institute, besides being headquartered, supports local policy makers on a regular basis.

In addition, at national level, the approach proposed in this chapter has been recently awarded the status of "methodological best practice" by the National Association of Italian Municipalities. For this reason, the methodological approach coined by authors has become part of the Italian Handbook of smart cities.

In the conclusive remarks it is crucial to discuss also some of the limitations that characterize this work, as they may represent an interesting starting point for future research. In spite of the methodological rigor, the framework proposed presents some limitations essentially related to the selection process of KPIs. As a matter of fact, the numerousness of "core" indicators being part of the framework appears—at least to some extent—exiguous if compared to the expectations of policy makers yearning for a rich palette of indicators that provides a full picture of their city in relation with similar/comparable cases. This is due to difficulties faced with respect to data availability, data granularity, and time relevance which represent Achilles' heels in comparability.

Finally, the research conducted provides food for thought for further studies. Trying to envisage future works based on this research endeavor, it will be worthwhile to leverage the present methodological framework as a foundation for developing a wide-ranging impact assessment model for smart city policy actions capable to evaluate the impacts of alternative options as net changes compared to a "no policy change" scenario.

References

Alawadhi, S., Aldama-Nalda, A., Chourabi, H., Gil-García, J., Leung, S., Mellouli, S., Nam, T., Pardo, T. A., Scholl, H. J., Walker, S. (2012). Building understanding of smart city initiatives. *Electronic Government, 7443,* 40–53.

Alford, J., & O'Flynn, J. (2009). Making sense of public value: Concepts, critiques and emergent meanings. *International Journal of Public Administration, 32*(3–4), 171–191.

Bannister, F., & Connolly, R. (2014). ICT, public values and transformative government: A framework and programme for research. *Government Information Quarterly, 31*(1), 119–128.

Bélissent, J. (2011). *The key to being a smart city is good governance: "Smart Governance".* http://blogs.forrester.com/jennifer_belissent_phd/11–05-15-the_key_to_being_a_smart_city_is_good_governance_smart_governance. Accessed 11 June 2014.

Benington, J., & Moore, M. (2011). *Public value: Theory and practice*. London: Palgrave Macmillan.
Berry, C. R., & Glaeser, E. L. (2005). The divergence of human capital levels across cities. *Regional Science, 84*(3), 407–444.
Bollier, D. (1998). *How smart growth can stop sprawl: A fledgling citizen movement expands*. Washington: Essential Books.
Brand, S. (2006). City Planet. *Strategy + Business,* Issue 42.
Caragliu, A., Del Bo, C., & Nijkamp, P. (2011). Smart cities in Europe. *Journal of Urban Technology, 18*(2), 65–82.
Chourabi, H., Nam, T., Walker, S., Gil-Garcia, J. R., Mellouli, S., Nahon, K., Pardo, T. A., Scholl, H. J. (2012). Understanding smart cities: An integrative framework. *Proceedings of 45th Hawaii International Conference on System Sciences* (pp. 2289–2297). Hawaii: IEEE.
Cole, M., & Parston, G. (2006). *Unlocking public value: A new model for achieving high performance in public service organizations*. Hoboken: Wiley.
Cordella, A., & Bonina, C. (2012). A public value perspective for ICT enabled public sector reforms: A theoretical reflection. *Government Information Quarterly, 29*(4), 512–520.
Cresswell, A. M., Burke, G. B., & Pardo, T. (2006). *Advancing return on investment, analysis for government IT: A public value framework*. Albany: Center for Technology in Government, University at Albany.
Danielsen, K. A., Lang, R. E., & Fulton, W. (1999). Retracting suburbia: Smart growth and the future of housing. *Housing Policy Debate, 10*(3), 513–540.
Dobbs, R., Smit, S., Remes, J., Manyika, J., Roxburgh, C., and Restrepo, A. (2011). *Urban world: Mapping the economic power of cities*. New York: McKinsey Global Institute.
Donath, J. (2011). Epilogue: The city as information organism. In *From social butterfly to engaged citizen: Urban informatics, social media, ubiquitous computing, and mobile technology to support citizen engagement* (pp. 485–490). Cambridge: MIT Press.
Dutton, W. H., Kraemer, K. L., & Blumler, J. G. (1987). *Wired cities: Shaping the future of communications*. Indianapolis: Macmillan.
Elkington, J. (1998). Partnerships from cannibals with forks: The triple bottom line of 21st-century business. *Environmental Quality Management, 8*(1), 37–51.
Etzkowitz, H., & Leydesdorff, L. (2000). The dynamics of innovation: From national systems and "Mode 2" to a Triple Helix of university–industry–government relations. *Research Policy, 29*(2), 109–123.
European Commission. (2001). *European governance: A white paper*. Brussels: Commission of the European Communities.
Ewing, R., Pendall, R., & Chen, D. (2002). *Measuring sprawl and its impact*. Washington: Smart Growth America.
Ferro, E., Caroleo, B., Leo, M., Osella, M., & Pautasso, E. (2013). The role of ICT in smart cities governance. *Proceedings of International Conference for E-Democracy and Open Government* (pp. 133–146). Krems: Edition Donau-Universität Krems.
Giffinger, R. (2007). *Smart cities—ranking of European medium-sized cities*. Vienna: Centre of Regional Science at Vienna University of Technology.
Habermas, J., & Mccarthy, T. (1973). *Legitimation crisis*. Boston: Beacon Press.
Hall, R. E., et al. (2000). The vision of a smart city. In *Proceedings of the 2nd International Life Extension Technology Workshop*. Paris.
Hallsmith, G. (2003). *The key to sustainable cities: Meeting human needs, transforming community systems*. Gabriola Island: New Society Publishers.
Harrison, C. (2010). Foundations for smarter cities. *IBM Journal of Research and Development, 54*(4), 1–16.
Harrison, C., & Donnelly, I. A. (2011). *A theory of smart cities*. Hull: ISSS.
Hills, D., & Sullivan, F. (2006). *Measuring public value 2: Practical approaches*. London: The Work Foundation.
Hollands, R. G. (2008). Will the real smart city please stand up? *City: Analysis of urban trends, culture, theory, policy, action, 12*(3), 303–320.
Hood, C. (1991). A public management for all seasons? *Public Administration, 69*(1), 3–19.

Inayatullah, S. (2011). City futures in transformation: Emerging issues and case studies. *Futures, 43*(7), 654–661.
Ishida, T., & Isbister, K. (2000). *Digital cities: Technologies, experiences, and future perspectives*. Berlin: Springer.
Johnston, E., & Hansen, D. (2011). Design lessons for smart governance infrastructures. In D. Ink, A. Balutis, T. F. Buss (Eds.), *Transforming American governance: Rebooting the public square?* (pp. 197–212.). Armonk: M.E. Sharpe.
Karunasena, K., & Deng, H. (2012). Critical factors for evaluating the public value of e-government in Sri Lanka. *Government Information Quarterly, 29*(1), 76–84.
Kelly, G., Mulgan, G., & Muers, S. (2002). *Creating public value: An analytical framework for public service reform*. London: UK Cabinet Office's Strategy Unit.
Komninos, N. (2009). Intelligent cities: Towards interactive and global innovation environments. *International Journal of Innovation and Regional Development, 1*(4), 337–355.
Kraay, A., Zoido-Lobaton, P., & Kaufmann, D. (1999). *Governance matters*. Washington: World Bank.
Lam, W. (2005). Barriers to e-government integration. *The Journal of Enterprise Information Management, 18*(5), 511–530.
Lee, J., & Lee, H. (2014). Developing and validating a citizen-centric typology for smart city services. *Government Information Quarterly, 31*(S1), S93–S105.
Lombardi, P., Giordano, S., Caragliu, A., Del Bo, C., Deakin, M., Nijkamp, P., & Kourtit, K. (2012). An advanced triple-helix network model for smart cities performance. In O. Y. Ercoskun (Ed.), *Green and ecological technologies for urban planning: Creating smart cities* (pp. 59–73). Hershey: IGI Global.
Maslow, A. (1943). A theory of human motivation. *Psychological Review, 50*(4), 430–437.
Max-Neef, M. (1989). *Human scale development conception application and further reflections*. New York: Apex Press.
Meadows, D. H., Meadows, D. L., & Randers, J. (2004). *Limits to growth: The 30-year update*. White River Junction: Chelsea Green Publishing.
Misuraca, G., Reid, A., & Deakin, M. (2011). *Exploring emerging ICT-enabled governance models in European cities*. Seville: JRC-IPTS.
Mooij, J. (2003). *Smart governance? Politics in the policy process in Andhra Pradesh, India*. London: Overseas Development Institute.
Moore, M. H. (1995). *Creating public value: Strategic management in government*. Cambridge: Harvard University Press.
Nam, T., & Pardo, T. (2011). *Conceptualizing smart city with dimensions of technology, people and institutions* (pp. 282–291). New York: ACM.
Nam, T., & Pardo, T. (2012). *Transforming city government: A case study of Philly311* (pp. 310–319). New York: ACM.
Nam, T., & Pardo, T. (2014). The changing face of a city government: A case study of philly311. *Government Information Quarterly*, April. Volume In Press, Corrected Proof.
Odendaal, N. (2003). Information and communication technology and local governance: Understanding the difference between cities in developed and emerging economies. *Computers, Environment and Urban Systems, 27*(6), 585–607.
OECD. (1997). *Final report of the Ad Hoc Working Group on participatory development and good governance*. Paris: OECD.
OECD. (2001). *Citizens as partners: Information consultation and public participation in policy making*. Paris: OECD.
OECD. (2006). *Competitive cities in the global economy*. Paris: OECD.
Parmenter, D. (2010). *Key performance indicators (KPI): Developing, implementing, and using winning KPIs*. Hoboken: Wiley.
Paskaleva, K. A. (2009). Enabling the smart city: The progress of city e-governance in Europe. *International Journal of Innovation and Regional Development, 1*(4), 405–422.
Rhodes, R. A., & Wanna, J. (2007). The limits to public value, or rescuing responsible government from the platonic guardians. *Australian Journal of Public Administration, 66*(4), 406–421.

Savoldelli, A., Misuraca, G., & Codagnone, C. (2013). Measuring the public value of e-Government: The eGEP2.0 model. *The Electronic Journal of e-Government, 11*(2), 373–388.
Schaffers, H. (2011). *Smart cities and the future internet: Towards cooperation frameworks for open innovation.* In *Future internet* (pp. 431–446). Berlin: Springer.
Smart Cities Stakeholder Platform. (2013). *Integrated action plan—Report process & guidelines for smart cities*. Brussels: European Commission.
Stern, N. (2007). *The economics of climate change: The stern review*. Cambridge: Cambridge University Press.
Stoker, G. (2003). *Public value management (PVM): A new resolution of the democracy/efficiency trade-off.* Manchester: Institute for Political and Economic Governance, University of Manchester.
Talbot, C. (2011). Paradoxes and prospects of 'public value'. *Public Money & Management, 31*(1), 27–34.
UNDP. (1997). *Governance for sustainable human development*. New York: UNDP Policy Document.
UN-HABITAT. (2011). *Cities and climate change: Global report on human settlements*. London: Earthscan.
Van der Hoeven, M. (2012). *Urban energy policy design*. Paris: International Energy Agency.
Wallace, J. 2013. *The rise of the enabling state: A review of policy and evidence across the UK and Ireland.* Dunfermline: Carnegie UK Trust
World Bank. (2006). *The current status of city indicators*. Washington: City Indicators Discussion Paper.
World Commission on Environment and Development. (1987). *Our common future*. Oxford: Oxford University Press.
World Health Organization & UN-HABITAT. (2010). *Hidden cities: Unmasking and overcoming health inequities in urban settings*. Geneva: WHO Publications.

Michele Osella is the head of Business Model and Policy Innovation Unit at the Istituto Superiore Mario Boella (ISMB), a multidisciplinary research team studying economic, social and policy implications of ICTs. In the academic sphere, he is an adjunct lecturer in Innovation Management and Business Modeling at the Polytechnic of Turin. Prominent strands of research on which he is focused are social computing, open data, smart city governance as well as Internet economics, with a keen interest for groundbreaking business models. Michele is currently involved in a number of international research projects, developed both in private and public sectors.

Enrico Ferro is the head of the Innovation Development Area at the Istituto Superiore Mario Boella (ISMB), a research team of 20 researchers working on the strategic, socioeconomic, and policy implications of information and communication technologies. Over the last 10 years Dr. Ferro has worked in many projects financed by the European Commission with roles ranging from scientific supervisor to senior expert. Dr. Ferro also covers an adjunct professor position at the Polytechnic of Turin and at the International Labor Bureau of the United Nations where he regularly lectures on innovation management in both the public and the private sector. His research work has produced over forty academic publications, one handbook of research and over 50 research reports.

Elisa Pautasso graduated in Industrial Engineering and Management at the Polytechnic of Turin. She has worked in the Business Model and Policy Innovation Unit at the Istituto Superiore Mario Boella (ISMB) since 2004. Her research activities are mainly focused on policy analysis and impact assessment. She has been involved in local and international projects, aimed to study the diffusion of ICTs between citizens, enterprises, and PAs, in order to create interregional benchmarking and provide support to decision makers for the definition of policy actions. More recently, her studies have concentrated on the integration of technologies in the urban context, examining the smart cities paradigm, and on social innovation.

Smart Cities and Resilience Plans: A Multi-Agent Based Simulation for Extreme Event Rescuing

Karam Mustapha, Hamid Mcheick and Sehl Mellouli

Abstract The concept of smart cities is one that relies on the use of new information and communication technologies in order to improve services that cities provide to their citizens. The resilience of a city is one of the services that it can provide to its citizens. Resilience is defined as its capacity to continue working normally by serving citizens when extreme events (EEs) occur. This chapter will propose a new framework based on multi-agent systems to help cities build simulation scenarios for rescuing citizens in the case of an EE. The main contribution of the framework will be a set of models, at different levels of abstraction, to reflect the organizational structure and policies within the simulation, which involves the integration of truly dynamic dimensions of this organization. The framework will also propose methods to go from one model to another (conceptual to simulation). This framework can be applied in different domains, such as smart cities, earthquakes and building fires.

Keywords Extreme events · City resilience · Agent based simulation · Multi-agent systems · Organization · Architecture · Modelling · Simulation

List of Abbreviations

AA	Agent Artefact
ABDiSE	Agent-Based Disaster Simulation Environment
ABS	Agent Based Simulation
ACL	Agent Communication Language

K. Mustapha (✉) · H. Mcheick
University of Quebec in Chicoutimi, 555 Boulevard Université,
Chicoutimi, QC G7H 2B1, Canada
e-mail: karam.mustapha1@uqac.ca

H. Mcheick
e-mail: hamid_Mcheick@uqac.ca

S. Mellouli
Laval University, 2325 Rue de l'Université, Québec, QC G1V 0A6, Canada
e-mail: sehl.mellouli@fsa.ulaval.ca

J. R. Gil-Garcia et al. (eds.), *Smarter as the New Urban Agenda A Comprehensive View of the 21st Century City,* Public Administration and Information Technology 11,
DOI 10.1007/978-3-319-17620-8_8

AUML	Agent Unified Modeling Language
BDI	Believe, Desire, Intention
CAOM	Conceptual Agent Organizational Model
CROM	Conceptual Role Organizational Model
D^4S^2	Dynamic Discrete Disaster Decision Simulation System:
EE	Extreme Events
FACL	Form-based ACL
FIPA	Foundation of Intelligent Physical Agents
GIS	Geographical Information System
JADE	Java Agent Development Environment
MAS	Multi Agent System
MDA	Model Driven Architecture
MDD	Model Driven Development
MOON	Mu1tiagent-Oriented Office Network
ND	Natural disaster
OMT	Object Modeling Template
OPAM	Operational Agent Model
PIM	Platform Independent Model
PSM	Platform Specific Model
RTI	Real Time Infrastructure
SAMoSAB	Software Architecture for Modeling and Simulation Agent-Based
UEML	Unified Enterprise Modeling Language

1 Introduction

The concept of smart cities is one that relies on the use of new information and communication technologies in order to improve services that cities provide to their citizens. One of the services that cities have to provide to their citizens is their resilience. The resilience of a city is defined as its capacity to continue working normally by serving citizens when extreme events (EEs) occur. The management of EEs is becoming more complex since EEs are becoming more frequent and more powerful. These events may be caused either by nature, like storms or earthquakes, or by humans, like wars or airplane crashes. When happening, these events require the intervention of different teams to rescue people such as police, firefighters, non-governmental organizations (NGOs) like the Red Cross, etc. In the meantime, these emergency teams have to collaborate and coordinate their activities to better rescue people. However, these teams have different skills, use different tools, adopt different strategies and play different roles. All this heterogeneity adds a lot of complexities to the rescuing activity. To this end, these teams will need tools to help them make efficient interventions. Simulations are one of the methods that can be used by these teams to predict their behaviour during an EE.

Over the past few decades, EEs such as droughts, floods, cyclones, earthquakes and volcanic eruptions have resulted in the mortality of approximately 3 million people and affected the lives of 800 million people worldwide. These have caused diseases as well as serious economic losses and homelessness. Therefore, modelling and simulating the rescue procedure may help to facilitate their management and limit their impact on society. These simulations may improve the efficiency of the teams on the field that may lead to reducing losses and damage of goods and saving lives. Multi-agent systems (MAS) are among the techniques used for modelling and simulating EE emergencies.

An MAS can model the behaviour of a set of entities expertly, more or less organized by respecting the laws governing their relations (Ferber and Perrot 1995). Agents have a degree of autonomy and are immersed in an environment in which and with which they interact (Erceau and Ferber 1991). There are several areas where MAS can be applied; they can act as a modelling paradigm or as a solution for software implementation. Therefore, the application of MAS in this area could help managers to experiment with all possible scenarios of a disaster and assist them in making decisions. This approach involves the simulation of systems in terms of models and their use.

Agent-based simulation (ABS; Russel and Norvig 2003) has spread out into many areas, including sociology, biology, economics, physics, chemistry, ecology, industrial applications and EE. ABS has the ability to capture different dynamic models which usually consist of simple entities (called reactive agent if simple behaviour is required) or more complex entities (called deliberative agent if decision-making and negotiation are needed). The global objective of this chapter is to provide a methodological framework that ranges from domain model analysis to running a simulation while considering the different entities that can make an intervention in the case of an EE. In Mustapha et al. (2010), we proposed a specific agent-based methodological framework allowing, from modelling to simulation, the production of observables at different levels of detail related to an EE rescuing activity. The proposed framework in this chapter is an extension of our previous work (Mustapha et al. 2010) with the objective to integrate dynamic organizational characteristics of an EE rescuing activity in the modelling and simulation procedures. It will also include the specification of the translation process from generic models to specific models, to ensure the transition between the proposed models.

This chapter is structured as follows: Sect. 2 defines the objectives of the research and related concerns of modelling and simulation of EE. Section 3 introduces an organizational-oriented methodological framework, which is capable of taking into account the organizational aspects at both the conceptual and the operational abstraction levels. Section 4 describes the dynamic EE organization. Section 5 introduces a model-driven architecture to transform the models proposed in the methodological framework. Section 6 details the agent-based software architecture in line with the proposed methodological framework, to simulate EE's organizational aspects. Section 7 presents an illustrative example of the proposed software architecture through the modelling of a building fire. Finally, conclusions and recommendations for future work are summarized in Sect. 8.

2 Smart Cities and Resilience Plans

The smart city concept is an emerging strategy, based on information and communication technologies that cities are adopting to mitigate the problems generated by the growth of the urban population and rapid urbanization. This emerging concept is a type of urban development able to meet the needs of institutions, businesses and citizens, both economically, environmentally and socially. We can describe a smart city when we invest in human capital, traditional communication, energy, social infrastructure and a high quality of life, with a wise organization of natural resources, and through participatory governance[1].

There is an increase in frequency of use of the phrase "smart city"; there is no clear and reliable understanding of the concept in academia. A limited number of studies have investigated and begun to systematically consider questions related to this new urban phenomenon of smart cities.

The concept is used all over the world with different nomenclatures, the context is still emerging and the research work of defining and conceptualizing it is evolving (Boulton et al. 2011; Hollands 2008). Several work definitions have been put forward and adopted in academic use. This discord of definitions is resulting in calls for conceptual research in this regard (Boulton et al. 2011).

- A city performing well in a forward-looking way in economy, people, mobility, environment, governance and living, built on the smart combination of endowments and activities of self-decisive, independent and aware citizens (Giffinger et al. 2007).
- A city that monitors and integrates conditions of all of its critical infrastructures, for example, roads, bridges, tunnels, rails, subways, airports, seaports, communications, water, power, even major buildings, can better optimize its resources, plan its preventive maintenance activities and monitor security aspects while maximizing services to its citizens (Hall 2000).
- A city "connecting the different infrastructure, like physical, IT, social and the business to leverage the collective intelligence of the city" (Harrison et al. 2010).
- A city striving to make itself "smarter" (more efficient, sustainable, equitable and livable) (Boulton et al. 2011).
- Use of smart computing technologies to make the critical infrastructure components and services of a city—which include city administration, education, healthcare, public safety, real estate, transportation and utilities—more intelligent, interconnected and efficient (Weber and Khademian 2008).

To become smart, existing cities should develop new efficient services in all areas such as:

- Intelligent transport and mobility: one of the challenges is to integrate different modes of transport—car, cycle and walking—in one system that is efficient, accessible, affordable, safe and environmentally sound. This integration allows a

[1] http://en.wikipedia.org/wiki/Smart_city

reduced environmental space, optimizing the use of urban space, and offers a diverse range of solutions of urban mobility that meet all their needs. In addition, the city will have to implement the latest technologies of transportation and electric mobility;

- Environmental sustainability: cities must act in two main areas: waste and energy. On waste, cities' mission will be to reduce or avoid their waste and put in place effective systems for recovery and recycling of waste (process by which material waste or a useless product is transformed into a new material or product quality of greater value).
- Responsible and intelligent urban habitat: the high value of property in city centres combined with the limited availability of land make the current urbanization complex. For example, buildings must be smarter to facilitate and improve the management of energy, or reduce consumption.

3 Agents and Organization Oriented EE Modelling and Simulation

A multi-agent based system is a powerful modelling technique for simulating individual interactions in a dynamic system and is distinctive in its ability to simulate situations with unpredictable behaviour (Lampert 2002).

Previous researches have focused on modelling of the rescue during natural disasters (NDs). However, current technological developments allow envisioning a systems approach that includes modelling of all aspects of an EE, from its impact on the resources and population to the required response by the involved agencies. The ABS approach can help to model and simulate these aspects, and it allows the simulation designer to model different levels of representations, such as individuals and groups of individuals. Hence, agent-based modelling allows capturing the dynamic nature of the EE and facilitates the study of numerous resource coordination associated with the interaction of multiple teams (Monteiro et al. 2008).

3.1 Agent-Based EE Frameworks

Even if agents are used in the simulation of EEs, a few researchers have proposed a framework to support both the design and the implementation of the EE simulation. Two studies are presented hereafter:

- ABDiSE (Agent-Based Disaster Simulation Environment) is a framework that provides model elements and tools to support the modelling and the simulation of different types of natural disasters such as fires, floods and debris flows. This tool describes how agents move, attach and interact with each other and with their environment (Hsu and Liu 2012);

- D^4S^2 (Dynamic Discrete Disaster Decision Simulation System) is a comprehensive decision support system to simulate large-scale disaster responses. This model has a specialized architecture designed for decision-makers who can be public safety service officials such as firefighters and the police (Wu et al. 2007c). More precisely, the proposed architecture integrates several models such as an agent-based simulation model, geographical information system (GIS) databases, a rule-based system for responders and optimization modules to create a hybrid system of agent-based and discrete simulation components.

3.2 Agent-Oriented Frameworks

The organizational modelling in multi-agent systems is based on the management of a process metaphor that underrates the organizational structure (Mustapha et al. 2010). A more general study of agent-oriented software engineering methodologies, undertaken in order to find conceptual and operational solutions, has confirmed that organizational issues were added to the actor approach. This approach is the basis of a methodological framework for helping the domain experts to design their models in their own language, as well as transitional agent-based models which are used to produce the distributed simulation model on which experiments are conducted (Labarthe et al. 2007). Methods like GAIA (Zambonelli et al. 2003), CRIO (Gaud et al. 2008), MOISE+ (Hübner et al. 2007) or Luis Antonio's work (Antonio et al. 2008) provide only a part of the solution for the required objectives. Most of these approaches use the notion of roles in order to promote flexibility in the design process, even with different abstraction or hierarchical levels. As an abstract view of the distributed organization, roles can be combined and associated to the agents' specific architecture, from complex information processing units (i.e. with deliberating capacities) to simple programmable units (reactive agents or state-machine like automata).

3.3 EE Specific Models

There exist other EE models that do not use the agent approach as a method for modelling and simulating complex problems such as emergency responses, evacuations, fires, traffic events, earthquakes and flooding. Among these approaches, we mention:

- Emergency response framework (Jain and McLean 2003): this framework allows the integration of modelling, simulation and visualization tools for emergency response. The development and implementation of this framework should significantly improve the nation's capability in the emergency response area.
- Buildings evacuation models: there are more than 26 models that have focused on simulating building evacuations. Many of these models are used to simulate

evacuation procedures from different types of structures. Featured models include: EVACNET4, WAYOUT, STEPS, PedGo, PEDROUTE, Simulex, Grid-Flow, ASERI, FDS+Evac, Pathfinder, SimWalk, PEDFLOW, buildingEXODUS, Legion, SpaceSensor, Evacuation Planning Tool (EPT), MassMotion, PathFinder, Myriad II, ALLSAFE, CRISP, EGRESS, SGEM, Evac/FDS, Massegress and Hidac (Kuligowski and Peacock 2006).
- Traffic models: there are three main approaches for the modelling and simulation of traffic:
  - The macro-simulation approach, also referred to as macroscopic (Helbing and Treiber 1999)
  - The micro-simulation approach also referred to as microscopy (Ben-Akiva 1994; Kosonen and Pursula 1991)
  - The mesoscopic approach (De Palma et al. 1996) that is widely used in economic research and studying patterns of movement

3.4 Limitations of the Presented Methods

From the above literature review, we found that the different presented models can be improved at different levels:

- The different presented methodologies do not take sufficiently into account the purely organizational aspects of an EE, that is, explicitly including the structure and organizational dynamics, particularly those related to the behaviour of actors or agents, behaviours generally associated with multiple roles.
- The presented methodologies do not take into account observables and indicators specific to the organization of the natural disaster. Observables and indicators are data and information used in ongoing decision processes, which need to be highlighted in the simulation results.
- The presented models lack aspects of evaluation. In evacuation modelling (in the EE case), validation refers to a systematic comparison of model predictions with reliable information (Galea and Gwynne 2005). Model predictions are dependent upon the data, codes and user of the evacuation model. The lack of suitable experimental data to feed the evacuation modelling causes a challenge.
- The fourth element to be improved is related to the presentation of occupants in the evacuation models (or EE). Accurate occupants' representation based on comprehensive anthropometric data and human performance and behaviour should be used in evacuation modelling to provide an additional level of validity to the models. Otherwise, building codes and standards should be reformed according to the dynamic changes of individuals' ages and sizes.
- The interoperability between emergency response modelling and simulation applications is currently extremely limited such as, for example, the interoperability between different models such as fire model, evacuation model).

- The cost of transferring data between emergency response simulation software applications is often very high.
- The emergency response organizations usually do not have the technical expertise or the time for building simulation models (Jain and McLean 2003).

Therefore, this study proposes a solution to overcome some limitations related to organizational dynamics, interoperability and transferring data in order to allow modelling/simulation of more "corporate" management after an EE occurs.

4 A Methodological Framework for EE Organizational Aspects, Modelling and Simulation

The proposed modelling approach is based on an incremental process to deal with the complexity in the modelling and implementation simulation process of EE. The proposed approach relies on a gradual increasing of the level of detail in the models. The real system is represented by a domain model of EE (e.g. an UEML model—Unified Enterprise Modeling Language—www.ueml.org) to represent the organizational aspects. In our previous work, an organizational methodological framework for modelling a complex system was proposed, which was according to two main abstraction levels: a conceptual and an operational level (Mustapha et al. 2010). Using the domain model provided by the domain expert, a simulation model is built step by step. The conceptual level proposes concepts and models, helping to grasp the complexity of the problem and its simulation objectives, whereas the operational level involves the implementation of the simulation model which includes the software integration issues. The different models and the transition to agent-oriented modelling and simulation in our methodological framework are presented in Fig. 1 (refer to Mustapha et al. 2010 for further details).

The modelling of the organizational concepts is engaged through a dialogue between the domain expert and an agent-knowledgeable modeller. Identifying the active entities and their organizations from the domain model produces an actor model. The modeller translates/abstracts the domain model into a conceptual organizational model (COM) based on (hierarchical) levels, actors, roles and groups named conceptual role organizational model (CROM) (Fig. 2; refer to Mustapha et al. 2010 for further detail). This stage highlights the organizational structure of the EE as well as the structural and dynamic relations between the entities comprising the EE. Following this conceptual model, an agent-based model is produced on the basis of observables that the user needs to obtain from the simulation, building up the route towards the implementation of the simulation.

The software designer details the Conceptual Agent Organizational Model (CAOM) by associating a conceptual agent with a software agent architecture (e.g. BDI—Believe, Desire, Intention—(Rao and Gorgeff 1991)) and specifying their behaviours (e.g. an UML—"Unified Modeling Language"—activity diagram for a reactive agent) and interactions (e.g. AUML—"Agent Unified Modeling Lan-

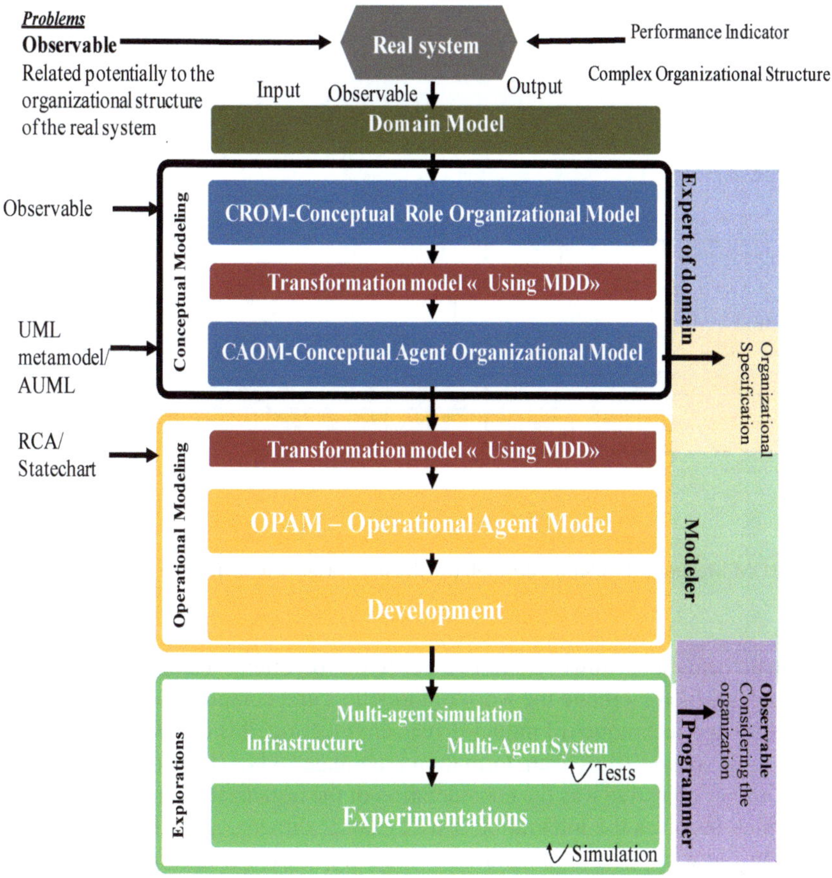

Fig. 1 A Methodological Framework for ND

guage"—sequence diagram (Odell et al. 2001)), resulting in an Operational Agent Model (OPAM). The implementation of these models in a simulation(s) environment results in an ABS system that can be executed. The observables that are related to the organizational structure of a real system are not described in the design model (Labarthe et al. 2007). They are only mentioned in the multi-agent system model, that is, one step before the implementation stage. It is necessary to describe them earlier in the modelling process (at a conceptual and operational level) as they may induce different modelling requirements.

Our work can be summarized as the enrichment of the proposed methodological framework, which would include the specification of the translation process to ensure the transition between the CROM models, CAOM and OPAM (proposed in the methodological framework models) (Mustapha et al. 2010) using MDD (Model Driven Development) (France et al. 2006) or MDA (Model Driven Architecture) (Mukerji and Miller 2003). Further, a decision-making criteria has been proposed

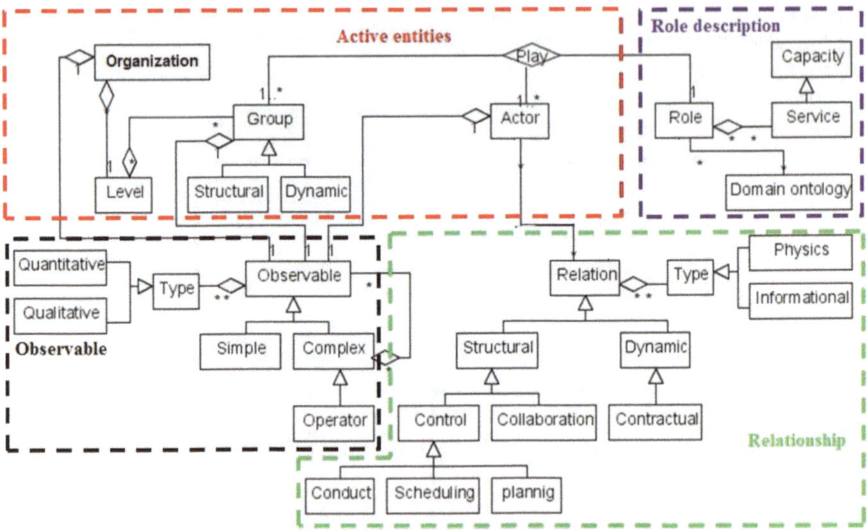

Fig. 2 CROM Metamodel (refer to Mustapha et al. 2010 for further detail)

for the transformations, however, more extensive experiments are required to define explicit rules which facilitate the work of modellers and computer scientists. As for the concepts and techniques of model-driven engineering, this could be implemented using tools such as Atlas Transformation Language (ATL) (Kosonen and Pursula 1991). Another objective was the consideration of the organization in the simulation of EE which involves the integration truly dynamic dimension of the organization by making more explicit organization's image that offers (Piunti et al. 2009) by combining A & A (Montagna et al. 2008) and Moise+ (Hübner et al. 2007).

5 Dynamic EE Organization

In this section, we solve one of the problems presented in Sect. 2.4, which involves the integration of truly dynamic dimensions of the organization in the methodological framework proposed in Mustapha et al. (2010). The dynamic organization is composed of levels that include one or several groups. Each group contains the actors of the CROM model (presented by the agent or a group of agents in the CAOM model) where each agent can play several roles and a role can be played by different agents. The organization, the group and the agent can generate observables (quantitative or qualitative). An observable is characterized by the activity it monitors (quality, cost), its quantitative or qualitative nature that requires defining its measuring units and the authorized values (whole or real number if quantitative, list of values if qualitative) and finally, its dated value. A role can provide services to other roles in the same group. Relationships can be developed between agents; every agent may receive an invitation to join a group such as police, firemen, first aid, evacuation or traffic. An agent can communicate with another agent when needed, to own business process (collaboration). In this case, agents are invited to join a group to meet some order. During the communication process, all agents compete with each other and that may orient their behaviour, their business volume and their capability. Also, during the modelling and simulation of EE, all the agents are able to ensure the success of the final management plan and collaborate to resolve problems. Agents can join or leave a group when the requested tasks are completed. Then, groups can be automatically created in case two agents intend to communicate in a new group and vice versa. In the case of a group that contains two agents that have completed their work, this group can be terminated. Therefore, in our methodological framework, the organization can be changed during the simulation including agents joining and leaving groups, changing the agents' behaviours and rules and removing/adding groups.

6 Model Transformation Using Model Driven Architecture: From CROM to CAOM

In our research, we propose to use Model Driven Architecture (MDA) to transform models in order to ensure the interoperability between the proposed models and to transfer data between these models without any loss of information (Limits presented in the Sect. 2.4). The MDA approach can be seen as a "pragmatic" framework for the implementation of the transformational approach. It advocates the development of software for model transformations. The basic idea is to separate the specification of the functions of a system for the implementation of these functions on a specific platform. Thus, during the development of a system, the desired functionality must be specified independently of the platform (i.e. business model). The platform must also be described by models. We choose a particular platform for the system and the platform independent model (PIM) is converted into a platform specific model (PSM). MDA proposes getting into abstraction by manipulating models and building applications by processing models. MDA offers different approaches to transform models.

The approach to transform the PIM to PSM is based on two steps: first, mark the PIM, and second, transform the PIM. A brand is a concept of a platform and it is used to mark a PIM to indicate how the element should be changed. These marks are defined in a model that describes the target platform. When a platform is chosen, the brands associated with it are then used to mark the elements of PIM to guide transformation to PSM. A marked PIM is then transformed to get the PSM.

In Fig. 3, we illustrate the transformations between CROM and CAOM models using MDA. Firstly, we marked the PIM model (CROM), and then transform it to PSM or CAOM model. The principle transformation task is to decide about the

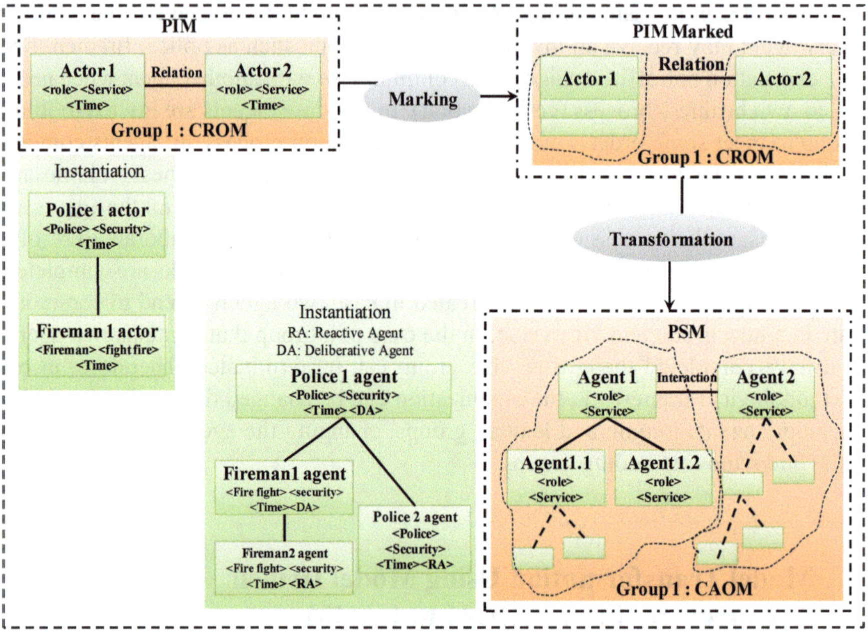

Fig. 3 Transformation model: CROM to CAOM

role that should be included in the CAOM model. Roles can be combined into one or several agents, according to the kind of behaviour that is expected to be studied (simple or complex). The second task of this transformation to the CAOM model is related to the transformation of the relationship between the CROM actors. Thus, CROM relationships are transposed into agents as interactions while keeping their classification; we used observables to filter the reactive agents and deliberative agents. Then, we transform the conceptual model (CAOM) to an operational model. The operational model provides a solution for implementing the conceptual model (CAOM). This step has led to the development of an operational model based on agents and that includes the choice of the agent architectures. For the representation of agents and their behaviour at the operational level, we propose a modelling approach that allows differentiation between reactive agents and deliberative agents.

The transformation between different models (CROM, CAOM, OPAM) can be considered as a new communication tool between modellers and developers. The objective of CAOM is to specify the behaviour of each CROM actor. It involves "filtering" the CROM model to only retain actors that have some interest for the simulation along with the desired level of detail of their behaviours. Thus, it highlights the quantitative/qualitative observable data relevant to the objectives of system representation. The following meta-model in Fig. 4 shows these concepts that form the building blocks of a CAOM model. The above meta-model is obtained after the transformation of the CAOM.

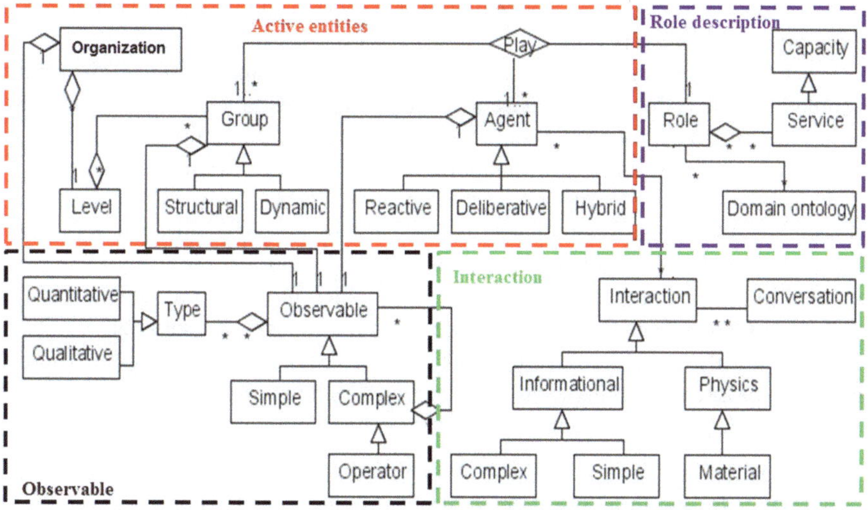

Fig. 4 CAOM Metamodel (refer to Mustapha et al. 2010 for further detail)

7 An Agent-Based Software Architecture

The simulation of the operational model, produced after several stages of model refinement, assumes the existence of a software infrastructure that supports heterogeneous simulation models. In addition, it should ensure the integrity of the distributed simulation (of two or more software environments) while providing the desired simulation data (observable). In this section, first we present what requirements rise up from these objectives, before introducing the general architecture of an agent and the organizational-oriented simulator.

7.1 Architectural Requirements

This section addresses simulation integration and interoperability issues, viewed as the management of data and event dependencies between simulators. Considering the complexity of such a task, we combine different integration approaches: FIPA (Foundation of Intelligent Physical Agents) specifications on agent-based software integration (DMSO 1998) and high-level architecture (HLA) specifications on distributed simulation integration (FIPA 2002), in order to redefine initial ad hoc actor simulation architecture (Labarthe et al. 2007).

FIPA proposes to agentify software services in order to separate the discovery and selection of services from the actual service call. Interaction protocols are defined to support the chain of actions that agents follow to track and execute the software distributed over an open environment. It is a general software integration approach that, however, does not deal with data sharing and time synchronization at a conceptual or software level.

HLA, an IEEE standard, is totally dedicated to the management of distributed simulations. HLA does not propose a software implementation or consider the internal structure of simulators (Federate). Its reckoning by the simulation community has resulted in numerous implementations and adaptations to different application domains, including ND simulations (Ounnar et al. 2008). A distributed simulation is seen as a federation of simulators, coordinated by a central unit—the RTI (Real Time Infrastructure)—exchanging data and instantiating an object modeling template (OMT) in respect with simulation rules which maintain the integrity of the global simulation (data format, time synchronization, events causality chain…).

There are also several research works that have been proposed to solve different problems of interoperability such as asynchronous communication and agent-based communication language. Below are several architectures we studied:

- An Agent Platform for Reliable Asynchronous Distributed (Bellissard et al. 1999): in this platform, the authors introduce a distributed communication model based on autonomous software agents. Agents act as software components that can migrate from node to node.
- Agent-Based Middleware for Web Service Dynamic (Lin and Maheshwari 2005): in this architecture, the author aims to construct an agent-based middleware for Web service dynamic integration based on peer-to-peer networks to facilitate the integration of optimal quality of Web services for application integration.
- XML-Based Mobile Agents (Steele et al. 2005): the authors present a mobile agent system based on the use of XML.
- An Agent-Based Distributed Smart Machine (Kao and Chen 2010): the authors present a software agent-based technology to enhance the remote tool service system by developing related remote service ontology to manage its smart and distributive characteristics.
- An Agent XML based Information Integration Platform (Li 2010): an Agent/XML based information integration platform is proposed in order to integrate process operation systems effectively. The subsystems are encapsulated as agent-based on XML. The encapsulation of different subsystems was implemented through agent technology.
- A Cross-Platform Agent-Based Implementation (Cervera 2005): the authors present a cross-platform, networked implementation of control tasks based on software agents and streaming technologies.
- FACL (Form-Based ACL; Chusho and Fujiwara 2000): the authors describe a multi-agent framework and an Agent Communication Language (ACL) for the Multi-Agent-Oriented Office Network (MOON) systems that are distributed systems of e-commerce. The multi-agent framework is a Java application framework and includes a form-based ACL (FACL) as a common protocol for passing application forms.

Looking at these different platforms, we found that there are several limitations: (1) these platforms support only a small number of executions and are limited to link with complex applications (Like JASON Platform (Bordini and Hubner 2006; Bordini et al. 2005)); (2) they are not open to no-agent software environments; (3) they cannot keep track of the execution; and (4) they have a low level of autonomy.

To overcome these limitations, our approach to EE simulation considers heterogeneity of agents and behaviour as the consequences of the domain which expert observable choices and not necessarily the nature of the EE entities. Therefore the simulation deals with heterogeneous complex behaviour which the simulation framework must integrate.

7.2 *Software Architecture for Modelling and Simulation Agent-Based (SAMoSAB)*

The objective of this section is to propose a software architecture that facilitates the production of these simulations by integrating the functional requirements and software related to the simulation of EE.

The different stages of the methodological framework presented in the previous section lead to an operational modelling crystallized by the OPAM model and characterized by:

- The refinement of the multi-agent organization split into two separate environments: one called cognitive (or deliberative) simulating decision-making process, and the other so-called reactive simulating single process or highly automated.
- Specification of the agent's behaviour: in languages appropriate to the granularity of the agents, it describes how agents should behave during the simulation without prejudice to the way they are actually implemented (programming language, simulation language and environment).
- Specification of the interactions between agents: This results in the dynamic simulation. These interactions will present issues set out according to different agents that are involved in the same society or two different societies (reactive and cognitive).

As a first step toward generality, we have considered two simulation environments integrated through a mediator. The basic idea was to identify and isolate the simulation functions that ensure the simulator integration. As shown in Fig. 5, the architecture is divided into the following component elements:

- *Simulators:* our generic architecture is open to different software platforms to simulate different behaviours (simple or complex behaviour).
- *Mediator:* the mediator realizes the transmission of information (message, signals, objects, data…) while keeping simulation-specific constraints respected (e.g. time synchronicity between both environments). The mediator role in the integration process is synthesized into six services (agent management, communication management, organizational model management, interoperability, time management, indicator management).

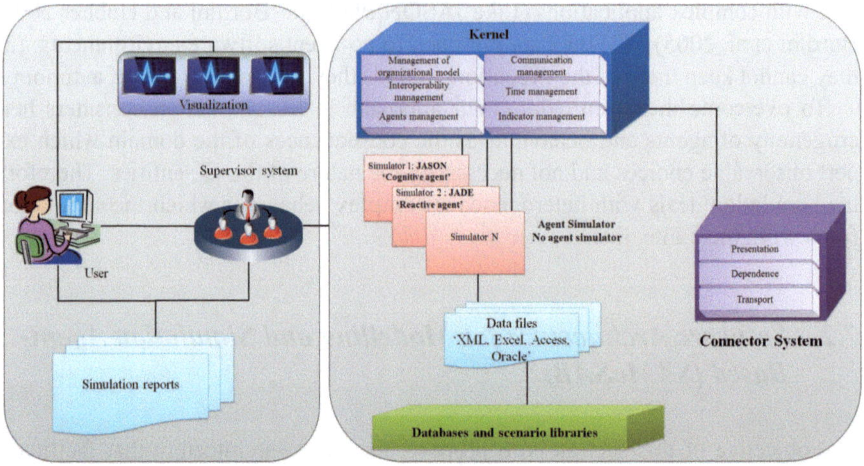

Fig. 5 SAMoSAB architecture

- *Connector system:* This element is used to ensure the interoperability between agent and non-agent platforms, which are situated in distributed systems; these must be designed to satisfy the requirements: transport of information, description of dependency, interoperability between different type of simulators and format of information (Mcheick and Qi 2011).
- *Database and libraries:* a number of different databases and libraries can be used to drive the simulation. The libraries will include the event scenarios and the related files required by the multiple simulators.
- *Data files:* These files are important for understanding the entire concept of the integrated simulators. The files predict a mechanism for configuring the simulator modules and sharing data between them. We have used different types of files such as XML, Excel and Oracle to encode the data. These files contain the executable data to be processed by the simulation. These files also contain the descriptive text that is intended only for human interpretation. We have also added the links between different types of data required for the interaction between different simulators.
- *Supervisor simulation:* this element is responsible for synchronization of all the simulators. It configures the scenario through synchronized initialization of all the simulators from data contained. It also coordinates the execution of the various simulators during a simulation run and outputs simulation reports.
- *User:* this element provides capabilities to create, modify the scenario data files, manage the display screens for configuring the system, observe the simulation runs and display results. It is also responsible for the generation of simulation results.
- *Visualization:* this element provides the graphic representation of the scenario and the events.

However, while keeping in mind such objectives, we have chosen to test our propositions by starting with two "specialized" simulation environments. SAMoSAB is thus presently composed of: (i) the JASON platform, (ii) the JADE platforms, (iii) a mediator and (iv) connector.

The JASON platform is adapted to the development of BDI (i.e. deliberative) agents (Bordini and Hubner 2006; Bordini et al. 2005). It is an extended interpreter (Vangheluwe et al. 2002) of AgentSpeak (Rao 1996), a BDI programming language allowing complex behaviour modelling. The JADE platform (Java Agent Development Framework) (Serment et al. 2007) is also a FIPA compliant agent-oriented software engineering tool implemented in Java. It proposes a framework for agent management (agent directories, communication management). Agent internal structure is open and left mostly to the programmer initiative. The mediator, see Serment et al. (2007), supports the simulation integration by proposing generic services as independent as possible of the simulator architecture. JASON is used to implement and simulate decision-making processes, whereas JADE deals with simple agent behaviours. Agents from both environments must interact; the mediator realizes the transmission of information (message, signals, objects and data) while keeping simulation-specific constraints respected (e.g. time synchronicity between both environments). A database is also included to capture the model parameters, record simulation data and results analysis. It is accessed by the simulators and the mediator. Figure 5 summarizes the general architecture of SAMoSAB.

7.3 Agent Modelling and Interoperability

The current SAMoSAB environment contains several type of agents: (i) deliberative agents, developed in JASON, that implement ND decision-making processes, that is, ND entities whose behaviours produce complex observables; (ii) reactive agents, developed in JADE, implementing basic behaviours and (iii) service agents, that is, agents not directly concerned by the simulation models but supporting the simulation process.

Table 1 summarizes the different types of agents and their roles in SAMoSAB, some of them provided by the JADE platform.

8 An Illustrative SAMoSAB Implementation

Figure 6 illustrates a software architecture supporting our methodological and "simulation-related" requirement. As exposed in the previous section, this SAMoSAB implementation contains JADE agents, JASON agents, non-agent platforms and a mediator in charge of the interactions.

The "simulation model" presented in Fig. 6, results from applying our methodological approach, that is, progressive translation of the CROM and CAOM models

Table 1 Agents description

Agent	Description
AMS	Manage agent life cycle, as well as “white pages” directory, i.e. the list of the agents’ name and their communication address
DF	Provide a “Yellow Page” service. It records agent roles, capabilities and may answer requests for other agent directory needs
ACC	Routes messages from one agent to another, independently of the platform of both agents. Implements for this purpose the IIOP protocol
IAg	Is associated to an indicator: It provides computational facilities to produce the value of aggregated indicators. Thus it agentifies the observables identified in the conceptual models. Indicator agents are also categorized depending on the type of indicator they represent (activity, productivity and quality)
DSA	DSA centralizes the source of data in a group of agents, and is responsible for finding the agents that have the required information. DSA then regroups and sends these data to the right indicator agent (IAg), an IAg is needed for exploiting these values
GMA	Manages a group, i.e. allows an agent to play a role in the group, as well as represent the agents in the group for specific requests; e.g. if an IAg needs a particular type of data, the group manager will locate the agents producing that data

AMS agent management system, *DF* directory facilitator, *ACC* agent communication channel, *IAg* indicator agent, *DSA* data source agent, *GMA* group manager agent

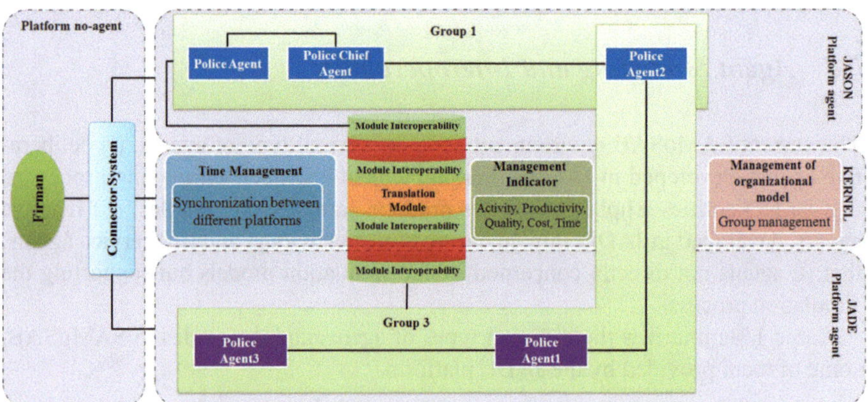

Fig. 6 SAMoSAB architecture illustration

of the case study presented in Mustapha et al. (2010). It is composed of two groups describing a simplified ND organization structure. Communication between three platforms is done through messages. Therefore, a mediator layer (denoted Kernel) and connector system ensures the communication link between different platforms (“physical” interoperability is simulated in this case as both are FIPA compliant environment). Note that the mediator is presently developed as a group of specialized agents.

9 Conclusion and Future Work

In an agent-based ND simulation, we have presented an organizational oriented methodological framework, which permits modelling and simulation of ND organizational aspects. It allows observables of different levels of detail while reproducing the ND behaviour according to desired observables. This methodological framework is structured according to conceptual and operational abstraction levels. At the conceptual level, the modelling is based on a conceptual role organizational model (CROM), which is then refined into a conceptual agent organizational model (CAOM). At the operational level, modelling is mainly based on the operational agent model (OPAM).

This framework allows the study of the impact of a specific ND organizational structure and its related management policies on ND performance. Based on a ND expert modelling of a particular ND, an organization/role oriented (CROM) and an agent-oriented (CAOM) conceptual model help in designing a simulation model, which will reproduce the ND global and local behaviours. These conceptual models are defined independently of particular agent architecture or even on specific software architecture, but propose transitional steps to guide their development.

In this chapter, we focused on the proposal of an open software architecture supporting the transformation of the conceptual model into an operational model by generalizing the previous "hard wired" architecture inspired by previous agent-based integration framework. This architecture can be seen as the interaction between different simulation platforms (agent platform and no-agent platform). We showed how different types of agents—deliberative and reactive—can interact during simulation, as well as the role of some service agents (group manager, indicator and DataSource agent) supporting this simulation. Development is currently based on the interaction between the JADE platform (for the reactive agent) and the JASON environment (for the deliberative agent).

In our future work, we account to work on several points: real data collection in order to have more accurate results (simulation) (example: data structure fire or other EE). These simulations are the first goal, a validation of the operationalization of the methodological framework for modelling and agent-oriented simulation, taking into account explicitly the organizational aspects of NDs. These simulations should also allow us to validate the software architecture proposed, architecture for the implementation of the previous methodological framework and the execution of simulations. To illustrate the interest of our approach to modelling and simulation-oriented agents for the management of EE, we also propose to explore different contextual scenario management of NDs, such as building fires, earthquakes, etc. The different results may well show the interest of our tool in understanding the behaviour of an EE.

References

Antonio, L., D'Amours, S., & Frayret, J. M. (2008). "A methodological framework for the analysis of agent-based supply chain planning simulations", SpringSim'08: Proceedings of the 2008 spring simulation multiconference, Society for Computer Simulation International San Diego, CA, USA.

Bellissard, L., Freyssinet, A., De Palma, N., Herrmann, M., & Lacourte, S. (1999). An agent platform for reliable asynchronous distributed programming. IEEE: FRANCE.

Ben-Akiva, M. E., Koutsopoulos, H., & Mukundan, A. (1994). A dynamic traffic model system for ATMS/ATIS operation. *Journal of Intelligent Transportation Systems, 2,* 1–19. (MITTNS).

Bordini, R., & Hubner, J. (2006). An overview of Jason. *Association for Logic Programming Newsletter, 9*(3), 1–6. http://www.cs.kuleuven.ac.be/~dtai/projects/ALP/newsletter/aug06/nav/articles/article5/fs.pdf.

Bordini, R., Hubner, J., & Vieira, R. (2005). Jason and the golden fleece of agent-oriented programming. In R. Bordini, J. Hubner, & R. Vieira (Eds.), *Multi-agent programming* (pp. 3–37). New York: Springer.

Boulton, A., Brunn, S. D., & Devriendt, L. (2011). Cyberinfrastructures and "smart" world cities: Physical, human, and soft infrastructures. In P. Taylor, B. Derudder, M. Hoyler, & F. Witlox (Eds.), *International handbook of globalization and world cities*. Cheltenham: Edward Elgar.

Cervera, E. (2005). A cross-platform agent-based implementation. IEEE.

Chusho, T., & Fujiwara, K. (2000). A formbased agent communication language for enduser-initiative agent-based application development. IEEE.

De Palma, A., Marchal, F., & Nesterov, Y. (1996). A modular system for dynamic traffic simulations. *Transportation Research Record Journal of the Transportation Research Board, 1607,* 178–184.

Defense Modeling and Simulation Office (DMSO). (1998). High Level Architecture Object Model Template, Version 1.3, dated 5 February 1998. Available online at the HLA Homepage.

Erceau, J., & Ferber, J. (1991). 'L'Intelligence Artificielle Distribuée'. *La recherche, 22,* 750–758.

Ferber, J., & Perrot, J. F. (1995). *Les Systèmes multi-agents, vers une intelligence collective*. Paris: InerEditions.

FIPA. (2002). FIPA Contract Net Interaction Protocol Specification, Foundation for Intelligent Physical Agents, www.fipa.org/specs/fipa00029/.

France, R., Ghosh, S., Dinh-Trong, T., & Solberg, A. (2006). Model-driven development using UML 2.0: Promises and pitfalls. *Computer, 39*(2), 59–66.

Galea, E., & Gwynne, S. (2005). *Principles and practices of evacuation modeling*. London: CMS Press.

Gaud, N., Galland, S., & Koukam, A. (2008). Towards a multilevel simulation approach based on holonic multi-agent. Published in the 10th International Conference on Computer Modeling and Simulation (EUROSIM/UKSiM"08), pp. 180–185, England.

Giffinger, R., Fertner, C., Kramar, H., Kalasek, R., Pichler-Milanovié, N., & Meijers, E. (2007). *Smart cities: Ranking of European medium-sized cities*. Vienna: Centre of Regional Science (SRF), Vienna University of Technology.

Hall, R. E. (2000). The vision of a smart city. In Proceedings of the 2nd International Life Extension Technology Workshop, Paris, France, September 28.

Harrison, C., Eckman, B., Hamilton, R., Hartswick, P., Kalagnanam, J., Paraszczak, J., & Williams, P. (2010). Foundations for smarter cities. *IBM Journal of Research and Development, 54*(4), 1–16.

Helbing, D., & Treiber, M. (1999). Numerical simulation of macroscopic traffic equations. *IMPACT of Computing in Science and Engineering, 1,* 89–99.

Hollands, R. G. (2008). Will the real smart city please stand up? *City, 12*(3), 303–320.

Hsu, T. L., & Liu, J. W. S. (2012). An agent-based disaster simulation environment. RITMAN Workshop 2012, Taipei, Taiwan.

Hübner, J. F, Sichman, J. S., & Boissier, O. (2007). Developing organised multi-agent systems using the Moise + Model: Programming issues at the system and agent levels. *International Journal of Agent-Oriented Software Engineering, 1*(3/4), 370–395

Jain, S., & McLean, C. R. (2003). "A Framework for modeling and simulation of emergency response," Proceedings of the 2003 Winter Simulation Conference, Dec. 7–10, New Orleans, Louisiana, pp. 1068–1076.

Kao, Y.-C., & Chen, M.-S. (2010). An agent-based distributed smart machine tool service system. In 3CA 2010. IEEE.

Kosonen, I., & Pursula, M. (1991). A simulation tool for traffic control planning. IEEE Conference Publication Number 320. Third international Conference on Road Traffic Control. Vol. 320, pp. 72–76.

Kuligowski, E. D., & Peacock, R. D. (2006). A review of building evacuation models, Fire Research Division Building and Fire Research Laboratory.

Labarthe, O., Espinasse, B., Ferrarini, A., & Montreuil, B. (2007). Toward a methodological framework for agent-based modeling and simulation of supply chains in a mass customization context. *Simulation Modeling Practice and Theory International Journal (SIMPAT), 15*(2), 113–136.

Lampert, R. (2002). Agent-based modeling as organizational and public policy simulators. *Proceedings of the National Academy of Sciences of the United States of America, 99,* 7195–196.

Li, X. (2010). An Agent/XML based information integration platform for process industry, in 2010 2nd International Conference on Computer Engineering and Technology. IEEE.

Lin, A., & Maheshwari, P. (2005). Agent-based middleware for web service dynamic integration on peer-to-peer networks. In S. Zhang & R. Jharvis (Eds.), *AI 2005: Advances in artificial intelligence* (pp. 405–414). Berlin: Springer Verlag. (LNAI 3809).

Mcheick, H., & Qi, Y. (2011). Dependency of components in MVC distributed architecture. 24th IEEE (ccece'2011). May 8–11, Ontario, Canada.

Montagna, S., Ricci, A., & Omicini, A. (2008). A & A for modeling and engineering simulations in systems biology. *International Journal of Agent-Oriented Software Engineering, 2*(2), 222–245.

Monteiro, T., Anciaux, D., Espinasse, B., Ferrarini, A., Labarthe, O., & Roy, D. (2008). The interest of agents for supply chain simulation. In C. Thierry, A. Thomas, & G. Bel (Eds.), *Simulation for supply chain management.* London: ISTE.

Mukerji, J., & Miller, J. (2003). MDA: Model Driven Architecture Guide Version, www.omg.org/cgibin/doc?omg/03-06-01, Juin (2003).

Mustapha, K., Tranvouez, E., Espinasse, B., & Ferrarini, A. (2010). An organization-oriented methodological framework for agent-based supply chain simulation. 4th International Conference on Research Challenges in Information Science, France.

Odell, J., Parunak, H. V. D., & Bauer, B. (2001). Representing agent interaction protocols in UML, Proceedings of the First International Workshop on Agent-Oriented Software Engineering, CIANCARINI, P. & WOOLDRIDGE, M.

Ounnar, F., Archimède, B., Pujo, P., & Charbonnaud, P. (2008). HLA Distributed Simulation Approaches for Supply Chain. In Hermès Science Europe Ltd (Ed.), Simulation for Supply Chain Management (pp. 257–294). Wiley, ISTE.

Piunti, M., Ricci, A., Boissier, O., & Hübner, J. F. (2009). Embodying organisations in multi-agent work environments, Web Intelligence and Intelligent Agent Technologies, 2009. WI-IAT'09. IEEE/WIC/ACM International Joint Conferences on 2, pp. 511–518.

Rao, A. S. (1996). AgentSpeak(L): BDI Agents speak out in a Logical Computable Language. In W. Van de Velde & J. Perram (Eds.), Proceedings of the Seventh Workshop on Modeling Autonomous Agents in a Multi-Agent World (MAAMAW'96), Jan. 22–25, Eindhoven, Netherlands, no. 1038 in LNAI, pp. 42–55. Springer-Verlag, London, U.K.

Rao, A. S., & Gorgeff, M. P. (1991). Modeling rational agents within BDI-Architecture. In J. Allen et al. (Ed.), Proceedings of the 2nd International Conference on Principles of Knowledge Representation and Reasoning. San Mateo, USA, Morgan Kaufmann, Pub, pp. 473–484.

Russell, S., & Norvig, P. (2003). *Artificial intelligence: A modern approach* (2nd ed.). Upper Saddle River: Pearson Education.

Serment, J., Espinasse, B., Tranvouez, E. (2007). An agent integration infrastructure for the development of environmental decision support systems based on simulation, AIS-CMS International modeling and simulation multiconference, Buenos Aires—Argentina. (ISBN 978-2-9520712-6-0).

Steele, R., et al. (2005). XML-based mobile agents. In Proceedings of the International Conference on Information Technology: Coding and computing (ITCC'05). IEEE.

Vangheluwe, H., et al. (2002). In Proceedings of the 2002 AI, Simulation and Planning in High Autonomy Systems Conference, AIS'2002, (Lisboa, Portugal, April 2002). 9–20.

Weber, E. P., & Khademian, A. M. (2008). Wicked problems, knowledge challenges, and collaborative capacity builders in network settings. *Public Administration Review, 68*(2), 334–349.

Wu, S., Shuman, L. J., Bidanda, B., Kelley, M., Sochats, K., & Balaban, C. (2007c). System implementation issues of dynamic discrete disaster decision simulation system (D4S2)—Phase I. In the Proceedings of the 2007 Winter Simulation Conference.

Zambonelli, F., Jennings N., & Wooldridge, M. (2003). Developing multi-agent systems: The GAIA methodology. *ACM Transactions on Software Engineering and Methodology, 12*(3), 317–370.

Karam Mustapha is a post-doc in computer sciences in the Polytechnic University, Montreal, Canada. He has done a post-doctorate degree of one and a half year in the Department of Computer Science at the University of Quebec at Chicoutimi, Canada. His topic research is related to multi-agent system, natural disaster, smart cities and tumor cancer evolution. Professor Mustapha has several publications in journals and conferences.

Hamid Mcheick is working as an associate professor in the Computer Science department at the University of Québec at Chicoutimi, Canada. He has more than 20 years of experience in both academic and industrial areas. He has done his PhD in Software Engineering and Distributed Systems in the University of Montreal, Canada. He has a master's degree in computer science from the University of Quebec At Montreal (UQAM), Canada and a master's degree in mathematics and computer science from Lebanese University, Lebanon.

Sehl Mellouli is a full professor at the Management Information Systems Department, Faculty of Business Administration, Laval University, Quebec, Canada. His main research interests are related to e-government, smart cities, intelligent systems and knowledge management. Professor Mellouli has several publications in well-known journals and conferences. He acts as a reviewer for different journals. He also acts for program committees and organization committees of different conferences. Professor Mellouli is currently secretary of the Digital Government Society. He is a former member of the North American Digital Government Group. He is currently a member of the eGovPolinet research consortium funded by the European Union. Professor Mellouli received his PhD in Computer Science from Laval University.

Smart City Implementation Framework for Developing Countries: The Case of Egypt

Karim Hamza

Abstract Current smart city frameworks and models are not enough to fulfill the requirements of developing countries in order to face their challenges in applying the smart city concept due to weak integration of social, economic, and political needs, and lack of a holistic and integrated approach to sustainable city development. In addition, most of the developing countries lack the proper means for implementation, proper infrastructures, enough funding, sufficient economic growth, and political stability. Moreover, the challenges that can prevent the success of such a concept are poverty, inequality, cultural barriers, and the continual rise of slums and unplanned immigration from rural areas to cities.

This chapter recommends a "strategic implementation framework for smart city" tailored for developing countries such as Egypt. This general framework aims to assist different successive governments in countries such as Egypt to develop and maintain smart city strategies that help sustainable development of the country instead of building separate isolated cities that cannot face different political, economic, social, and environmental challenges.

Keywords Smart city · Developing countries · Egypt · Government · Urban planning · Sustainability framework · Technology

1 Introduction to Smart City

The value and importance of smart city concepts are still under debate, but the more important question today is how we can build a sustainable smart city and not why we need a smart city. The concept is still emerging, and the work of defining and conceptualizing it is in progress (Boulton et al. 2011; Hollands 2008; Chourabi et al. 2012). Some view smart city as a strategy to mitigate the problems generated by the urban population growth and rapid urbanization (Chourabi et al. 2012; Natural Resources Defense Council 2014, October 11).

K. Hamza (✉)
Institute for European Studies (IES), Vrije Universiteit Brussel,
21 Oksheed st. Manial al Rowda, Cairo 11451, Egypt
e-mail: khamza@vub.ac.be

J. R. Gil-Garcia et al. (eds.), *Smarter as the New Urban Agenda A Comprehensive View of the 21st Century City,* Public Administration and Information Technology 11,
DOI 10.1007/978-3-319-17620-8_9

The concept of smart city is evolving between technological innovation and sustainable urban planning, but limiting smart city concept to these disciplines only can reveal a limited vision and result in a fragile concept. The concept of smart city needs to have dimensions other than technology and urban planning in consideration, including economic development, social development, ecological development, and political mechanisms.

In addition to the limited vision of current smart city concept, developing countries face the biggest challenges in implementing and building a smart city since most of them lack the proper means of implementation, proper infrastructure, enough funding, sufficient economic growth, and political stability. There are also a lot of other risks and challenges that can prevent the success of this concept, such as poverty, inequality, cultural barriers, a continual rise of slums, and unplanned immigration of rural areas to cities.

2 Why Developing Countries Need Smart City?

There is a continuous shift from rural to urban population (see www.unfpa.org); currently more than half of the world's population lives in urban areas (Dirks et al. 2010; Chourabi et al. 2012; Dirks et al. 2009). Unfortunately, such enormous and complex growth of urban population in developing countries is unplanned, which means that such growth will be inevitably messy, unsustainable, and would not match the minimum requirement for reasonable quality of life.

The cities that develop in such conditions will face three categories of problems: first, operation problems such as difficulty in waste management, air pollution, traffic congestions, and inadequate and deteriorating infrastructures of cities; second, the social and political complexity problems such as inequality, societal turmoil, rising human health concerns, increasing repercussions due to slums, and growth of public institutions that are ineffective and unable to satisfy the needs of communities; and third, economic problems such as unsustainable economic development, unemployment, scarcity of resources due to depletion, misuse and wasting of the state resources, and the efflux of foreign investments (Johnson 2008; Rittel and Webber 1973; Chourabi et al. 2012).

Despite the fact that operational; social; political and economic problems constantly increase urban poverty and put more pressure on urban planning, challenges such as how to address the urbanization of poverty and increasing inequality, how to address rapid and chaotic emergence of unplanned areas, and how to meet the needs of the youth who constitute the majority of the urban population (UN-HABITAT 1998).

The ability of smart cities in developing countries to respond for operational challenges; and the accumulated problems either socially or economically with inadequate financial resources, constrained time frame, and other different social, economic, and political challenges. It is a difficult challenge for any government to face these problems alone (UN-HABITAT 1998). Therefore, developing countries need to develop a strategic vision for smart city approach to overcome operational, political, social, and economic challenges, and maintain sustainable development.

3 Methodology

The strategic implementation framework for smart city in developing countries is based on multidiscipline literature review. It analyzes the current smart city frameworks and models and proposes broad recommendations on how to improve the existing models to match the developing country's needs. The analysis (a) takes sample of frameworks developed by technology companies such as CISCO and IBM from information technology practitioner's point of view, (b) understands the factors and characteristics of smart city developed by academics, and finally (c) takes the European Union standards for smart city and sustainable development recommendation by UN-Habitat.

The aim of this chapter is to combine different disciplines to enrich the current frameworks with different views and tailor a general framework that can assist developing countries in deploying smart city to overcome the sustainability challenges, since most of the frameworks and strategies are based more on information technology point of view and do not take into account the challenges that developing countries face in implementing a smart city. Finally, the framework tries to simulate a smart city strategy for Egypt based on the strategic implementation framework for smart city.

The present chapter aims to answer these questions: (a) Why developing countries need smart cities?, (b) What are the main frameworks for a smart city?, (c) What is the recommended framework for developing countries to implement a smart city?, and (d) What is the strategic implementation framework for a smart city in a country like Egypt?

4 Smart City Frameworks

Current smart city frameworks and models are not enough to fulfill the requirements of developing countries to face the challenges in applying smart city concept, due to lack of integration of local systems and regional systems, lack of a holistic and integrated approach to sustainable city development, a lack of consideration of human factors and human–environment interactions, and weak integration of social, economic, and political needs (Kim and Steenkamp 2013).

Going through the literature of the smart city frameworks and models, we find that they can be grouped into two main segments: (a) *segment I—smart city structure:* this segment addresses the amount of layers the smart city concept is based on and the relation between these layers, with more elaboration on the role of technology and (b) *segment II—smart city factors and characteristics:* this segment aims to address the main factors and characteristics attached with smart city concepts that allow us to call a city "smart", with more emphasis on UN-Habitat recommendation for the human settlement, which is a major concern for developing countries, in addition to the European Commission standards for a smart city.

4.1 Segment I: Smart City Structure

Going through the literature focusing on smart city structure, we find a basic approach like the "Urban Information Model" which is more common in urban planning and geographic information system (GIS) disciplines; also there is another approach which is more modern like the "Smart City Framework" and is developed by the Cisco Internet Business Solutions Group (IBSG). These types of models and frameworks are more appealing for technology companies such as Cisco, IBM, and Microsoft to promote their views and products in smart cities, since information technology can be the backbone for developing infrastructure in a smart city.

The *Urban Information Model* is often applied by the geographic information system (GIS); this model includes five main groups of layers: (1) natural environment, including topography, natural resources, geology, etc.; (2) infrastructure, including roads, bridges, tunnels, buildings, telecommunication, etc.; (3) resources, representing materials necessary for city operation and industry including water, air, oil, minerals, etc.; (4) services, including transportation, energy, commerce, health care, etc.; (5) social systems, including locations and actions of people, such as commerce and culture, laws, regulations, governance, etc. (Harrison and Donnelly 2011).

On the other hand, the "Smart City Framework" developed by Cisco Internet Business Solutions Group (IBSG) tries to describe the process of developing smart city by: (1) understanding how cities operate, (2) defining city objectives and stakeholder roles, and (3) understanding the role of information and communication technology (ICT) within physical city assets. Additionally, the framework defined four main layers: (1) city objectives which aim to improve social, environmental, and economic pillars by linking the city's objectives to projects, policies, and initiatives; (2) city indicators which aim to match international indicators and benchmarks to city objectives; (3) city components which aim to describe the physical components of a city like utilities, transportation, real estate, and services; (4) city content which aims to map the city objectives with best practices and policies (Falconer and Mitchell 2012).

Additionally this framework highlighted the importance of emphasizing the stakeholder roles in developing smart city strategy. It identified five key stakeholders groups: governments, regulators, developers, owners, and operators (Falconer and Mitchell 2012).

4.2 Segment II: Smart City Factors and Characteristics

This segment aims to address the main factors and characteristics attached with smart city concepts, the weight and description of the factors and characteristics may differ from city to city or from one country to another, depending on the economic, political, and social condition of every society. This segment is of interest to academicians, United Nation organizations, European Union, and institutes interested in community development. The UN-Habitat recommendation for human settlement is focused here, since it is a major concern for developing countries.

Going through the literature related to this segment, we find an integrative framework based on extensive literature review, funded by the social sciences and humanities research council (SSHRC). This framework resulted in identification of eight critical factors to form the basis of an integrative framework that can be used to examine how local governments look for smart city initiatives: (1) management and organization, (2) technology, (3) governance, (4) policy, (5) people and communities, (6) the economy, (7) built infrastructure, and (8) the natural environment (Chourabi et al. 2012).

Another literature was the European Union project for smart city (see www.smart-cities.eu) which also tried to identify the characteristics and factors for smart city to form indicators and assessment framework to assess city's performance as smart city. The characteristics of a smart city identified are: (1) smart economy, including factors all around economic competitiveness such as innovation, entrepreneurship, trademarks, productivity, and flexibility of the labor market; (2) smart people: it is described not only by the level of qualification or education of the citizens but also by the quality of social interactions regarding integration and public life and the openness towards the "outer" world; (3) smart governance: comprises aspects of political participation, services for citizens, as well as the functioning of the administration; (4) smart mobility: local and international accessibility are important aspects of mobility as well as the availability of information and communication technologies and modern and sustainable transport systems; (5) smart environment: is described by attractive natural conditions (climate, green space, etc.), pollution, resource management, and also by efforts towards environmental protection; finally, (6) smart living: comprises various aspects of quality of life such as culture, health, safety, housing, tourism (Giffinger et al. 2007).

Finally, the UN-Habitat recommends global goals for human settlement that aims to develop an adequate shelter and sustainable human settlement in an urbanizing world. Such recommendations are necessary for most of the developing countries. The main principles of these goals are: (1) equality: planning should promote human settlements where all people, especially women, children, and youth have equal access to basic housing, infrastructure, health care, green and open spaces; equal opportunity for education, for productive and freely chosen livelihood and for personal, spiritual, cultural, and social development; equal rights and obligations with regard to the conservation and use of natural and cultural resources; equal opportunity to participate in public decision-making; (2) eradication of poverty:

planning should aim at meeting the basic needs of the low-income groups within human settlements and provide for full productive and freely chosen employment; (3) sustainable economic development: guarantee employment opportunities and social progress; (4) livability: physical conditions and spatial characteristics of towns, villages, and cities have to be taken into account. In addition, city layout and district land-use patterns, population and building densities, and ease of access to adequate public amenities have to be accounted for; (5) civil engagement and government responsibility: generate a sense of citizenship and identity, a spirit of volunteerism and civic engagement. It should take into account that governments have the responsibility to protect their citizens' health, safety, and general welfare, and ensure all rights are protected under law (UN-HABITAT 1998).

Additionally, the UN-Habitat highlighted that sustainable development requires maintaining balance among the three distinct, everyday development processes: economic development, community development, and ecological development. But the pressure that local communities face make it extremely difficult for any one institution to maintain this balance alone (UN-HABITAT 1998).

5 Strategic Implementation Framework for Smart City

Based on the smart city frameworks and models literature review, we recommend a generic framework that can be tailored by developing countries in order to set a smart city strategy of implementation. The framework is divided into three main parts: (1) smart city structure; (2) smart city factors; and finally (3) smart city strategy. The sequences of developing these parts may differ depending on the type of city or cities required to be developed. But in order to have a sustainable smart city, these framework parts need to be prepared and integrated (see Table 1).

5.1 Part I: Smart City Structure

This part is derived from the "Urban Information Model", and it tries to divide the smart city structure into six main groups of layers instead of five: (1) environment layer, including topography, natural resources, geology, etc.; (2) infrastructure layer, including roads, bridges, tunnels, buildings, telecommunication, etc.; (3) resources layer, representing materials necessary for city operation and industry including water, air, oil, minerals, etc.; (4) services layer, representing different kinds of services, including transportation, energy, health care …, etc.; (5) social systems layer, such as culture, laws, regulations, governance, etc.; (6) city economy layer, which focuses on revenue generating resources to the city, includes commerce, trade, factories, harbors, tourism, mining, employment. This layer is the most important one for developing countries because it can stimulate investments and increase population attraction.

Table 1 Smart city strategies framework for developing countries

Part I Smart city structure	Part II Smart city factors	Part III Smart city strategy
Environment layer: topography, natural resources, and geology	Smart economy: innovation, entrepreneurship, productivity, trade, sustainable development	City assessment—where are we now? (current)
Infrastructure layer: roads, bridges, tunnels, buildings	Smart people: education, social interactions, human settlement, equality, eradication of poverty	City objectives—where do we want to go? (future)
Resources layer: water, air, oil, minerals	Smart governance: policy, political participation, services for citizens, administration	City indicators—how to assess our achievement? (measurement)
Services layer: transportation, energy, commerce, health care	Smart mobility and connectivity: Local and international accessibility, ICT, modern and sustainable transport systems	City stakeholders—who is responsible for smart city development? (people): governments, regulators, business, owners, operators, and users
		City strategy—how are we going to achieve the smart cities objectives? (strategy)
		(a) First strategic stream—developing existing cities
		(b) Second strategic stream—developing new smart cities
		(c) Third strategic stream—integration of smart cities
Social systems layer: culture, laws, regulations, governance	Smart environment: attractive natural conditions (climate, green space, etc.), pollution, resource management	
City economy: revenue generating resources to the city, which includes commerce, trade, factories, harbors, tourism, mining, employment	Smart quality of life: health, safety, housing, tourism	
	Smart institutions: management and organization, integration, participation	
	Smart infrastructure: integration, technology, eco-friendly	

The structure of a smart city can be the foundation of urban planning but it is necessary to develop and apply the smart city factors and strategies. The sequence of these layers is not significant but in the end it should be integrated.

5.2 Part II: Smart City Factors

This part uses the main factors and characteristics attached with smart city concepts, developed in the (a) integrative framework by the social sciences and humanities research council (SSHRC), (b) standards for smart city by European Union, and (c) human settlement goals by UN Habitat. The weight and description of these characteristics may differ from city to city or from one country to another, depending on the economy, political, and social condition of a society. These factors and characteristics of a smart city include but not limited to:

1. Smart economy: includes factors of revenue generation such as innovation, entrepreneurship, productivity, employment; the main aim is to maintain sustainable economic development to guarantee employment opportunities and social progress.
2. Smart people: adequate education system; equality in access to basic housing, infrastructure, health care, green and open spaces; equal opportunity for education, for productive and freely chosen livelihood and for personal, spiritual, cultural, and social development; equal rights and obligations with regard to the conservation and use of natural and cultural resources; equal opportunity to participate in public decision-making; eradication of poverty by meeting the basic needs of the low-income groups within human settlements and provide productive employment opportunities.
3. Smart governance: policy; regulations; political participation, services for citizens as well as the functioning of the administration; civil engagement by generating a sense of citizenship and identity, a spirit of volunteerism, and civic engagement; government responsibility since governments have responsibility to protect their citizens' health, safety, and general welfare, and to ensure all rights are protected under law.
4. Smart mobility and connectivity: local and international accessibility are important aspects of mobility as well as of the availability of information and communication technologies and modern and sustainable transport systems.
5. Smart environment: is described by attractive natural conditions such as climate, green space, pollution, resource management, and also by the efforts towards environmental protection.
6. Smart quality of life: comprises various aspects of quality of life such as culture, health, safety, housing, and spatial characteristics of towns, villages, and cities that have to be taken into account. In addition, it also includes city layout and district land-use patterns, population and building densities and ease of access to adequate public facilities.

7. Smart institutions: development of city management and organization, integration, participation on all parts of the city either private or public sector.
8. Smart infrastructure: infrastructure will be based more on the technology front to increase cost control, efficiency, and integration; also these infrastructures need to be eco-friendly to maintain sustainable development.

Finally, to maintain a sustainable development for smart city it is essential to establish a balance among the four development processes: economic development, community development, political development, and ecological development.

5.3 Part III: Smart City Strategy

This part is developed from the "smart city framework" developed by Cisco Internet Business Solutions Group (IBSG). It aims to divide the smart city strategy setting process into main elements that should be considered to make sure that these strategies can be implemented successfully. The strategy elements include:

1. City assessment—Where are we now? (current): the aim is to identify the current situation of the city, understand how cities operate, identify strength and weakness, such as location, population, demography, resources, economy, attractions, size, risks, slums, etc.
2. City objectives—Where do we want to go? (future): mainly define city objectives and aim to improve social, environmental, and economic environment.
3. City indicators—How to assess our achievement? (measurement): aim to match international indicators and benchmarks to city objectives.
4. City stakeholders—Who is responsible for smart city development? (people): identify the roles and impact of different stakeholder groups in developing smart city strategy, the key stakeholders groups are: governments, regulators, businesses, owners, operators, users.
5. City strategy—How are we going to achieve the smart city objectives? (strategy): (a) *first strategic stream—developing existing cities:* developing the existing cities, so that it can be smart sustainable cities from economic, social, ecological, and political dimensions, and this strategic stream will focus on mega cities, medium cities, and small cities; (b) *second strategic stream—developing new smart cities:* the aim of these new cities is to attract population from existing cities with high population density, redistribute the population across the whole country, and stimulate the economic growth of the country. These cities can have different sizes and should be connected to allow mobility and efficient supply chain; and (c) *third strategic stream—integration of smart cities:* integrate the existing smart cities and the new smart cities with federal or regional objectives, enhance supply chain, and increase sustainable development on the regional scale of a country.

6 Smart City in Egypt

Since 2010, Egypt suffers major political turbulences, plus social and economic challenges. In addition to continuous enormous population growth rate (approximately 1.7%), Egyptian population reached approximately 85 million in 2013, and it is projected that Egyptian population will reach approximately 125 million by 2050 (United Nations 2012). This is why Egypt faces a high poverty rate; around 20% of the population is below the poverty line (unfpa 2009). The average urban population growth rate is 2.0%, and average rural population growth rate is 1.4% (UNdata 2012). There are large developmental disparities between rural and urban areas, Upper and Lower Egypt, as well as between and within governorates and cities. Nearly 40% of all investments are concentrated in Cairo Governorate. Nevertheless, the public expenditure on social services remains low (Central Agency for Public Mobilization And Statistics 2014).

6.1 Background of Smart City in Egypt

The concept of smart city in Egypt is not well defined, since most of the trials presented by the Egyptian government were in the form of establishing industry- or services-specific focused cities in specialized cities or zones. There were different types of cities such as industrial zones as "10th of Ramadan city"; new cities like "6th of October city"; and technology-based cities like "Smart Village in Cairo."

Smart Village was established in 2001 by the Egyptian government and private sector investors. The "Smart Village" (see www.smart-villages.com) can be considered as successful model for applying smart city standards from technology prospective. This village was developed to host the information technology companies and provide essential services standards such as networks security, high standards building, landscape, and other facilities. But smart village was not able to sustain as a successful model by its own, since it depended on the surrounding cities for supply; it did not include housing, and also it was not well connected with the transportation facilities. It lacked the coordination and integration with rest of the cities. Therefore, after the change of the government's officials who promoted this idea, the interest in such type of cities has become negligible, and the investors have started to withdraw their investments and reallocate in other locations.

Other efforts by former Egyptian governments to enhance citizens' quality of life were building new urban areas surrounding Cairo, the capital city of Egypt, in order to absorb population from the capital and encourage business and industrial investments. The cities such as "6th of October city", "Al Obur city" (the Crossing), "AL Tagamou Al Khames" (Fifth District area), and "AL Shrouk city" (sunrise), were built between 20 km and 60 km far from Cairo city downtown. The main problem was that these new cities were not sufficient enough for its citizens to dispense the need to be in Cairo.

The new cities that developed surrounding Cairo caused the citizens and residents to go daily or frequently to Cairo downtown to meet their needs, which increased the traffic problem in Cairo and the connecting roads to new cities like

"26th of July Axis" and "Cairo Ring Road". By the end of 2010, after almost 10 years of developing these new cities, these became an extension to Cairo city and the connecting roads were overwhelmed by unplanned areas and slums. This also resulted that life in greater Cairo became almost impossible due to increased pollution and extreme traffic congestion (World Bank 2010; Japan International Cooperation Agency (JICA) 2008; United Nation Development Program 2007).

Other developing cities like Sharm Al Shake and Hurgadah are ideal for quality of life but lack economic sustainability, since the economy of these cities is focused on tourism only and not on diversity of business and industries. Despite that the tourism sector in Egypt almost generates an average of 11.3 % of the gross domestic product (GDP) with receipts in 2011 valued at US$8.7 billion; it dropped to almost US$5.6 billion in 2013. Nevertheless, the tourism industry is seasonal and fragile; it also depends on global economic and political conditions plus the security condition in the country. Since 2011 revolution, the tourism sector is almost drained due to the political and security unrest in Egypt. As a result, these cities were impacted as unemployment had increased and cost of living had become unaffordable (State Information Services 2014; Central Agency for Public Mobilization And Statistics 2014).

The main authorities in Egypt responsible for planning and developing urban cities, are Ministry of Housing, Utilities and Urban Development (www.moh.gov.eg); New Urban Communities Authority (www.newcities.gov.eg); General Organization for Physical Planning (www.gopp.gov.eg); and Ministry of Planning (www.mop.gov.eg). In addition, different international organizations that support Egypt in development are the United Nation Development Programme (UNDP), United Nations Human Settlements Programme (UN-HABITAT), and Japan International Cooperation Agency (JICA).

Urban planning in Egypt does not lack planning but lacks the coordination between the plans and strategies, integrated sustainable development vision, and there is a need for more coordination between different authorities and ministries in planning and implementing the strategy of urban planning. Also the biggest fault is that implementation of cities is constrained by solving problems in existing cities, instead of setting a strategy for redistribution of the population congestions. Also, most of the services and investments are centralized in Cairo governorate and delta area, which reduce the attention of development in other territories such as Red Sea area and west of Egypt (United Nation Development Program 2007).

Therefore, countries like Egypt need to develop more smart cities to ensure sustainable development and benefit from technology, since technology will open more employment and investment opportunities, as well as increase efficiency and cost control. One of the major challenges Egypt faces is integration of the development of smart cites in a strategic framework in order to achieve the goals and objective of sustainable development for different territories in Egypt.

6.2 *Smart City Strategic Framework—Prototype on Egypt*

This section tries to simulate Egypt's requirements for smart cities as a prototype on the suggested generic framework "smart city strategic framework" illustrated earlier. Such simulation helps in setting smart city implementation strategies in Egypt. The framework is divided into three main parts: (1) smart city structure, (2) smart city factors, and (3) smart city strategy and recommendation for Egypt. The sequences of developing these parts may differ depending on the type of city or cities required to be developed. But in order to have a sustainable smart city, the frameworks parts should be in order together and integrated.

6.2.1 Part I: Smart City Structure

This part is mainly the responsibility of local level, infrastructure planners, and urban planners, and considers the smart city structure in six main groups of layers: (1) environment, including topography, natural resources, geology; (2) infrastructure, including roads, bridges, tunnels, buildings, telecommunication; (3) resources, representing materials necessary for city operation and industry including water, air, oil, minerals; (4) services, representing different kinds of services, including transportation, energy, health care.; (5) social systems: culture, laws, regulations, governance, etc.; (6) city economy, which focuses on revenue generating resources to the city, including commerce, trade, factories, harbors, tourism, mining, and employment. This layer can be the most important one for developing countries because it can stimulate investments and increase population attraction.

The structure of smart city requirements will be different depending on its size, population density, new or existing city, and type of economy focus. The Egyptian government will need to stimulate the private sector and foreign investments to support the development of these layers in different smart cities. Also it is necessary for Egypt to invest and utilize (a) desert renewable energy technology such as solar energy and wind energy; this will help expand vast areas in the country, reduce the impact of power shortage in Egypt, and reduce the cost of connecting networks of infrastructure for new cities; (b) desert architecture and urban planning to help reduce energy consumption and utilization of landscape; and, (c) coastal areas development, which will generate opportunities for alternative power and water resources, and provide different means of transportation.

6.2.2 Part II: Smart City Factors

The setting of the smart city factors is the responsibility of central government or regional level, and generalized on local level. The main aim is to maintain a sustainable development for smart city by assuring balance among the three development processes: economic development, community development, and ecological development. The factors of smart city should include smart economy, smart people, smart governance, smart mobility and connectivity, smart environment, smart quality of life, smart institutions, and smart infrastructure.

Egypt government will have to focus specially on smart people factor to consider: (1) equality by giving equal access to basic housing, infrastructure, health care, green and open spaces; equal opportunity for education; social development; equal rights and obligations with regard to the conservation and use of natural and cultural resources; equal opportunity to participate in public decision-making; (2) eradication of poverty by aiming to meet the basic needs of the low-income groups in the society and provide employment opportunities.

6.2.3 Smart City Strategies—A Recommendation for Egypt

This part aims to divide the smart city strategy setting process into main elements that should be considered to make sure that these strategies can be implemented successfully. The strategy elements include city assessment, city objectives, city indicators, city stakeholders, and city strategy.

Egyptian government needs to develop three streams of strategies: *First strategic stream—developing the existing cities*: so it can be smart sustainable cities from economic, social, ecological, and political dimensions, this strategic stream focuses on (a) mega cities like Cairo and Alexandria, (b) medium cities like Port Said and Damietta, (c) small cities like Sharm El Sheikh and Hurghada, and (d) developing villages to be more sustainable and encouraging for farmers and agriculture investment, by renovation and development of existing villages to match at least minimum requirement of quality of life, such as, health, education, electricity, etc. *Second strategic stream—developing new smart cities*: the aim of these new cities is to attract population from Nile delta and cities with high population density and redistribute it across the country to stimulate the economic growth of Egypt. These cities can be in different sizes and should be connected to allow mobility and efficient supply chain. These cities should allow population expansion, usage of deserted areas, utilization of renewable energy and maintain contentious sustainability. Finally, *third strategic stream—integration of smart cities*: this stream aims to integrate different smart cities on the whole country in order to achieve the national objectives, and maintain sustainable development political, economic, social, and environmental dimensions (Table 2).

Table 2 Smart city strategies—a recommendation for Egypt

First strategic stream developing existing cities	Second strategic stream developing new smart cities	Third strategic stream integration of smart cities
Mega cities like Cairo and Alexandria	Attract population from Nile delta and cities with high population density	Integrate different smart cities on the whole country in order to achieve the national objectives
Medium cities like Port Said and Damietta	Redistribute the population across the whole country	Maintain sustainable economic development
Small cities like Sharm El Sheikh and Hurghada	Stimulate economic growth	Maintain sustainable social development
Developing villages to be more sustainable and encouraging for farmers and agriculture investment	Allow mobility and efficient supply chain	Maintain sustainable political development
	Consider population expansion	Maintain sustainable environmental development
	Utilize usage of deserted areas	
	Utilize renewable energy	
	Maintain contentious sustainability	

7 Discussion and Conclusion

Current smart city frameworks and models are not enough to fulfill the requirements of developing countries to face the challenges in applying smart city concept, as most of these frameworks and models are based on information technology point of view, and do not take into account the challenges that developing countries face in implementing smart cities, such as the weak integration of social, economic, and political needs and lack of a holistic and integrated approach to sustainable city development. Nevertheless, most of the developing countries lack proper infrastructure, adequate means of implementation, enough funding, sufficient economic growth, and political stability. Moreover, the challenges that can prevent the success of such a concept are poverty, inequality, cultural barriers, and the most difficult ones are the continual rise of slums and unplanned immigration from rural areas to cities.

Developing countries are still in need to deploy smart city concept, as there is a continuous shift from rural to urban population throughout the world, but unfortunately such enormous and complex growth of urban population in developing countries is unplanned, which means that such a growth will be inevitably messy, unsustainable, and does not match minimum requirement for reasonable quality of life. Most of the cities developed recently face different types of problems such as difficulty in waste management, air pollution, traffic congestion, and inadequate and deteriorating city infrastructure, and also the social and political complexity problems such as inequality, societal turmoil, rising human health concerns, increasing repercussions due to slums, and growth of public institutions that are ineffective and unable to satisfy the needs of communities. In addition to increasing unemployment and unsustainable economic development, these problems and others can be mitigated through the adoption of scalable solutions that take advantage of ICT to increase efficiency, reduce cost, and enhance quality of life.

The literature of the smart city frameworks and models were grouped into two main segments: (a) *Segment I—Smart city structure*: this segment addresses the amount of layers the smart city concept is based on, and the relation between these layers, with an elaborate discussion on the role of technology. (b) *Segment II—Smart cities factors and characteristics*: this segment aims to address the main factors and characteristics attached with smart cities concepts, and consider the UN-Habitat recommendation for the human settlement, which is a major concern for developing countries, in addition to European Commission standards for smart cities.

A generic framework called "strategic implementation framework for smart city" was developed, and used in order to set a smart city strategy of implementation. The framework is divided into three main parts. *Part I—smart city structure*: that includes six main layers (1) environment layer, (2) infrastructure layer, (3) resources layer, (4) services layer, (5) social systems layer, and (6) city economy layer. *Part II—smart city factors*: that includes eight main factors (1) smart economy, (2) smart people, (3) smart governance, (4) smart mobility and connectivity, (5) smart environment, (6) smart quality of life, (7) smart institutions, and (8) smart infrastructure.

Finally, *part III—smart city strategy*: that includes five main steps (1) city assessment (current), (2) city objectives (future), (3) city indicators (measurement), (4) city stakeholders (people), and (5) city strategy (strategy). The framework was used to set a prototype smart city strategy framework for Egypt to implement and develop smart cities strategy.

The Egyptian government in order to implement a smart city strategy needs to have three streams of strategies: (1) developing existing cities, (2) developing new smart cities, and (3) integrating smart cities. The strategic streams aim to integrate different smart cities, on the whole the country, to achieve the national objectives, maintain sustainable development and political, economic, social, and environmental dimensions. In addition, Egypt government needs to give special attention to smart people factor to consider: (1) equality by giving equal access to basic housing, infrastructure, health care, equal opportunity for education, and sustainable social development; (2) eradication of poverty by aiming to meet the basic needs of the low-income groups in the society and provide employment opportunities.

It is also essential for Egypt to invest and utilize (a) desert renewable energy technologies, such as solar energy and wind energy, mainly to reduce the impact of power shortage in Egypt and the cost of connecting networks of infrastructure for new cities; (b) desert architecture and urban planning to help reduce energy consumption and utilization of landscape; and, (c) coastal areas development, which will generate opportunities for alternative power and water resources and provide different means of transportation.

Strategic implementation framework for smart city aims to develop a generic framework for implementing smart city strategies in developing countries, by considering integration of local systems and regional systems, considering human factors, integrating social, economic and political needs, and maintaining an integrated sustainable development. The main benefit of this framework is its ability to consider other dimensions in addition to technology and urban planning; these dimensions include economic development, social development, ecological development, and development of political mechanisms.

References

Boulton, A., Brunn, S., & Devriendt, L. (2011). Cyberinfrastructures and "smart" world cities. In P. Taylor, B. Derudder, M. Hoyler, & F. Witlox, *International handbook of globalization and world*. UK: Edward Elgar.

Central Agency for Public Mobilization And Statistics. (2014). *Egyptian population report 2013*. Cairo: Central Agency for Public Mobilization And Statistics.

Chourabi, H., Nam, T., Walker, S., Gil-Garcia, J., Mellouli, S., Nahon, K., et al. (2012). Understanding smart cities: An integrative framework. *45th Hawaii international conference on system sciences*. Hawaii.

Dirks, S., Keeling, M., & Dencik, J. (2009). How smart is your city?: Helping cities measure progress. (IBM Global Business Services). ftp://public.dhe.ibm.com/common/ssi/ecm/en/gbe03248usen/GBE03248USEN.PDF. Accessed 1 Jan 2014.

Dirks, S., Gurdgiev, C., & Keeling, M. (2010). Smarter cities for smarter growth: How cities can optimize their systems for the talent-based economy. IBM Global Business Services. ftp://public.dhe.ibm.com/common/ssi/ecm/en/gbe03348usen/GBE03348USEN.PDF. Accessed 1 Jan 2014.
Falconer, G., & Mitchell, S. (2012). Smart city framework a systematic process for enabling smart + connected communities. Cisco Internet Business Solutions Group (IBSG). https://www.cisco.com/web/about/ac79/.../Smart-City-Framework.pdf. Accessed 1 Jan 2014.
Giffinger, R., Fertner, C., Kramar, H., Kalasek, R., Pichler-Milanović, N., & Meijers, E. (2007). *Smart cities ranking of European medium-sized cities*. Vienna: Centre of Regional Science (SRF), Vienna University of Technology.
Harrison, C., & Donnelly, I. (2011). *A theory of smart cities. 55th Annual meeting of the international society for the systems sciences*. New York: IBM Corporation.
Hollands, R. (2008). Will the real smart city please stand up? *City, 12*(3), 303–320.
Japan International Cooperation Agency (JICA). (2008). *Strategic development master plan for sustainable development of the greater Cairo region*. Cairo: Japan International Cooperation Agency.
Johnson, B. (2008). Cities, systems of innovation and economic development. *Innovation: Management, Policy & Practice, 10*(2–3), 146–155.
Kim, J., & Steenkamp, A. (2013). Analysis of smart city models and the four-foci taxonomy for smart city design. *Architectural research conference—Urbanism: Technology, connectedness and the urban environment* (pp. 638–649). Southfield: Lawrence Technological University.
Natural Resources Defense Council. (2014, October 11). Sustainable Communities. Natural Resources Defense Council. http://www.nrdc.org/sustainable-communities/. Accessed 1 Sept 2014.
Rittel, H., & Webber, M. (1973). Dilemmas in a general theory of planning. *Policy Sciences, 4,* 155–169.
State Information Services. (2014). State information services Tourism importance. State information services. http://www.sis.gov.eg/En/Templates/Articles/tmpArticles.aspx?ArtID=67693. Accessed 6 March 2014.
Toppeta, D. (2010). The smart city vision: How innovation and ICT can build smart, "Livable", sustainable cities. (The Innovation Knowledge). http://www.thinkinnovation.org/file/research/23/en/Toppeta_Report_005_2010.pdf. Accessed 1 March 2014.
UNdata. (2012). UNdata | country profile | Egypt. http://data.un.org/CountryProfile.aspx?crName=EGYPT. Accessed 8 Jan 2014.
UN-HABITAT. (1998). *Strategic urban development planning in Lake Victoria region: Lessons of experience*. Toronto: International Council for Local Environmental Initiative (ICLEI).
United Nations Population Fund. (2009). UNFPA Annual Report 2008. United Nations Population Fund. http://www.unfpa.org/publications/unfpa-annual-report-2008. Accessed 1 Oct 2014.
United Nation Development Program. (2007). *Strategic urban development plan for greater Cairo*. Cairo: UNDP.
United Nations. (2012). World population prospects: The 2012 revision. http://esa.un.org/wpp/unpp/panel_indicators.htm. Accessed 1 Dec 2013.
World Bank. (2010). *Egypt—Cairo traffic congestion study—phase 1*. Washington, DC: World Bank.

Karim Hamza has 15 years of professional experience in government technology and project management with leading international organizations such as European Union, Oracle, and Mahindra Satyam. He has worked on developing several strategic portfolios and programs to the Egyptian government. His research and publications focus on government's development, governance technology, policy analysis and modeling, and smart city, with a special focus on developing and unstable countries such as the Arab spring countries. He participates in different international conferences as a guest speaker, jury member, and review committee. Currently, he is engaged in the European Union project "EGOVPOLINET" which focuses on standardizing ICT Technology standards and methodology for policy analysis and governance for public sector in European Union countries. Also he is a senior associate researcher at the Institute for European Studies in Brussels and a fellow researcher at the Center for Technology in Government in New York. He was awarded his PhD from Vrije University Brussels.

Part II
Smart Cities Around the World

How Do Southern European Cities Foster Innovation? Lessons from the Experience of the Smart City Approaches of Barcelona and Milan

Mila Gascó, Benedetta Trivellato and Dario Cavenago

Abstract Innovation, and technological innovation in particular, can help city governments meet the challenges of urban governance, improve urban environments, become more competitive, and address sustainability concerns. To prevent and manage these challenges, cities need to operate in an innovative way. In this context, the smart city approach is emerging as a way of solving tangled and difficult problems. However, there is not a unique and right strategy to develop a smart city. By drawing on a comparison between the experiences of Barcelona (Spain) and Milan (Italy), this chapter aims to explore similarities and differences in the way these two Southern European cities, both being the second largest in their respective countries, are building their smart city agenda. The ultimate aim is to identify the main features of two still developing approaches, which appear to be influenced by the increasing integration of smart dimensions and initiatives in the cities' strategic agendas and the related opportunities and challenges.

Keywords Smart cities · Barcelona · Milan · Urban planning · Innovation

Abbreviations

BIE	Bureau International des Expositions
EU	European Union
ICT	Information and Communication technologies
IT	Information Technology
LGP	Local Government Plan

M. Gascó (✉)
Institute of Public Governance and Management, ESADE-Ramon Llull University, Av. Pedralbes 60-62, 08034 Barcelona, Spain
e-mail: mila.gasco@esade.edu

B. Trivellato · D. Cavenago
Department of Sociology and Social Research, University of Milano-Bicocca, Via Bicocca degli Arcimboldi, 8, 20126 Milano, Italy
e-mail: benedetta.trivellato@unimib.it

D. Cavenago
e-mail: dario.cavenago@unimib.it

J. R. Gil-Garcia et al. (eds.), *Smarter as the New Urban Agenda A Comprehensive View of the 21st Century City*, Public Administration and Information Technology 11,
DOI 10.1007/978-3-319-17620-8_10

1 Introduction

Over the past few decades, the challenges faced by municipal administrations, such as urban growth or migration, have become increasingly complex and interrelated. In addition to the traditional land use regulation, urban maintenance, production, and management of services, governments are required to meet new demands from different actors regarding water supply, natural resources' sustainability, education, safety, or transportation (Naphade et al. 2011; Albrechts 2006). Furthermore, today, cities are in strong competition for companies, tourists, and most of all talents (Zenker et al. 2013), and they are also experiencing unprecedented socioeconomic crises.

Innovation, and technological innovation in particular, can help city governments meet the challenges of urban governance, improve urban environments, become more competitive, and address sustainability concerns. To prevent and manage these challenges, cities need to operate in an innovative way. In this context, the smart city approach is emerging as a way of solving tangled and difficult problems (Nam and Pardo 2011a).

Although the literature is rich in references to the smart city, it is also fragmented: this is still a fuzzy concept that is not being consistently used (Meijer and Rodríguez-Bolívar 2014). Nevertheless, agreement exists on the fact that "smart cities" is a construct in which to frame local government transformation by using innovative technologies. In this respect, a smart city is one with a strong commitment to innovation not only in technology but also in management and policy (Nam and Pardo 2011a).

This fragmentation is also reproduced in terms of the strategies that different cities follow to become smarter. There is no single route to becoming smart, and different cities have adopted different approaches that reflect their particular circumstances. According to the Centre for Cities (2014), this is dependent on a number of factors, ranging from the financial and managerial capacity, private sector offerings, and what citizens and businesses want.

Given this context, this chapter aims at comparing how two Southern European cities, Barcelona and Milan, both being the second largest in their respective countries, are building their smart city agenda. The ultimate goal is to identify the main features of two still developing approaches, which appear to be influenced by the increasing integration of smart dimensions and initiatives in cities' strategic agendas and the related opportunities and challenges.

The two cases, which are of great relevance for Spain and Italy have been particularly hit by the economic crisis. Their cities have a strong need to be (more) innovative in their pursuing of new engines of growth and competitiveness.

The remainder of this chapter is organized as follows. First, we present the literature on smart cities. The data and methods used to analyze and compare the strategic plans and processes of the two selected European cities are then explained. Subsequently, we describe the cases of Barcelona and Milan as well as the results of their comparison and discuss our findings. Second, we bring to a close, drawing some conclusions.

2 Smart Cities: A Conceptual Framework

Various attempts have been made to academically define and conceptually describe a smart city (AlAwadhi and Scholl 2013). Although no generally accepted academic definition has emerged so far, several works have identified certain urban attributes that may characterize what a smart city is.

Giffinger et al. (2007) rank 70 European cities using six dimensions: smart economy (competitiveness), smart people (human and social capital), smart governance (participation), smart mobility (transport and Information and Communication Technologies (ICT)), smart environment (natural resources), and smart living (quality of life). As a result, they define a smart city as "a city well performing in a forward-looking way in these six characteristics, built on the "smart" combination of endowments and activities of self-decisive, independent, and aware citizens" (p. 11). According to Lombardi (2011), these dimensions, which have been used by several authors, connect with traditional regional and neoclassical theories of urban growth and economic development.

As a result of the review of multiple smart city's definitions and related terms, Nam and Pardo (2011b) suggest three conceptual dimensions of a smart city: technology, people, and community. For them, technology is the key ecause of the use of ICT to transform life and work within a city in significant and fundamental ways. However, a smart city cannot be built simply through the use of technology. Hence, the role of human infrastructure, human capital and education, on one hand, and the support of government and policy, on the other, also become important factors. Considering these three variables, the authors conclude that "a city is smart when investments in human/social capital and IT infrastructure fuel sustainable growth and enhance a quality of life, through participatory governance" (p. 286).

In turn, Leydesdorff and Deakin (2011) introduce a triple helix model of smart cities. They argue that cities can be considered as densities in networks among three relevant dynamics: the intellectual capital of universities, the wealth creation of industries, and the democratic government of civil society. Lombardi et al. (2011) build on this model and refer to the involvement of the civil society as one of the key actors, alongside the university, the industry and the government. In Lombardi's words (2011): "this advanced model presupposes that the four helices operate in a complex urban environment, where civic involvement, along with cultural and social capital endowments, shape the relationships between the traditional helices of university, industry and government. The interplay between these actors and forces determines the success of a city in moving on a smart development path" (p. 8).

Yet, one of the most comprehensive and integrative frameworks for analyzing smart city projects has been presented by Chourabi et al. (2012). The model has already been tested, and several cities' smartness strategies and initiatives have been assessed using them. Chourabi et al. (2012) present a set of eight dimensions, both internal and external, that affect the design, implementation, and use of smart cities initiatives: management and organization, technology, governance, policy context, people and communities, economy, built infrastructure, and natural environment (see Table 1).

Table 1 Smart cities integrative framework. (Source: Chourabi et al. (2012))

Dimensions	Description
Management and organization	Managerial and organizational factors, such as project size, managers' attitudes and behaviors or organizational diversity, influence projects broadly
Technology	A smart city relies on a collection of smart computing technologies applied to critical infrastructure components and services. However, ICT impact is unclear and it can either increase the quality of life of citizens or the digital divide
Governance	Processes, norms, practices that rule the exchange of information among different stakeholders. It involves several factors such as leadership, collaboration, communication, data exchange, partnership, or service integration
Policy context	Political (e.g., policy agendas) and institutional (e.g., regulatory barriers) components of the environment
People and communities	Individual persons and communities of the city affecting and affected by the implementation of smart city initiatives. It involves several factors such as participation and partnership, accessibility, quality of life, or education
Economy	Economic inputs to and economic outcomes from smart cities initiatives such as innovation, productivity, or flexibility
Built infrastructure	Availability and quality of ICT infrastructure such as wireless infrastructure and service-oriented information systems
Natural environment	Sustainability and good management of natural resources

3 Research Methods

Focusing on the two smart city cases (Barcelona and Milan), this empirical study aimed at exploring similarities and differences between the two cities. Rich data were collected through semi-structured interviews with individuals in charge of the smart city strategy of the cities, with people in charge of specific smart city projects, and with experts on smart cities. In the case of Milan, four interviews took place. Also, the authors attended four out of the seven public events organized by the municipality and had the opportunity to talk to participants and organizers. In the case of Barcelona, 19 interviews were conducted. Many documents were analyzed as well. Some of them were official documents and presentations by the respective city councils, while many others had been written by researchers or academic authors. This was particularly in the case of Barcelona, a city whose smartness strategy has already been widely studied.

4 Case Descriptions

4.1 Barcelona

After the local elections of June 2007, the organizational structure and the unit in charge of promoting ICT adoption in the city of Barcelona and within the Barcelona

City Council experienced some changes. As a result, an ICT master plan was defined and implemented in September 2008. The plan pursued: (1) to improve the municipal functions and services by focusing on citizens and quality, (2) to reinforce proximity by developing a new territorial model based on 73 neighborhoods instead of 10 districts, (3) to achieve processes innovation by intensively using new technologies, and (4) to measure management. In sum, the new model was intended to make Barcelona a smarter city by means of having a simple and effective, closer to citizens, connected, ubiquitous, and innovative public (local) administration. As a result, three main areas were prioritized: infrastructures, smart services, and citizens' interaction.

Although 2011 was witness to a change of government, the vision on the smart city strategy did not change because it was already shared by the two main parties since the beginning (Gavaldà and Ribera 2012). Thus, during the new term a phase of relative policy continuity began in which already started projects kept developing while new ones were planned.

The new government approach is based on the will of reinforcing the smart city brand of Barcelona as a promoter of a new economy of urban services. The goal is to show Barcelona as an essential reference for all those cities which seek to redirect its economy and its external promotion following this paradigm (Gavaldà and Ribera 2012). The Smart City Expo and World Congress, held for the very first time in 2011, was the starting point of this policy, which has evolved around two main projects based on public–private partnerships: the Smart City Campus at 22@ and the development of the City Protocol.

Other ICT-related initiatives, particularly regarding the urban transformation and the development of city-wide ICT infrastructures, have also played an important role in the city governance model that has tried to change Barcelona into the ICT metropolis of the Mediterranean. It has been the case of the 22@ Barcelona project.

22@ Barcelona is the result of an urban strategy, that dates back to 2000, aimed at transforming 200 ha of industrial land of Poblenou into an innovative district offering modern spaces for the strategic concentration of intensive knowledge-based activities in five strategic fields: media, ICT, energy, medical technologies, and design. In this sense, this is a project of urban, economic, and social refurbishment.

According to the last available data (June 2012), 4500 new companies have moved to the district since 2000 (an average of 545 per year and 1.2 per day). Of them, 47.3 % are new start-ups and 31 % are technological or knowledge-based companies. As a result, the number of people working in Poblenou has risen significantly. There are currently more than 56,000 new workers (about 71 % with University studies).

From an urban point of view, since the beginning of the project, approximately 70 % of the refurbishment of the industrial area has already been accomplished, under 141 plans for urban amelioration. The said projects will result in obtaining 3031.510 m^2 of floor plant for new production facilities, social housing facilities, and technical services. The real estate sector has decisively supported the project: 85 out of the 141 plans approved are promoted by the private sector.

However, assessments of the 22@ project have not always been that positive, particularly, when such assessments have been related to the contribution of 22@ to the development of a smart city. Leon (2008) and Charnock and Ribera-Fumaz

(2011), for example, state that the district hosts a poorly educated capital, a low level of local entrepreneurship, venture capital scarce resources, little presence of large international firms, and little connectivity to businesses located in other European and Latin American cities. In this regard, companies based in 22@ with a turnover of over €15 million are only 8.1 %, 68 % are microenterprises with up to 10 employees, and less than a quarter part export goods or services to other countries.

From a social perspective, according to Gavaldà and Ribera (2012), 22@ has given rise to functional specialization, which has favored the concentration of talent and the advent of a new type of residents: the so-called creative class. However, at the same time, another phenomenon has taken place: the displacement of part of the population residing in the district—mostly low-middle socioeconomic class, vulnerable in terms of housing—and its replacement by a new middle class.

Since 2011, ICT-enabled urban development has clearly become part of the Barcelona smart city strategy, although, it is still very much related to the development of 22@, it has surpassed it. So, among other initiatives, the Barcelona City Council has promoted the setup of the Institute of Technology for Urban Habitat (BIT for the Habitat), a new foundation that fosters innovation in new urban services (such as planning and infrastructures, housing, and environment) through new forms of collaboration with private companies, and which is supported by Cisco. The yearly appointment of the Smart City Expo World Congress and the official naming of Barcelona as the resident place of the World Mobile Congress have also contributed to the development of the current smart city strategy.

As a result of its smart city strategy, Barcelona has designed 24 programs for 2014, which are driven by five values: an efficient, sustainable, productive, social, and free Barcelona. Three of them (the new telecommunications network, the urban platform, and the intelligent data project) are transversal. The rest of them have to do with diverse areas such as smart lighting, smart parking and transportation, smart water, optimized waste collection, smart regulation, or open government.

Most of them are underway and results are not clear yet. In this respect, Gavaldà and Ribera (2012) declare that probably the projects lack an environmental sustainability perspective as well as bottom-up approaches, which give users a more important role in shaping the city and its services. Having said this, Barcelona is worldwide perceived as a leading smart city. Several studies have ranked the city among the smartest in Spain, Europe, and even at the international level. Just recently, in March 2014, the European Commission awarded the European Capital of Innovation ("iCapital") prize to Barcelona for introducing the use of new technologies to bring the city closer to citizens.

In sum, although a lot remains to be done, it seems that the city has taken a unique position of not only advancing its own initiatives but also trying to provide support for the smart cities movement around the world.

4.2 Milan

Following a reform introduced by Lombardy's Regional Law 12/2005, in 2011 Milan's City Council approved its first Local Government Plan (LGP), which implied a wider and more strategic approach to local development and required citizens' participation from the early stages of the process. By focusing on issues such as greening, infrastructures, and public services, the plan aimed at simplifying access to public services and promoting the contribution of the private sector (both for profit and nonprofit) in the pursuit of public sector interests (Mingione et al. 2010). Different from the past was the method through which the plan itself was built: rather than relying exclusively on the internal department responsible for urban planning, the administration also involved a group of external experts and contributors drawn from the design and environmental planning sectors, academia, communication, and program management.

Although the LGP featured a much more strategic approach relative to past planning efforts, there was no specific focus on ICT use to support local development or internal reorganization within the municipality. However, several programs aimed at enhancing citizens' quality of life, services' provision, business support, and local development have recently been explicitly reframed and grouped within a more comprehensive smart city strategy: this involved increased investment in ICT instruments and infrastructures, with an acceleration related to the upcoming Expo 2015. The main driver for the municipality at the beginning of the process was the EU's Smart Cities and Communities Initiative, which provided the opportunity to obtain funding for environmental sustainability and energy projects. In the year 2012, the municipality, therefore, devised a strategy for the construction of its smart city agenda based on coordination rather than implementation. Responsibility for such coordination effort was given to the Councilor for Employment Policies, Economic Development, University and Research, the member of the municipal government who is responsible for these areas, as well as to a municipal manager, the head of the department in charge of Economic Innovation, Smart City and University.

The method is based essentially on coordination, rather than implementation, because it implies a critical coordination effort on the part of the department both internally (within the municipality) and externally (in the municipality's interaction with external stakeholders and with the citizens). Internally, the department became quickly aware that Milan had already started to develop or implement a number of projects or programs with "smart" features. By grouping them under a common smart city development plan, they could be better appreciated and developed. This effort was not easy due to the difficulties of involving different departments, but the focus on a "cross-sectional" theme such as the smart city, sustained internal motivation. As a subsequent step, the following macro-areas were identified as priorities for the development of Milan's smart city strategy: digital city, mobility, environment, inclusion and cohesion, and services to citizens.

In terms of the process for the construction of the smart city strategy, the coproduction of such strategy with the citizens and selected categories of stakeholders (firms, universities, financial institutions, the third sector, and other public administrations) assumes a particularly high value for the municipal administration. The consultation process with these categories of stakeholders involved the creation of six working groups linked to the six pillars of a smart city (smart economy, smart living, smart environment, smart mobility, smart people, and smart governance), in addition to a seventh one linked to Expo 2015, and by organizing over the course of 1 year—from mid 2013 to mid 2014—one large public and participatory event related to each pillar. This process was constructed and carried out with strong co-operation and involvement on the part of the Chamber of Commerce of Milano, also with the aim of identifying an appropriate governance structure. A website for Milano Smart City was built only in April 2013, but as of March 2014 (when this Chapter is being written) it is far from being comparable with other similar web-sites, as it only acts as repository for the documents and presentations of the public events mentioned above. As of March 2014, no independent or autonomous agency for the management of the smart city strategy has yet been established in Milan, though there are plans for the establishment of a shared governance body involving other stakeholders in addition to the local municipality. The municipality of Milan is also heavily relying on the collaboration of large private high-tech firms in the construction of Milano Smart City: this is evident both in the consortia that were established to apply for national or EU-based financing, and in the partnerships that were developed especially, though not exclusively, for Expo 2015.

From the perspective of specific smart city programs and mechanisms, several new projects have been introduced over the past few years, while previous or existing services and instruments have often been restructured with ICT-enabled new features and/or a more citizen-centered approach. This includes, for instance, the establishment of an incubator for social enterprises, or the EU-financed My Neighborhood-My City project (implemented in a formerly run-down neighborhood) that, building on open innovation, moves from citizens' needs to renew the area also through the ICT. Numerous specific projects were also developed that relate to the above mentioned six pillars of a smart city and which leverage the opportunities provided by the ICT to provide better services to citizens as well as to public and private institutions.

The smart city drive provides an opportunity for greater coordination relative to the past. Over the past decade, the municipality has focused itsattention on large events, such as Expo 2015, interventions such as the Fashion City Project (Citylife) and the resolution of specific problems, such as pollution, through instruments such as the congestion charge. However, according to Mingione et al. (2010), these interventions have rarely been interrelated and have hardly communicated with each other: sectorial focus and separation are the main characteristics of these actions. On the other hand, precisely because of its scope and international visibility, Expo 2015 will be an opportunity to accelerate and drive to convergence a number of trends and developments which will contribute to wider recognition of Milan as a smart city.

The implementation process of Expo 2015 has not only experienced various governance-related difficulties but also drastic reductions in the available resources as a result of the economic crisis that broke out shortly after the approval of the project by the BIE in 2008. For this reason, the initial design has undergone many changes, with the gradual elimination of entire works. Starting in 2012, the project and its implementation have experienced a significant rethinking in a smart city perspective (Gallione 2012; Morandi et al. 2013), with three technology platforms especially developed to support the event. While obviously maintaining its overarching theme—Feeding the Planet, Energy for Life—Expo 2015 is now taking advantage of a number of partnerships and collaborations with technological partners, who not only provide much needed funds but also an acceleration of technological developments that will benefit Milano Smart City in the longer run. At the same time, this allows to link the development of the municipality's smart city strategy to Expo 2015, thereby creating a mutually reinforcing process.

5 Comparing Barcelona and Milan

This section compares Barcelona and Milan in terms of their smart city strategies using Chourabi et al.'s integrative framework (2012).

Table 2 lists both the similarities and differences of Barcelona and Milan's efforts to become smart cities.

Table 2 Comparing Barcelona and Milan

Dimensions	Barcelona	Milano
Management and organization	New public management approach, planning tools, definitions of priorities, change of the organizational structure	Coordination by different bodies (internal and external)
Technology	Core	Residual in the beginning, more important nowadays
Governance	Leadership by the city council. Participation of private actors and collaboration with, mainly, the Autonomous Government of Catalonia	Coordination by specific department within the municipal administration. Leadership by large events' organizers. Difficult collaboration among different levels of government. Participation of private actors (including Expo 2015 SpA, the company in charge of Expo 2015 implementation)
Policy context	The smart city within a broader ICT policy/strategy	The smart city within a broader urban planning policy/strategy
People and communities	Citizens hardly participate	Citizen participation is important in the process of building the smart city
Economy	Economic competiveness (lower than Milan)	Economic competitiveness (higher than Barcelona)
Built infrastructure	Very important. The smart city project started with a smart district project, so to speak (22@) based on building IT infrastructure. The city has one transversal project around building/strengthening IT infrastructure, which is key	Not relevant. Infrastructure development was left to the private sector although, recently, more has been done within the smart city drive and the acceleration prompted by Expo 2015 implementation
Natural environment	Projects designed under the 2014 smart city strategy lack an environmental sustainability perspective	Environment matters

Regarding management and organizational issues, the approaches adopted by the two cities differ. It seems that Barcelona's management and organization are part of a broader management model based on the new public management stream of theory and, among other, on territorial decentralization, service externalization, and the adoption of managerial tools (such as strategic plans). The smart city strategy has been led politically and executively. The Computer Municipal Institute has played a key role in the development of the strategy. The Institute has adapted to the evolution of the strategy. Currently, the city has created a Smart City Personal Management Office within the Institute in which the projects belong, which coordinates all the projects in the city that are classified under the smart city tag. This has meant transitioning from siloed work to transversal work. Milan has not set up a separate organization in charge of the smart city strategy implementation, although it seems a Milano Smart City Association will be announced soon. So far, the traditional structure has taken on the responsibilities on the smart city strategy by coordinating internal and external efforts: the main actors are two individuals within the municipal administration, and their respective teams: the *Assessore* (member of the municipal executive) for Employment Policies, Economic Development, University and Research, and the Head of the Department for Economic Innovation, Smart City, and University. All activities are carried out by these two units in strict cooperation with the Milan Chamber of Commerce.

Although technology can be said to be the key for both cities nowadays, this has not always been the case, which gives rise to a new difference between Barcelona and Milan. For the former, technology has been at the core of its urban development model. In this respect, ICT have been an essential crosswise tool that has supported the multi-faceted innovation process. In the case of Milan, technology has been less important. The smart city has not been built around ICT although their relevance is actually growing: at the moment, many smart city projects are clearly based on ICT adoption. The differences between the city's most important projects also show this contrast: on one hand, Barcelona's key smart project, 22@, is clearly one that shows the strategic use of ICT by the city; on the other, Expo 2015 in Milan is an umbrella event, with not only links to the different dimensions of the smart city strategy (and therefore, to technology) but also an event that goes beyond that.

Both Barcelona and Milan have involved several stakeholders in the definition and implementation of their smart city's strategies. But there is also divergence regarding their respective governance approaches in terms of leadership and involvement of third actors. Clearly, it has been the Barcelona City Council the one to lead the efforts in the city. On the contrary, only recently the Municipality of Milan has devoted increasing efforts toward the development of the smart city strategy. In a way, the Barcelona City Council has led in isolation, that is, there has been participation of other actors, as will be explained at once, but the city council has explicitly given direction. The Municipality of Milan has coled the process. In this respect, large events' organizers and, particularly, the Milan Chamber of Commerce have taken an important leading role. It is interesting to note that this collaboration has materialized in Expo 2015 S.p.A., the public company in charge of Expo 2015 implementation, whose shareholders are the four administrations at different levels (national, regional, provincial, and municipal), together with the Chamber of

Commerce. Public–private partnerships have proved central in the implementation of the smart city's strategies of both cities. So has collaboration with other public administrations, although it has taken place differently. Thus, in the case of Barcelona, support/collaboration has come from the Autonomous Government of Catalonia, particularly in relation to wider projects that have included other territories (other than strictly the city of Barcelona). Regarding Milan, collaboration has taken place among local administrations at different levels (such as the Lombardy region or the Milan Metropolitan Area), but these relationships have often been difficult and forced by necessity (e.g., Expo 2015 implementation in the face of significant delays). Finally, European Union (EU) funding programs have been very important for both cases.

In terms of the policy context, the Barcelona's smart city strategy has clearly expanded within the framework of a broader ICT policy/strategy that has developed throughout time always having in mind the four dimensions of the so-called Barcelona model. This model gave a lot of importance to territorial decentralization and, therefore, to urban development in districts and neighborhoods, resulting in the prioritization of urban transformation projects, such as 22@. Although urban planning has also been a driving force in the case of Milan (actually, urban innovation in the city dates back to the 1980s), more traditional perspectives, accompanied by laws and regulations, have predominated: territorial governance, integrated planning, or urban and environmental policies. Also, in the case of Milan, urban governance and planning are much less structured and ordered compared to other Italian and European cities: strategic planning at the urban and metropolitan level not only tends to be driven by large events, such as the Expo 2015 but also includes a number of regular international exhibitions which have become very important for Milan's economy (such as the International Home Furnishing Exhibition or Milan Fashion Week).

Smart cities require smart citizens, individuals, and groups (communities) that participate in the building of the city. Barcelona and Milan are not alike in this respect. The former has hardly implemented participation projects and, generally speaking, in Barcelona, there is a lack of bottom-up approaches. Although there have been slight changes in the past few years, the unsuccessful electronic consultation on the transformation of the Diagonal Avenue, the city's main street, carried out in 2010 has forced the city council to be cautious regarding participation. Milan has adopted a totally different approach: the smart city strategy is being developed with the active involvement of citizens. Actually, the people in charge refer to the coproduction of the smart city with citizens (and other stakeholders as well).

Despite the differences listed so far, both cities have a dynamic economic environment, which has been recognized worldwide. Barcelona was awarded in March 2014 the European Capital of Innovation. Milan has also implemented an innovative approach prize due to its efforts to promote a broad-based innovative culture. Other factors also make it a competitive economy, such as its entrepreneurial culture (several events, many led by Barcelona Activa, take place to promote entrepreneurship) or the promotion of ICT-based economic activities located in the 22@ district. Milan is also a vibrant city from an economic perspective. It has stimulated innovation not only in creative and knowledge sectors but also in traditional core industries. Actually, Milan is the most important economic center of Italy and a major financial and business center. Interestingly enough, the process by which these cities have become economically important is quite different, for Barcelona's has been particularly led by the public sector (both the city council and the Autonomous Government of Catalonia) and Milan's has been characterized by private efforts, although, some public attempts to coordinate collaboration among local industries have been made at the regional level.

Regarding built infrastructures, Barcelona and Milan differ; although the latter is given growing importance to this dimension within the new smart city strategy. Barcelona has always prioritized the development of ICT infrastructure. In particular, within the Barcelona 2.0 model framework, several projects were put in place such as Barcelona Wi-Fi (a service that allows the citizen to connect to the Internet through Wi-Fi access points) and the Wi-Fi mesh network (a municipal network for ubiquitous services). On the contrary, Milan has paid much more attention to physical infrastructures (roads, highways, subway, and railway) and only, recently and shyly, there has been some investment in terms of wireless infrastructure and service-oriented information systems. Actually, the smart city in Barcelona is developing around the main smart project the city has—22@, which has given rise to infrastructures plans aimed at building a modern network of energy, telecommunications, district heating and pneumatic refuse, and waste collection systems, whereas Milan's smart city project is focusing on delivering smart services, that is ICT-based public services, such as the Readt application (to access the services of municipal libraries), the new tourism portal, or Infoalert (a service that aims at reducing road congestion by means of information shared in real time). A recent acceleration both in terms of coverage and intensity is linked to Expo 2015.

Finally, dissimilarities also arise for the last dimension: natural environment. Concerning Barcelona, it was already advanced that, according to Gavaldà and Ribera (2012), generally speaking, smart city projects lack an environmental sustainability perspective: "the effects of the economic crisis have put in standby the achievement of higher levels of quality of life and have paralyzed investments in environmental sustainability at macro level" (p. 24). Quite the reverse, in Milan, sustainability and the better management of natural resources matter. Lombardy's regional law 12/2005 clearly shows so, as it includes elements with a renewed emphasis on the environmental sustainability of territorial developments. Also, in Milan, the start of the smart city strategy was linked to the EU's Smart Cities and Communities Initiative, which provided the opportunity to obtain funding for environmental sustainability and energy projects, showing an important emphasis on the natural environment. Finally, nowadays, one of the core dimensions of such strategy has to do with the smart environment.

6 Conclusions

Two common trends seem to arise from the previous comparison. On one hand, the relationship between innovation and the development of a smart city is not obvious. It may seem that Barcelona is more innovative than Milan, not only because of the iCapital award but also because many of its smart cities projects are technology driven. But Milan has invested in the open innovation approach, emphasizing the need to cocreate and coproduce with citizens. Thus, there is not enough evidence to draw a conclusion on innovativeness. On the other hand, from a general point of view, objective results are still unclear in both cases. So taking these two aspects into consideration, it is not possible to say that Milan's strategy is working better than Barcelona's, or the other way round. Still, we can timidly refer to the following:

- There are important differences between the two cities. Yet, none of them is throwing clear positive results, at least, in terms of outcomes (citizens' quality of life improvement, economic development, and sustainability). However Barcelona's reputation is much better than Milan's in this area. As already stated, the former is worldwide perceived as a leading smart city to which many other cities turn to in order to "copy" some of its ideas. Barcelona is building a smart city and, at the same time, is successfully managing a brand: the Barcelona Smart City. It is doing so simultaneously, following a city management model that has already proved successful in the past regarding other topics. In sum, real projects matter but image is also important. In this respect, Barcelona has taken better advantage of its assets than Milan.
- It may be inferred from our findings that there are different ways to develop a smart city depending on the wider institutional/policy/governance context as well as the economic environment. It seems that Barcelona and Milan are developing their smart cities strategies according to their wider context and, therefore, to their traditional way of doing things. What is not clear is their city model and the type of city they want to become. It is about being faster, more intelligent, or more technology oriented. These tools only shape the city according to the city model that should guide all the efforts.
- Building on the above conclusion, it seems that there is not only a single strategy but also, by definition, there is no single way to carry out a strategy. Mintzberg's (1994) concept of a realized strategy being the result of two components is very useful in this respect: a strategy that was actually planned by the relevant actor (the "deliberate" strategy) and the "emergent" strategy that results from learning and interaction with other stakeholders. Taking this distinction into account, the planned component seems more important in the Barcelona case while the emergent component seems more relevant in the Milan case. This may explain the important differences in terms of leadership. Regarding Barcelona, the public sector, and particularly the city council, has always been a strong leader. It is the city council the one which has conceptualized the smart city strategy and implemented it. The city council had a clear direction, a clear concept of what a smart city should be ("something" around the 22@ project). It has involved other actors in terms of what the city council thought it was best. In the case of Milan, leadership is not obvious. Even, coordination by the municipality is weak. The city has been developed in collaboration. In the beginning, there was no clear idea of what to achieve but the city has taken advantage of different opportunities that has come along the way, which have been the result of cooperation.

Finally, although our intention was not to test Chourabi et al.'s model (2012), implementing it has raised a few issues worth taking into account for its future improvement. We have found that the model is useful to describe a smart city strategy's components and, probably, to organize the various perspectives and focuses of the smart city literature. But it is not enough to understand results. It is not an analytical model. Therefore, it cannot be used to give an answer to questions related to success or failure of smart city strategies. The model does not clearly state the direct effect of the dimensions onto results. It does not take into account cross-dimensional issues either.

References

AlAwadhi, S., & Scholl, H. J. (2013). *Aspirations and realizations: The smart city of Seattle*. Paper presented at the 46th Hawaii International Conference on System Sciences. Maui, HI, January 7–10.

Albrechts, L. (2006). Shifts in strategic spatial planning? Some evidence from Europe and Australia. *Environment and Planning A, 38*(6), 1149–1170.

Centre for Cities. (2014). *Smart cities*. London: Centre for Cities.

Charnock, G., & Ribera-Fumaz, R. (2011). A new space for knowledge and people? Henri Lefebvre, representations of space, and the production of 22@Barcelona. *Environment and Planning D: Society and Space, 29*(4): 613–632.

Chourabi, H., Nam, T., Walker, S., Gil-Garcia, J. R., Mellouli, S., Nahon, K., Pardo, T., & Scholl, H. J. (2012). *Understanding smart cities: An integrative framework*. Paper presented at the 45th Hawaii International Conference on System Sciences. Maui, HI, January 4–7.

Gallione, A. (2012) *Dossier expo*. Milan: Rizzoli.

Gavaldà, J., & Ribera, R. (2012). *Barcelona 5.0: From knowledge to smartness?. Working Paper Series, WP12-002*. Barcelona: Universitat Oberta de Catalunya.

Giffinger, R., Fertner, C., Kramar, H., Kalasek, R., Pichler-Milanovic, N., & Meijers, E. (2007). *Smart cities: Ranking of European medium-sized cities*. Vienna: Center of Regional Science.

Leon, N. (2008). Attract and connect: The 22@Barcelona innovation district and the internationalisation of Barcelona business. *Innovation: Management, Policy and Practice, 10*(2/3), 235–246.

Leydesdorff, L., & Deakin, M. (2011). The triple-helix model of smart cities: A new-evolutionary perspective. *Journal of Urban Technology, 18*(2): 53–63.

Lombardi, P. (2011). New challenges in the evaluation of smart cities. *Network Industries Quarterly, 13*(3), 8–10.

Lombardi, P., Giordano, S., Farouh, H., & Wael, Y. (2011). *An analytic network model for smart cities*. Paper presented at the 11th International Symposium on the Analytic Hierarchy Process. Sorrento, June 15–18.

Meijer, A., & Rodríguez-Bolívar, P. (2014). Governing the smart city: A review of the literature on smart urban governance. *International Review of Administrative Science* (forthcoming).

Memo, F., Rancati, S., & Zajczyk, F. (2011). Il governo del territorio in Lombardia. Un "modello" di deregolamentazione?. In A. Mazzette (Ed.), *Esperienze di governo del territorio* (pp. 56–101). Roma & Bari: Editori Laterza.

Mingione, E., Dell'Agnese, E., Mugnano, S., D'Ovidio, M., Niessen, B., & Sedini, C. (2007). *Milan city-region: Is it still competitive and charming? Pathways to creative and knowledge-based regions. ACRE report 2.12*. Amsterdam: AMIDSt.

Mingione, E., Zajczyk, F., Dell'Agnese, E., Mugnano, S., D'Ovidio, M., Sedini, C., & Parrino, L. (2010). *Policies and strategies in the Metropolitan Region of Milan. How to enhance the city's competitiveness. ACRE report 10.12*. Amsterdam: AISSR.

Minzberg, H. (1994). *The rise and fall of strategic planning*. New York: Free Press.

Morandi, C., Andrea, R., & Di Vita, S. (2013). ICT: Interfacce tra persone e luoghi. Sperimentazioni in corso per una smart (city-)region del Nord Italia: Il territorio tra Torino e Milano verso l'Expo 2015 e oltre. *TeMA: Journal of Land Use, Mobility and Environment, 1*(2013), 19–33.
Nam, T., & Pardo, T. (2011a). *Smart city as urban innovation: Focusing on management, policy, and context.* Paper presented at ICEGOV 2011. Tallinn, September 26–28.
Nam, T., & Pardo, T. (2011b). *Conceptualizing smart city dimensions of technology, people and institutions.* Paper presented at the 12th Annual international Conference on Digital Government Research. College Park, MD, June 12–15.
Naphade, M., Banavar, G., Harrison, C., Paraszczak, J., & Morris, R. (2011). Smarter cities and their innovation challenges. *Computer, 44*(6), 32–39.
Pasqui, G., Armondi, S. & Fedeli, V. (2010). I piani strategici alla prova. In R. Florio (Ed.). *10 anni di pianificazione strategica in Italia. Ragioni, esiti, criticità* (pp. 9–118). Florence: ReCS—Rete delle Città Strategiche.
Pyrozhenko, V. (2011). *Implementing open government: Exploring the ideological links between open government and the free and open source software movement.* Paper presented at the 11th Annual Public Management Research Conference. Syracuse, NY, June 2–4.
Sartorio, F. S. (2005). Strategic spatial planning: A historical review of approaches, its recent revival, and an overview of the state of the art in Italy. *DISP, 162*(3), 26–40.
Schaffers, H., Komninos, N., Pallot, M., Trousse, B., Nilsson, M., & Oliveira, A. (2011). Smart cities and the future Internet: Towards cooperation frameworks for open innovation. In J. Domingue et al. (Eds.). *Future Internet Assembly 2011. Achievements and technological promises* (pp. 431–446). New York: Springer.
Townsend, A. (2013). *Smart cities: Big data, civic hackers, and the quest for a new utopia.* New York: W. W. Norton & Company, Inc.
Trivellato, B., Cavenago, D., & Beltrami, G. (2013). *Is strategic urban planning becoming 'smarter'? Reflections on a selection of European cities.* Paper presented at the 2013 EGPA Annual Conference. Edinburgh, September 11–13.
Wang, H., Song, Y., Hamilton, A., & Curwell, S. (2007). Urban information integration for advanced e-planning in Europe. *Government Information Quarterly, 24*(4), 736–754.
Zenker, S., Eggers, F., & Farsky, M. (2013). Putting a price tag on cities: Insights into the competitive environment of places. *Cities, 30*, 133–139.

Mila Gascó holds an MBA and a Ph.D. in public policy evaluation. She is a senior researcher at both the Institute of Innovation and Knowledge Management and the Institute of Public Governance and Management, both at ESADE. For 7 years, she was a senior analyst at the International Institute on Governance of Catalonia where she was in charge of the local studies and the information society areas of research. She has also taught at different universities and done supervising work on information society-related research and Ph.D. thesis. She has consulting experience as well. She has worked for a wide variety of organizations such as the UNDP, the City Council and the Provincial Council of Barcelona, the Latin American Centre on Management for Development, the World E-Governments Organization of Cities and Local Governments, the Inter-American Development Bank, and Google.

Benedetta Trivellato is assistant professor of Management and Business Administration at the University of Milano-Bicocca. She holds a Ph.D. in Information Society from the University of Milano-Bicocca, an MSc in Economics for Development from the University of Oxford (UK), and a BSc in Economics from Bocconi University, Milano. She was previously a researcher at the Interuniversity Research Centre on Public Services (CRISP) at the University of Milano-Bicocca, where she conducted extensive consulting and re-search activities in cooperation with several public administrations, especially at the municipal level. Prior to her academic experience, she was a consultant at the UK-based consulting firm Oxford Analytica, where she was involved in major projects for institutional and corporate clients such as the International Monetary Fund, the European Commission, Telecom Italia, and others. Her main research interests include: services analysis, design and management (with a focus on public services), urban governance and strategic planning, and social innovation and sustainability.

Dario Cavenago is full professor of Management and Business Administration at the University of Milano-Bicocca, where he is also deputy director of the Interuniversity Research Centre on Public Services (CRISP). He is also president of Tabulaex, a spin-off of the University of Milano-Bicocca that provides business intelligence, knowledge management and consulting services, and carries out applied research mainly in the field of public services. He coordinates the 2-year Master of Science program in Service Management and Design, and is director of the 1-year master program in Management for Human Capital Development. Among his main research interests are the analysis and management of change within public and nonprofit organizations, and city management. He has recently initiated a research and training program on social business models and social innovation.

Smart Cities in a Digital Nation: Are Swedish Cities Enough Innovative?

Christophe Premat

Abstract Sweden is known to be a well-connected society. There is a will to transform the nation into a digital nation with a specific place granted to the smart cities. These smart cities began to map a digital agenda for the next years where e-business, universities, smart transportation, and green and cultural policies have to be combined. The aim of this chapter is to analyze a few municipal digital agendas (Stockholm, Göteborg, Malmö, Jönköping, Umeå) to describe the Swedish model of smart cities. The question is to know whether the digital agenda is linked to a strong development of smart cities.

Keywords Digitization · E-economy · Sustainable development · Metropolitan areas · E-business · Post-materialistic attitudes

1 Introduction[1]

The chapter focuses on the relation between smart cities and digital growth in Sweden. In 2011, Anne-Karin Hatt, the IT and regions minister[2], created a national digitization council. The idea was to highlight the IT-innovations that make Sweden one of the first digital nations of the world. In the meantime, three Swedish metropolitan areas (Malmö, Stockholm, and Göteborg) as well as other middle-sized cities (Umeå, Jönköping) initiated different e-policies to become smart cities. New services were provided and new taxes were introduced to create a stimulating environment. The main goal is to build up a sustainable development with well-balanced policies. Stockholm is an interesting example of a city that promoted such policies. Stockholm has been organizing a European prize for green cities for several years to encourage the combination of modern and green policies at the

[1] This study was possible thanks to the help of the Lars Hierta Minne Foundation in Sweden. This foundation helps me to plan a research on digital communication between 2013 and 2015.

[2] http://www.regeringen.se/sb/d/14479/a/162868 *[Last visit 21 February 2014].*

C. Premat (✉)
Badhusgatan 1A, tr. III 151 72, 151 72 Södertälje, Sweden
e-mail: christophe.premat@su.se

J. R. Gil-Garcia et al. (eds.), *Smarter as the New Urban Agenda A Comprehensive View of the 21st Century City,* Public Administration and Information Technology 11,
DOI 10.1007/978-3-319-17620-8_11

local level. The e-policies aimed at having a flexible work organization to avoid an urban congestion during rush hours (VTM 2012). E-policies refer to public policies that are based upon digital data to facilitate the emergence of new networks.

The criteria adopted by the Swedish smart cities to meet the challenges of the sustainable development (Ercoskun 2012) are the following ones: efficient local transportation, environmental policies, job flexibility, and e-business (Campbell 2012). In the first part, I present the Swedish digital plan for the next 30 years and the outlines of the digitization council. In the second part, I compare the different Swedish smart cities. What is the profile of these cities and how smart are they? Do we have a strong difference between the three metropolitan cities and the other smart cities in Sweden? Do we have a real digital growth (Wiewel and Knaap 2005) that can support the development of smart cities in Sweden?

2 The Digital Dream

2.1 State of the Art

The topic of smart cities is very recent in the public debate, it has been discussed for a few years with the access to a more interactive form of Internet, World Wide Web 2.0. It supposes a new type of governance where Internet supports decentralized and interactive relations between e-business, administration, and citizens (Johnston and Hansen 2011). The smart governance implies a green growth where digital economy avoids unnecessary movements in the city. According to Giffinger, there are six criteria that help us to identify what a smart city is: smart economy, smart governance, smart people, smart environment, smart living, and smart mobility (Giffinger et al., 2007, p. 12). Table 1 explains all these factors. A smart city is characterized by a collective network of citizens, local administrations, and companies.

Smart cities mix a spirit of innovation and a participatory governance based upon a high level of education. They are not isolated, they have a very efficient transportation system. They are attractive not only for business and companies but also for people who are interested in local quality of life. People are well-integrated into their communities, they create an efficient network. At the same time, the concept of smart cities can have different meanings (Sinkiene et al. 2014): innovative city, digital city, green city, and so on. It means that it depends whether the local government emphasizes one particular public policy (transportation, digital policy, or environmental policy). The concept of smartness is a transfer from technological sciences. It indicates that cities use a knowledge to reach a better quality of life (Mancilla-Amaya et al. 2010). Some authors insist more on the technological performance (Mishra 2013), whereas others prefer to underline the creativity of such a concept (Rios 2012). The multidimensional aspect of smart cities is prevailing as it combines a green and digital economy, an effective system of local transportation, an innovative environment linked to research, and a new architectural culture. People can have access to all these different components and see how they can take part

Table 1 Characteristics of a smart city. (Source: Giffinger et al. 2007, p. 12)

Smart economy (competitiveness)	Smart people (social and human capital)	Smart governance (participation)	Smart mobility (Transport and Information and Communications technology)	Smart environment (natural resources)	Smart living (quality of life)
Innovative spirit	Level of qualification	Participation in decision-making	Local accessibility	Attractivity of natural conditions	Cultural facilities
Entrepreneurship	Affinity to life long learning	Public and social services	International accessibility	Pollution	Health conditions
Economic image and trademarks productivity	Social and ethnic plurality	Transparent governance	Availability of Information and communications technology infrastructure)	Environmental protection	Individual safety
Flexibility of labour market	Flexibility	Political strategies and perspectives	Sustainable, innovative and safe transport systems	Sustainable resource management	Housing quality
International embeddedness	Creativity				Education facilities
Ability to transform	Cosmopolitanism/open-mindedness				Tourist attractivity
	Participation in public life				Social cohesion

in local economy. There is a new type of regulation in those cities and I would like to analyze some examples to see whether it is possible to strengthen the indicators.

In the beginning of 2000s, some researchers focused on the development of metropolitan areas and their implications in terms of public services, private actors, and job opportunities. Thad Williamson, David Imbroscio, and Gar Alperovitz dealt with the "urban sprawl": sprawl refers both to the fact of continuing outward development on the perimeters of metropolitan areas and to the specific form such development has taken, namely, construction of freeways, strip malls, and other car-centered uses of space" (Williamson et al. 2002, p. 71). The metropolitan effect was studied in different contexts (Hoffmann-Martinot 2006, p. 231) and some problems were enlightened such as dysfunctional phenomena: "absence of control or guidance in territorial development," "insufficient resources," "weak management and expertise capacities," "lack of structural cooperation in common problem solving," "social, ethnic, and fiscal segregation" (Hoffmann-Martinot 2006, p. 242). The urban sprawl can have disastrous effects if it is out of control. The concept of smart cities inheres a flexible way of seeing the urban development: The digital growth implies a better use of resources and has effects on the transportation system. In our opinion, the scientific discussions on the concept of smart cities can be seen as a response to the different challenges due to the metropolitan effect. Smart cities emerge in a core of postindustrial values (Inglehart 2008) where a well-balanced and sustainable development is preferable to other values (job opportunities).

2.2 Post-Materialistic Attitudes and Digital Agenda

Smart cities imply specific attitudes from citizens. They must have post-materialistic values to reach a better quality of life. There are no smart cities without smart citizens. Sweden is characterized by a modern social system and a strong use of digital resources. Citizens are connected to Internet and the whole social system is efficient (Skovdahl 2008). Each citizen has a personal identity number *(personnummer)* that is required for every transaction. It was introduced in 1947 and could cover the population of the country. The personal identity number is issued by the national tax office *(Skattverket)*. The administrative tasks are facilitated by the use of this social security number. Furthermore, Swedes are used to pay taxes on their iPhones, they can for instance download applications to pay the parking and so on. They always need this identity number for all the transactions they make (Ludvigsson et al. 2009). As a matter of fact, Swedes are used to have a very performant administration. Citizens are also very sensitive to environmental problems as the country has a strong recycling system: newspapers are turned into paper mass, food is composted, and plastic containers are transformed into plastic raw material[3]. The municipalities reuse the waste to create new energy. This is also why this special care for environmental issues is reflected by all the international surveys.

In the past studies of *World Values Surveys,* we also observe that post-materialistic attitudes grow (Taniguchi 2006, p. 416). In those surveys, an index of autonomy based on a selected core of question on perceptions of life was created[4]. The ques-

Table 2 Comparison of independent attitudes among respondents in Sweden, the USA, and Spain between 1995 and 2007. (Source: World Values Surveys, http://www.wvsevsdb.com/wvs/WVSIntegratedEVSWVSvariables.jsp?Idioma=I)

	1996 (Sweden)	2006 (Sweden)	1995 (USA)	2006 (USA)	1995 (Spain)	2007 (Spain)
Non independent values	229 (22.7%)	167 (16.7%)	510 (33.1%)	383 (30.7%)	363 (30%)	469 (39.1%)
Mixed values	476 (47.2%)	422 (42.1%)	324 (21%)	325 (26%)	282 (23.3%)	326 (27.2%)
Independent values	185 (18.3%)	352 (35.1%)	150 (9.7%)	164 (13.1%)	51 (4.2%)	92 (7.7%)
Total	1009 (100%)	1013 (100%)	1541 (100%)	1249 (100%)	1211 (100%)	1200 (100%)

[3] https://sweden.se/nature/99-recycling-thats-the-swedish-way/#start [Last visit 21 July 2014].

[4] http://www.wvsevsdb.com/wvs/WVSIntegratedEVSWVSvariables.jsp?Idioma=I [Last visit, 21 February 2014]. The autonomy index is built thanks to a set of questions regarding the independence and the perseverance.

tions are linked to the perceptions of the citizens: Are they autonomous? Do they have prejudices due to religious beliefs? The post-materialistic attitudes show the preference for values that highlight the quality of life. Table 2 synthetizes the results of the comparison. I compared the results with the surveys made in USA and in Spain. I took Spain as a core example for southern Europe.

In 1996, 22.7 % of Swedish respondents expressed heteronomous values, 47.2 % had mixed values (half independent) and 18.3 % expressed autonomous values (independence and perseverance), the rest was unknown. In 2006, we had a diminution of the first category (people who are not independent) with 16.7 %, a diminution of the second category with 42.1 % but the number of independent people was multiplied by two (35.1 %). We have much more people who feel independent in Sweden compared to Spain and the USA where religious beliefs are very strong. If we measure the post-materialistic attitudes (interest in environmental issues and quality of life, team work rather than a sterile hierarchy), then we see that they increased in Sweden (25 % in 2006 and 20 % in 1996)[5]. Digital policies are initiated in a country where citizens are used to have facilities (no bureaucracy, security of the relation between citizens and administration). In *World Values Surveys,* it is possible to find out a world map of happiness. There is an index measuring whether citizens feel happy or not[6]. Sweden is one of the countries where citizens feel really happy (the score is 192.4). Only Norway and Iceland have higher scores in 2008. If we analyze these results with the help of smart cities criteria (Giffinger 2007), then we have a smart environment and a smart governance which facilitate the life of the individuals in Sweden. The level of education is high in Sweden. According to the Organisation for Economic Co-operation and Development (OECD) reports in 2012, 40 % of the 30–34-year-olds hold a tertiary degree in the country (OECD 2012).

When it comes to Internet access, according to a study made in 2012, in 1995 2 % of the Swedes were connected to Internet, they were 89 % in 2012 (Findahl 2013, p. 8). Half of the Swedish population had the access to the Internet, thanks to the smartphones between 2011 and 2012. The Swedish Post and Telecom Authority (PTS) reported that mobile traffic in Sweden grew 32% from mid-2013 to mid-2014 (Global Mobile Data Traffic 2015).

The Swedish government built-up a digital agenda with some priorities for Information and communications technology (ICT) users: "easy and safe to use, services that create benefit, the need for infrastructure, and the role of ICT in societal development" (ICT for everyone, a digital agenda for Sweden 2011). In other words, if we refer to the terminology of Giffinger (Giffinger 2007), then we have a smart environment and a real digital agenda for the future in Sweden.

[5] http://www.wvsevsdb.com/wvs/WVSIntegratedEVSWVSvariables.jsp?Idioma=I [Last visit, 21 February 2014].

[6] http://www.jdsurvey.net/jds/jdsurveyMaps.jsp?Idioma=I&SeccionTexto=0404&NOID=103 [Last visit, 21 February 2014].

2.3 The Digitization Committee

There is a digital agenda for Europe with a wish to share many data by the end of 2020. The goal is to have standard systems and a high access to Internet for all European citizens. The broadband system should be extended to all the member states and the European Commission would like to have a more homogeneous market in open data societies. The European Commission defined concrete actions in order to reach this single market. It adopted new Broadband State aid Guidelines in December 2012. The "action 3" aims at opening up data resources for reuse: "public authorities produce large amounts of data that could become the raw material for new, innovative cross-border applications and services. Examples of products and services based on the reuse of public sector information (PSI) are Global Positioning System (GPS), weather forecasts, financial, and insurance services."[7] The Commission sees the digital agenda as a way of creating new jobs for the future and new possibilities for e-commerce.

Under the Swedish presidency of European Union in 2009, a declaration was made in Malmö, the 18th of November 2009 regarding a digital agenda in Europe. A few priorities were set up such as transparency, open government data, collaboration, privacy, and open source[8]. The idea was to have a real digital agenda to appreciate the evolution towards a digital Europe. "Citizens and businesses are empowered by eGovernment services designed around users' needs and developed in collaboration with third parties, as well as by increased access to public information, strengthened transparency, and effective means for involvement of stakeholders in the policy process."[9] There is an obvious link between open data society, transparency, and citizen participation. Different member states created national structures to think about special digital agendas. This was the case for Sweden in 2011, France in 2012, and other European countries[10]. All European governments see e-economy as a resource for the future and this is why they created commissions to follow up the digital agenda.

In 2011, under the initiative of the minister of regions and ICT, Ann-Karin Hatt, a national digitization committee was created in Sweden. The goal was to make Sweden a top high tech nation. The committee aims at encouraging digital policies in every field. The committee's tasks were defined in the directive of the 7th of June 2012. The committee has to make recommendations on the digital agenda by the end of December 2015. According to the directive, the committee has to publish annual reports (*Statens offentliga utredningar,* SOU 2013, p. 31). It defined 22 fields where digital policies could be measured: Internet connections, e-services, digital com-

---

[7] http://ec.europa.eu/digital-agenda/en/pillar-i-digital-single-market/action-3-open-public-data-resources-re-use [Last visit 21 February 2014].

[8] http://itlaw.wikia.com/wiki/Ministerial_Declaration_on_eGovernment [Last visit, 21 February 2014].

[9] http://www.intellitics.com/blog/2009/11/19/malmo-2009-ministerial-declaration-on-egovernment/ [Last visit, 21 February 2014].

[10] http://www.cnnumerique.fr/home-2/ [Last visit 21 February 2014].

petency, digital security, e-administration, e-business, e-care system, school and teaching, Democracy (e-consultations), digital culture (access to e-books), Internet in Sweden and in the world, security of the information (protection against viruses), infrastructures (e-invoices), geographic information (maps), robust electronic communication, broadband, research and innovation, ICT for environment, gender problematic, freedom on the net, copyright, and ICT for global development (international cooperation) (*En rapport för Digitaliseringskommissionen*, 23 January 2012). The committee defined a set of questions for all these indicators to measure the digital level of Sweden. According to the Global Information Report 2013 *(Growth and Jobs in a Hyperconnected world),* "broadband, 3G and the intelligent use of big data could also revitalize economic growth. Governments play a crucial role in supporting this digital development, from funding broadband networks to addressing complex issues such as privacy and security. The economy as a whole will eventually reap the benefits as remote rural areas are tied into the national network, resulting in new jobs and broader educational opportunities" (Bilbao-Osorio et al. 2013). Sweden is the third country in the world in terms of digital economy.

In a nutshell, many citizens share post-materialistic attitudes in Sweden, preferring the quality of life, the easy relation with public administration and the interest for environmental issues. These values create the conditions for a smart environment. The country adapted very well to the digital innovations with 89% of the population that is connected to Internet. The government created a digitization commission to set the digital agenda. At the same time, the government did not use the word of "smart cities" in its official publications. Some local governments use the word to show how the cities can transform themselves in the future.

3 Smart Cities in a Digital Environment

3.1 Metropolitan Challenges

We live in a world where most of 50% the population lives in metropolitan areas (Metropolis 2009). The big cities have to develop efficient local policies to answer to all different challenges that they face (Campbell 2012, p. 18). The municipalities grow and have to develop a smart way of thinking to satisfy the high social demand. Stockholm is often seen as a city where future challenges were defined very early. Stockholm Metropolitan Agglomeration has a population of 1,500,000 inhabitants, it includes 12 communes such as Danderyd, Huddinge, Järfälla, Lidingö, Nacka, Sollentuna, Solna, Stockholm, Sundbyberg, Tyresö, Täby, and Upplands Väsby (Metropolis 2009, p. 190).

The urban development is a constant preoccupation as the site of Stockholm is geographically limited by water. It is not possible to have new buildings everywhere, it is therefore important to think about a rational development. This is why the suburban areas grew very fast in Stockholm such as Solna and Sollentuna in the north part of the city and Nacka and Huddinge in the south part. Stockholm is known for early environmental policies. In 1972, the city welcomed an international conference on environmental issues. The participants shared ideas and recommendations that could be implemented all around the world to have a new movement of green cities. The principle 19 dealt with the necessity of having a solid education. "Education in environmental matters, for the younger generation as well as adults, giving due consideration to the underprivileged, is essential to broaden the basis for an enlightened opinion and responsible conduct by individuals, enterprises, and communities in protecting and improving the environment in its full human dimension. It is also essential that mass media of communications avoid contributing to the deterioration of the environment, but, on the contrary, disseminates information of an educational nature on the need to project and improve the environment to enable mal to develop in every respect."[11] Three social actors have to work together, individuals, enterprises, and communities to have efficient environmental policies. In Stockholm, some policies were adopted to limit the pollution such as the Congestion Tax Act.

On 2nd June 2003, the Stockholm City Council adopted a proposal implementing a congestion charging. The Swedish Parliament implemented the Congestion Tax Act (2004, p. 629) to limit the pollution inside the city. The implementation of a congestion tax began in January 2006[12]. A local referendum was held on 17th September 2006, the same day as the parliamentary election. The turnout was high with 76.4 % of voters; 51.3 % of the voters approved the tax. Fourteen other municipalities held local referendums on the implementation of congestion charges. Table 3 summarizes the results of these referendums in Stockholm and in all these 14 municipalities.

The idea was to have a green tax to limit the car traffic downtown[13]. There is a difference between the citizen attitudes in a big city such as Stockholm and the citizen attitudes outside Stockholm. The difference is around 12 % and is explained by the fact that the metropolitan area has a lot of commuters coming from suburban areas. Citizens can accept to pay a congestion tax to come downtown but they refuse to pay a congestion tax in their own suburban area. Some rich communes such as Lidingö, Danderyd, and Täby had a low proportion of yes-voters because most of the inhabitants had a personal car. The congestion tax means that those inhabitants would have had to pay a lot.

Automobilists who come downtown have to pay a tax that can be higher during rush hours (from 10 to 20 Swedish crowns). The municipality of Stockholm wanted to limit the car traffic, many parking lots were created in suburban areas to encourage the use of communal transportation. Even if the consequences of such taxes are criticized, they have a dissuasive effect on the population. The local transportation has to be better to attract more passengers. The statistics show that the commu-

[11] http://www.unep.org/Documents.Multilingual/Default.asp?documentid=97&articleid=1503 [Last visit, 21 February 2014].

[12] http://www.stockholmsforsoket.se/templates/page.aspx?id=183 [Last visit, 21 February 2014].

[13] http://www.stockholmsforsoket.se/templates/page.aspx?id=197 [Last visit, 21 February 2014].

Table 3 Results of the referendums held on the implementation of congestion taxes in Sweden in 2006. (Source: http://www.stockholmsforsoket.se/templates/page.aspx?id=10215)

Municipality	Number of cast votes	Yes(%)	No(%)
Stockholm city	243.055	51.3	45.5
All 14 municipalities combined	324.786	39.8	60.2
Danderyd	16.962	32.5	67.5
Ekerö	13.528	39.9	60.1
Haninge	37.548	40.8	59.2
Lidingö	24.926	29.6	70.4
Nacka	44.785	42.9	57.1
Nynäshamn	12.588	41.2	58.8
Salem	7.563	39.6	60.4
Sollentuna	32.409	40.8	59.2
Solna	35.598	43.9	56.1
Tyresö	22.526	44.3	55.7
Täby	35.630	34.2	65.8
Vallentuna	14.884	42.5	57.5
Vaxholm	5.699	45.9	54.1
Österåker	20.140	40.9	59.1

nal transportation system was preferred during the initial years but the high prices tend to make local transportation less and less attractive for commuters. There are current political debates in Sweden about the necessity of having lower prices[14]. The city of Göteborg implemented the same tax in the beginning of January[15]. The debate is current in Malmö as many local politicians would like to implement this tax[16]. The 2014 election is a test for the municipality of Göteborg as there were some reactions against the implementation of this tax[17]. The current debate for the congestion tax concerns the effects of the tax as the opponents claim that the tax limits growth opportunities and has discriminating consequences.

The big challenge of the municipality of Stockholm concerns housing issues. The prices are very high and it is almost impossible for a middle-class family to buy an apartment in Stockholm. The municipality did not build enough buildings and it is really difficult to find a place to live in the city. Even renting an apartment is a problem. This situation can affect the local economy as students find it very difficult to have a room in Stockholm. As a matter of fact, the mobility and the local economy can be affected. Stockholm is one of the European cities growing very

[14] http://www.dn.se/debatt/stockholmsdebatt/lagtrafiktaxa-kan-sanka-sl-kortet-med-hundralap-par/ [Last visit, 21 February 2014].

[15] http://www.transportstyrelsen.se/sv/vag/trangselskatt/trangselskatt-i-goteborg/ [Last visit, 5 March 2014].

[16] Alexander Kuprijanko, "Trängelskatt föreslås", *Sydsvenskan*, http://www.sydsvenskan.se/sverige/trangselskatt-foreslas/, 27 June 2012 [Last visit, 5 March 2014].

[17] http://www.nejtrangselskattgbg.se [Last visit, 5 March 2014].

fast. The economic activity is well diversified. According to the national agency for growth, 55 % of people between 18 and 70-years-old declare that they could possibly be entrepreneurs (*Tillväxtverket* 2012, p. 29). Thirty-eight percent of people between 18 and 70-years-old want to be entrepreneurs (*Tillväxtverket* 2012, p. 39).

The municipality of Stockholm has proposed, since 2007, a plan for Stockholm by the horizon of 2030[18]. The vision is described in the following way: "a creative city, diverse companies, a knowledge region, a research impact with the Nobel prize, a top city in IT, environment and techniques, a place for international conferences and congresses, a place which attracts people, companies, and international investments."[19] The municipality of Stockholm decided on 29th April 2013 to have a digital renewal *(Ett program för digital förnyelse 2013–2018)*[20]. This decision points out different fields to improve: the quality of e-services, the necessary education to encourage IT skills, and the capacity to handle the e-information in the right way. The municipality initiated a call for tenders to select the actors that can guarantee this evolution for Stockholm. In other documents provided by the municipality of Stockholm, the objectives are clear: a citizen e-book, a platform for mobile applications, a new research motor, an efficient digital communication with secured data between companies and the city of Stockholm, a client platform, and a geodata platform.

The municipality also initiated a method to promote open data[21]. On the 13th of March 2014, a series of data were available for citizens (information on the traffic, the environment, the population, and so on)[22]. A few questions were addressed to citizens so that they could take part in this open data era. How can the municipality grow? How can the city look like in 20 years? The Swedish population has more than 9.2 million inhabitants and almost one Swede out of five lives in Stockholm. By the year of 2030, around 3.5 million inhabitants are expected to live in the region of Stockholm. This concentration illustrates the necessity of dealing with all these issues to avoid an explosion of social inequalities. The upheavals that occurred in May 2013 in Husby illustrate that some areas of Stockholm feel excluded[23]. Many social debaters and politicians dealt with the word segregation[24].

Other big metropolitan areas such as Malmö and Göteborg also follow up a digital agenda but it is the result of a regional initiative. In the case of Malmö, the region

[18] http://www.stockholm.se/OmStockholm/Vision-2030/ [Last visit, 21 February 2014].

[19] http://www.stockholm.se/OmStockholm/Vision-2030/Innovativ-och-vaxande/ [Last visit, 21 February 2014].

[20] http://www.stockholm.se/OmStockholm/Vision-2030/Exempel-pa-olika-satsningar-inom-ICT/ [Last visit, 21 February 2014].

[21] http://open.stockholm.se/ [Last visit, 21 February 2014].

[22] http://open.stockholm.se/handlingsplan-oppna-data [Last visit, 21 February 2014].

[23] http://one-europe.info/stockholm-riots-reflect-european-immigration-policy-failures [Last visit, 21 February 2014].

[24] http://www.dn.se/nyheter/sverige/segregationen-skolans-storsta-utmaning/ [Last visit, 21 February 2014].

Table 4 Profile of the big metropolitan areas in Sweden. (Source: http://www.ekonomifakta.se/sv/Om-Ekonomifakta/)

	Malmö	Göteborg	Stockholm
Population	307.758	526.089	881.235
Average age of citizens	38 years old	39 years old	39 years old
Proportion of entrepreneurs	5.4%	5.3%	7.6%
Proportion of high educated people	30.3%	32.9%	38.3%
Unemployment rate	14.9%	9.4%	7.1%

of Scania adopted a digital agenda in February 2012[25] where all partners (companies, public sectors) agreed to share data to facilitate job opportunities. The region of Västra Götaland had a similar method but in this last case the digital agenda was described in details (Table 4).

The metropolitan areas face big social and economical challenges. They attempt to be real smart cities but they have to go over some hurdles such as the housing possibilities and the social differentiation between some areas. The congestion tax is an efficient weapon to tackle the environmental issues but the social cohesion is maybe the hardest challenge for the metropolitan cities.

3.2 *Middle-Sized Cities have Efficient Policies*

According to the study made by the website smart-cities.eu, Jönköping and Umeå are two smart cities well ranked[26]. If we apply the criteria of the team of researchers composed of Rudolf Giffinger, Christian Fertner, Hans Kramar, Nataša Pichler-Milanovic, and Evert Meijers (Giffinger 2007), then the cities of Jönköping and Umeå are well ranked. Jönköping is 19 and Umeå 23[27]. Jönköping obtains a high score for the smart economy whereas Umeå has a good score in the smart environment.

Jönköping is a city of Småland, a region in the south part of Sweden. It is located in the south part of the lake Vattern and has a population of 130,000 inhabitants[28]. It is the 10th city of Sweden. The region of Jönköping has initiated a digital agenda in 2013 to plan a better development in the future[29]. The region of Jönköping created a council for e-development to set the digital agenda. The objectives are to

[25] https://www.skane.se/sv/Nyheter/Omvarld/Skane-pa-vag-mot-en-digital-agenda/ [Last visit, 21 February 2014].

[26] http://www.smart-cities.eu/index2.html [Last visit, 21 February 2014].

[27] http://www.smart-cities.eu/model.html [Last visit, 21 February 2014].

[28] http://www.jonkoping.se/omkommunen/kommunfakta/statistikochutredningar/befolkningsstatistik.4.176db646136e9421c4942c.html [Last visit, 21 February 2014].

[29] http://www.regionjonkoping.se/web/Visa_reda.aspx?p=Ny_nationell_guide_for_regional_e-utveckling [Last visit, 21 February 2014].

have a better e-administration, a closer contact between public services, citizens, and private actors such as companies and define a common strategy to strengthen digital policies[30]. In the municipality of Jönköping, the concrete policies concern the university, e-services and schools with a national platform for pupils[31]. The objective is to build up an e-academy with different courses that can fit the needs for the economic development of the region. If a system of communication between companies, schools, and universities is created, it will be easier to have a better education system and enroll new students[32]. At the same time, we have an analysis of social needs in the region of Jönköping. The creation of all these platforms has a cost and an ineffective way would be to have different models. The project is based upon recommendations that clearly avoid the multiplication of local projects that are disconnected to each other. I asked the project manager whether it was possible to evaluate the first concrete policies and here's her answer: "I can unfortunately not say that there is a concrete development of digital culture in Jönköping even if the regional cooperation has begun."[33] This message shows us that there is a digital agenda, but there is a lack of concrete policies. This is a typical case for all the digital projects that have an evasive agenda (Schlegel 2009, pp. 190–191).

The communes cannot try a platform without being sure that it is worth doing it. The costs are too high and local politicians do not want to take this risk. Furthermore, it seems that there is a national attempt to share data and e-services. What would be the added value for a commune to launch a specific platform? Is there not a contradiction between the experimentation of smart cities and the existence of national e-data? Jönköping has a big university that attracts people from Småland but it has less international cooperation than the University of Umeå.

The problem is quite similar to Jönköping in Umeå. The region of Västerbotten launched a campaign named Digidel2013 in 2013 to strengthen the possibilities of connection to Internet in the region. The campaign was a reaction against the social exclusion of around 1.2 million people in Sweden who do not use Internet[34]. Different organizations and libraries took part in the project in Västerbotten. Umeå is a northern city with a population of 117,394 inhabitants, 37,000 students, and 11,000 registered companies[35]. The population has doubled since 1960 and Umeå is one of the ten first municipalities in Sweden.

---

[30] http://www.regionjonkoping.se/web/Vad_ar_E-utvecklingsradet.aspx [Last visit, 21 February 2014].

[31] https://www.jonkoping.se/kommunpolitik/nationelltregionaltochinternationelltsamarbete/regionaltsamarbete/jonkopingslanseutvecklingsrad/eutvecklingsradetsprogramforklaring.4.29487f7c137b6fc1ade1c95.html [Last visit, 21 February 2014].

[32] https://www.jonkoping.se/kommunpolitik/nationelltregionaltochinternationelltsamarbete/regionaltsamarbete/jonkopingslanseutvecklingsrad/regionaleutvecklingsdag.4.75b33e88137748dd6ec70d.html [Last visit, 21 February 2014].

[33] E-mail with the project manager of e-development council in Jönköping, 26 February 2014.

[34] http://www.digidel.se/ [Last visit, 21 February 2014].

[35] http://www.umea.se/mer/otherlanguages/inenglish/umeainbrief.4.1255481e123d7d67aaa800010679.html [Last visit, 21 February 2014].

In the presentation of Umeå, we have an insistence on the impact of the universities with all the students. "Umeå is a young, smart city. Young because the average of its citizen is 38. Smart because knowledge is our most valuable cost. The city's young population is one of the main reasons for the dynamics and city pulse associated with Umeå. The 37,000 students at Umeå University are an important reason why cultural events of both broad and more specialized appeal always find an inquisitive and interested audience in Umeå. The more well-known festivals include Kulturnatta, the Umeå International Jazz Festival, Umeå open, She's got the beat, Holmön Folk Song Festival, Sami week, MADE in Umeå, Umeå Football Festival, Edge, and the Täfteå Festival."[36] It is obvious that the word smart refers to the relation between universities and companies. Umeå developed an IT strategy very early, employment in private trade and industry has grown by more than 30% in 10 years. There is a smart economy and a smart mobility with a lot of international students. "The number of broadband connections per capita is among the highest in the world, not least thanks to the fact that Umeå built a fibre optic metropolitan area network very early on. Umeå was named IT Municipality of the Year in 2007."[37]

Umeå has a national center for scientific and parallel computing, the High-Performance Computing Center North (HPC2N). This is a cooperation between universities with a special work on virtual reality and scientific visualization[38]. The use of high-performance computing includes compute-intensive and communication-intensive applications. We found some concrete examples with Akka cluster, one of the fastest academic computer system in Nordic countries. "Akka was ranked 16 on the Green500 list of the most energy-efficient supercomputers in the world, which serve as a complementary view to the TOP500 list."[39] The focus is more on digital research in Umeå than in Jönköping that develops a platform of digital economy. If we now compare Umeå with Jönköping, we have the following indicators summarized in Table 5.

Table 5 Profile of middle-sized cities in Sweden. (Source: http://www.ekonomifakta.se/sv/Om-Ekonomifakta/)

	Jönköping	Umeå	Sweden (average)
Population	129,478	117,294	33,000
Average age of citizens	40.2	38.4	41.2
Proportion of entrepreneurs	5%	4%	6.8%
Proportion of high-educated people	25%	36.9%	24.8%
Unemployment rate	6.9%	7.2%	8.5%

[36] http://www.umea.se/mer/otherlanguages/inenglish/youngumea.4.1255481e123d7d67aaa800013017.html [Last visit, 21 February 2014].

[37] http://www.umea.se/mer/otherlanguages/inenglish/umeaanexpandingitcity.4.1255481e123d7d67aaa800012992.html [Last visit, 21 February 2014].

[38] http://www.hpc2n.umu.se/ [Last visit, 21 February 2014].

[39] http://www.hpc2n.umu.se/resources/akka [Last visit, 21 February 2014].

Umeå is a younger city with a big university. This is why the proportion of high-educated people is much significant in Umeå. The proportion of entrepreneurs is a little bit larger in Jönköping. At the same time, the communes still are the big employers both in Umeå and Jönköping. In Jönköping, the commune employs 12,675 persons whereas the commune of Umeå has 9775 employers. In Jönköping, Husqvarna is the first private employer (1575 persons) for Jönköping and Volvo is the one for Umeå (1975 persons). The profile of both communes is limited to a big company and a university. The project of smart cities can help these communes to reach diversity in economy. Umeå is the cultural capital of Europe in 2014, it has the possibility to highlight its effort in the cultural sector[40]. The municipality deals with the concept of cultural growth to qualify this trend in Umeå[41]. The cultural heritage of Umeå is preserved and several companies compete to take part in this cultural effort. Different researches from the University (Digital Humanities) are presented but the smart city is more a goal than a strong reality. There are research units which focus on digital innovation with companies using these results but there is no systematic connection between companies, the university, and the commune. E-business is not really included in a communal intranet such as in Karlstad and Kristianstad (Christiansson 2001). Both Umeå and Jönköping use an open data portal[42]. These middle-sized cities have a growth potential, they invest in education and culture and stimulate the local economy but they need to improve the system of open data. The challenge is to invest in an adequate platform of digital data. It can be too expensive for those municipalities.

4 Conclusion

There is a contrast between the national and the local levels concerning the development of e-services. If smart cities can be seen as interesting experimentations, Sweden has chosen to invest in a digitization council to be a high-tech nation. The council wants to strengthen the security of open data and encourage the growth of e-services. It is easier as many data are national whereas the local experimentations remain limited because of the cost. No communes want to take the risk of initiating a digital platform that could be useless. The local experimentations can go further if and only if the national authorities support it. In other words, we are in front of a dilemma: Are smart cities autonomous or do they use national data to have a sustainable development based upon a digital growth? It is maybe something adapted to metropolitan areas that can take this risk or cities that are part of a regional development.

It is surprising to see that the big municipalities do not necessarily link the digital agenda to the green policies. The congestion tax is a good example of that

[40] http://umea2014.se/sv/ [Last visit, 5 March 2014].

[41] http://umea2014.se/sv/om-umea2014/kulturdriven-tillvaxt/ [Last visit, 5 March 2014].

[42] http://www.lansstyrelsen.se/jonkoping/Sv/psi-data/Pages/psi-data.aspx [Last visit 5 March 2014].

but it illustrates the necessity of controlling the local transportation. A smart city cannot work without a smart transportation. If the municipalities of Stockholm and Göteborg implemented those policies, it could be interesting to see how the municipalities used the incomings. A smart city includes the ecological, cultural, and digital aspects to create an environment adapted to e-business.

Our conclusion is that the model of smart city is at its very beginning with a lot of different agendas (the European one, the national one, and the municipal experimentations). There is a need of having concrete policies in all these municipalities to see whether smart cities can improve the urban life with e-business, less transportation, better services, and so on (Coskun et al. 2011). There is a trend toward smart cities when citizens share common values based upon post-materialistic beliefs. It is the case in Sweden as the scientific surveys reflect a post-materialistic attitude and a high-educated population.

We are still in a prophetic period where a lot of announcements are made but where it is difficult to understand if people talk about the same thing. As more than 85 % of the Swedish population is connected to Internet, the concept of "smart city" can sound a little bit theoretical and sometimes elusive. In the language of public policies, the word "smart cities" is like a slogan (Morin 1999), a reference (Muller 2009) but the diversity of the concept makes it hard to define a real and coherent digital public policy. Do we have several concurrent plans for smart cities in the future? That could be a concrete way to discuss the alternative scenarios to reach an efficient growth for all these cities. A consensus conference on this topic could help politicians to make the right decisions for the future. The consensus conferences in Sweden were held on medical questions but the future of smart cities could be a prospective way to make concrete public policies emerge.

There is a need to see the effects of e-economy thanks to very specific and smart policies at the local level. The digital agenda requires a new digital culture where citizens, administrations, and companies accept to share more data. In other words, smart cities can be creative cities if these connections are strengthened. The philosopher Bernard Stiegler deals with a new digital education where citizens know from the beginning how they can work and live thanks to an adequate use of new technologies (Stiegler 2010)[43]. Can smart cities contribute to sustainable metropolitan areas? This is too early to answer this question but the Swedish case study shows a strong step forward in this direction.

5 Recommendations

This case study shows that there are both a regional effort to build up a digital agenda and a national interest in harmonizing the existing practices. At the same time, the system is decentralized as Swedish communes have strong competencies.

[43] http://reseaux.blog.lemonde.fr/2013/09/29/blues-net-bernard-stiegler/ [Last visit, 4 March 2014].

The risk for a decentralized system is to create a central platform of data for all these smart cities. Hence, I would like to address a few recommendations for these smart cities in Sweden.

1. The creation of a common platform of smart cities in Sweden where the experiences could be shared. A network could be created with all the project managers and a member of the digitization committee. We already had a symposium but the structure should be more official to have a better follow-up.
2. The digitization committee should also have contacts with the Swedish association of communes and regions[44] to edit guidelines considering the costs, the bugs and the hurdles of such projects.
3. The diversity of jobs should be encouraged in order to secure the future of middle-sized smart cities.
4. The agendas should not be multiplied so that the concrete steps can be tested by the concerned actors (companies, communes, and citizens). They should be replaced by a business plan that connects all the different actors.
5. Middle-sized smart cities should be helped in their attempts to encourage a digital economy. They do not have the same platforms as big cities and their efforts can be limited. An adapted business plan should be conducted for these cities.
6. The congestion tax should be connected to the efforts for having a green and smart city. The idea would be to communicate on a smart and effective transportation system. Smart cities in Sweden do not include this tax as a possibility of reaching a smart and well-balanced development.
7. Some alternative scenarios should be presented so that institutions, experts, citizens, and politicians can test the best digital public policies to develop smart cities.
8. Some consensus conferences should be organized on this topic to have an accurate consciousness of the challenges that smart cities face. These consensus conferences would help to develop a Swedish model of smart cities.

Key Terms and Definitions

Congestion tax: Tax for cars and tracks coming in Stockholm. The tax was implemented in Stockholm in 2006.

Digidel2013: Campaign launched in the region of Umeå to strengthen the internet connections of the inhabitants in Västerbotten.

Digitization committee: The committee's tasks were defined in the directive of the 7th of June 2012. The committee has to make recommendations on the digital agenda by the end of December 2015. According to the directive, the committee has to publish annual reports (*Statens offentliga utredningar,* SOU 2013, p. 31). It defined 22 fields where digital policies could be measured.

[44] http://www.skl.se/ [Last visit, 4 March 2014].

High-Performance Computing Center: This is a cooperation between universities with a special work on virtual reality and scientific visualization in Umeå. The use of high-performance computing includes compute-intensive and communication-intensive applications.

Metropolitan effect: The cities that absorb surrounding communes. The metropolitan areas need a clear governance and a territorial plan to think about better public policies connected to the social needs of citizens.

Open data: The data and statistics that are available for citizens. Some Swedish municipalities chose to publish statistics on the local population.

Post-materialistic attitudes: are the values that highlight a care for environmental issues, individual autonomy rather than beliefs in job and growth. These attitudes privilege team work rather than hierarchy.

References

Bilbao-Osorio, B., Duta, S., & Lanvin, B. (2013). *The global information technology report 2013, growth and jobs in a hyperconnected world*. Geneva: World Economic Forum.

Campbell, T. (2012). *Beyond smart cities: How cities network, learn and innovate*. New-York: Earthscan.

Christiansson, M.-T. (2001). Process orientation in inter-organisational co-operation—By which strategy? In A. G. Nilsson & J. S. Pettersson (Eds.), *On methods for systems development in professional organisations—The karlstad university approach to information systems and its role in society*. Lund: Studentlitteratur.

Coskun, V., Ok, K., & Ozdenizci, B. (2011). *Near field communication: From theory to practice*. Hoboken: Wiley.

En rapport för digitaliseringskommissionen, kartläggning av indikatorer för den svenska digitala agendan, 23 januari 2012.

Ercoskun, O. Y. (Ed.). (2012). *Green and ecological technologies for urban planning: Creating smart cities*. Hershey: Information Science Reference.

Findahl, O. (2013). *Svenskarna och internet 2013*. Stockholm: SE. https://www.iis.se/docs/SOI2013.pdf.

Giffinger, R., Meijers, E., Fertner, C., Kramar, H., Kalasek, R., & Pichler-Milanović, N. (2007). *Smart cities—Ranking of European medium-sized cities*. University of Vienna: Centre of Regional Science.

Global Mobile Data Traffic. (2015). Cisco Visual Networking Index: Global Mobile Data Traffic Forecast Update, 2014–2019. http://www.cisco.com/c/en/us/solutions/collateral/service-provider/visual-networking-index-vni/white_paper_c11-520862.html.

Government Offices of Sweden. (2011). ICT for everyone, A Digital Agenda for Sweden. http://www.government.se/reports/2011/12/ict-for-everyone-a-digital-agenda-for-sweden/.

Hoffmann-Martinot, V. (2006). Reform and modernization of urban government in France. In V. Hoffmann-Martinot & H. Wollmann (Eds.), *State and local government reforms in France and Germany* (pp. 231–251). Wiesbaden: Verlag für Sozialwissenschaften.

Inglehart, R. (2008). Changing values among western publics from 1970 to 2006. *West European Politics, 31,* 131–146.

Johnston, E. W., & Hansen, D. L. (2011) Design lessons for smart governance infrastructures. In D. Ink, A. Balutis, & T. Buss (Eds.), *Transforming American governance: Rebooting the public square* (pp. 197–212). London: Routledge.

Ludvigsson, J. F., Otterblad-Olausson, P., Pettersson, B. U., & Ekbom, A. (2009). The Swedish personal identity number: Possibilities and pitfalls in healthcare and medical research. *European Journal of Epidemiology, 24*(11), 659–667.

Mancilla-Amaya, L., Sanín, C., & Szczerbicki, E. (2010). Smart knowledge-sharing platforms for e-decisional community. *Cybernetics and Systems: An International Journal, 41*(1), 17–30.

Metropolis. (2009). *Metropolitan regions*, working paper (2nd ed.). http://www.metropolis.org/file/metropolitan-regions-working-papers.

Mishra, M. K. (2013). *Role of technology in SMART governance: 'Smart city, safe city'*. http://ssrn.com/abstract=2310465. Accessed 15 Aug 2013.

Morin, E. (1999). *Les sept savoirs nécessaires à l'éducation du futur*. Paris: Seuil.

Muller, P. (2009). *Les politiques publiques*. Paris: Presses universitaires de France.

OECD. (2012). *Education at a Glance 2012*. OECD indicators. OECD Publication. http://www.oecd.org/edu/EAG%202012_e-book_EN_200912.pdf.

Rios, P. (2012). Creating "the Smart City". https://archive.udmercy.edu/handle/10429/393.

Schlegel, H. (2009). Marketing événementiel et design d'aujourd'hui. Petites mythologies de la technologie de masse. *Les Cahiers Sens Public, 11–12,* 189–203.

Sinkiene, J., Grumadaite, K., & Liugailaite-Radzvickiene, L. (2014). *Diversity of theoretical approaches to the concept of smart city*. 8th International Scientific Conference, Business and management 2014.

Skovdahl, B. (2008). *Den digitala framtiden*. Stockholm: Institutet för Framtidsstudier.

Statens offentliga utredningar SOU. (2013:31). En digital agenda i människans tjänst—Sveriges digitala ekosystem, dess aktörer och drivkrafter. 253 p. http://www.regeringen.se/rattsdokument/statens-offentliga-utredningar/2013/05/sou-201331/.

Stiegler, B. (2010). *Taking care of youth and the generations*. Stanford: Stanford University Press.

Taniguchi, M. (2006). A time machine: New evidence of post-materialist value change. *International Political Science Review, 27*(4), 416.

Tillväxtverket. (2012). *Entreprenörskapsbarometern 2012*. Stockholm: Tillväxtverket.

Vehicular Traffic Management for Smart Cities (VTM). (2012). First international workshop. http://ieeexplore.ieee.org/xpl/mostRecentIssue.jsp?punumber=6384921.

Wiewel, W., & Knaap, G. (2005). *Partnerships for smart growth university-community collaboration for better public places*. Cambridge: M.E. Sharpe.

Williamson, T., Imbroscio, D., & Alperovitz, G. (2002). *Making a place for community, local democracy in a global era*. New York: Routledge.

Christophe Premat is an associate professor in cultural studies at the Department of Romance languages of Stockholm University. He is also an associate researcher at Centre Émile Durkheim (Institute for political studies of Bordeaux) and he is currently a MP for French residents overseas at the National Assembly in France. He published recently a book on the principle of political representation (L'idée d'une nouvelle représentation politique, Edilivre 2014) and a few articles on participatory democracy such as "Initiatives and referendums," Gabriel, Keil, Kerrouche (eds.), *Political Participation in France and Germany,* Essex, EPCR Studies, 2012, pp. 161–188 and "How do French municipalities communicate on citizen involvement?," in *E-Government Success Factors and Measures: Concepts, Theories, Experiences, and Practical Recommendations,* Jose Ramon Gil-Garcia (ed.), 2013, Pennsylvania, IGI Global, pp. 321–340.

Implementing Smart Services in Moscow: The Integrated Mobile Platform

Evgeny Styrin and Artem Kostyrko

Abstract The city of Moscow is a leading implementer of information and communication technologies (ICT) in public services. The high level of ICT and mobile penetration among Muscovites creates a strong demand for mobile and electronic services.

This chapter will explore a case study involving the creation and development of Moscow's integrated mobile platform (IMP). The case study illustrates ICT usage policy and the Moscow Government's priorities in terms of delivering and providing access to mobile public services. The case study takes a framework approach to mobile platform development and is also based on the lean government concept. Key success factors in IMP development as well as challenges involved in the collaboration and coordination of various IMP stakeholders are also examined. The case study provides examples of mobile applications developed on the basis of the IMP. The governance decision-making process and regulatory framework for IMP management are examined as well. The IMP is interconnected with innovative front office systems such as the Moscow Public Services Portal and Open Data Portal.

Keywords Mobile government · Mobile public services · Integrated mobile platform · Public services portal · Open data portal · Smart city

Abbreviations

DIT	Department of Information Technologies
HTTP	Hyper Text Transfer Protocol
ICT	Information and communication technologies
IMP	Integrated mobile platform
IT	Information technologies
J2EE	Java 2 Enterprise Edition
MMS	Multimedia Messaging Service

E. Styrin (✉) · A. Kostyrko
National Research University "Higher School of Economics", Myasnitskaya 20, Moscow, Russia 101000
e-mail: estyrin@hse.ru

A. Kostyrko
e-mail: akostyrko@hse.ru

J. R. Gil-Garcia et al. (eds.), *Smarter as the New Urban Agenda A Comprehensive View of the 21st Century City,* Public Administration and Information Technology 11,
DOI 10.1007/978-3-319-17620-8_12

OECD	Organisation for Economic Co-operation and Development
PSP	Public Services Portal
SMS	Short Messaging Service
USSD	Unstructured Supplementary Service Data

1 Introduction

Moscow, Russia's capital, with a population of over 10 million, is also the leading Russian city in terms of information and communication technologies (ICT). According to statistics provided by the Moscow Department of Information Technologies (DIT), Moscow has the highest levels of ICT penetration. For example, 3G network coverage in 2013 reached 90 %, 4G/LTE penetration was 75 %, and Internet connection with a speed of over 100 Mbit was more than 50 %. The 200 % penetration of mobile phones in Moscow means that each Muscovite owns at least two mobile phones or other mobile devices. The Moscow Government has more than 3000 information systems which need to be integrated or interconnected to provide the required quality of public services.

Since 2012, the Moscow DIT has administered over USD 1 billion each year for ICT infrastructure development, e-services, and support for other Moscow Government departments. Private companies invest over USD 0.6 billion annually.

At the same time, there is a high demand on the part of Muscovites for better quality services from the Moscow authorities. Citizens' needs, problems, complaints, and proposals cannot be responded to on a personal basis because of the city's size, population density, and functional load..

Unfortunately, it is very hard to plan an autonomous city strategy-based exclusively on Muscovites' interests, because Moscow, as the capital, is partly administered by federal authorities. Moscow is also the largest financial and transport hub in Russia and Eastern Europe.

Technological solutions, such as integrative technological platforms for effective interagency information sharing and decision-making, together with ongoing improvement in public services are the only possible means of controlling and regulating complex city systems of public transportation, urban planning and development, utilities and housing management, and predicting city development trends (Washburn et al. 2010).

This chapter is structured as follows. First, we discuss the Smart City concept as applied to Moscow, then propose a research question and method. Next we analyze government policy based on ICT usage. After that, we review the literature in order to analyze the mobile government development framework and its applicability to Moscow. Then, we present the case study of the Moscow's integrated mobile platform (IMP) for citizens and businesses as an example of a smart city solution and finally discuss problems and obstacles involved in implementing smart solutions in the context of governance in Moscow.

2 Moscow's "Smart City" Concept

The Moscow Government regards the role of ICT in smart city formation as twofold: first, as a means of communicating and establishing a feedback loop with citizens; second, as a way of reconfiguring the information system and establishing an effective information-sharing infrastructure for Moscow Government departments and public organizations founded or cofounded by the Moscow Government.

Therefore, integrative or umbrella front office projects, such as the Moscow Open Data Portal (www.data.mos.ru), IMP for services, Moscow Public Services Portal (www.pgu.mos.ru), Citizen Participation, Control and Feedback Portal (www.gorod.mos.ru), Integrated Medical Analytical System (http://emias.info), and Data Storage and Processing Center, have appeared as new front office means of communication, control, and transparency between the authorities and citizens (UN 2012).

Strategic ICT development in Moscow shows that the smart city concept for the Moscow Government means a secure, open, comfortable, manageable, and responsive city. For each of these factors, the Moscow Government has launched a separate project and information system (see Fig. 1). All the data gathered through each of these systems is to be linked and processed to provide extra social and economic value to citizens and the authorities. Moscow's approach to smart city implementation is close to the concept of an "information city," where the most convenient conditions are provided to people who are willing to communicate, work, and interact with the authorities via the Internet (Sairamesh et al. 2004; Sproull and Patterson 2004).

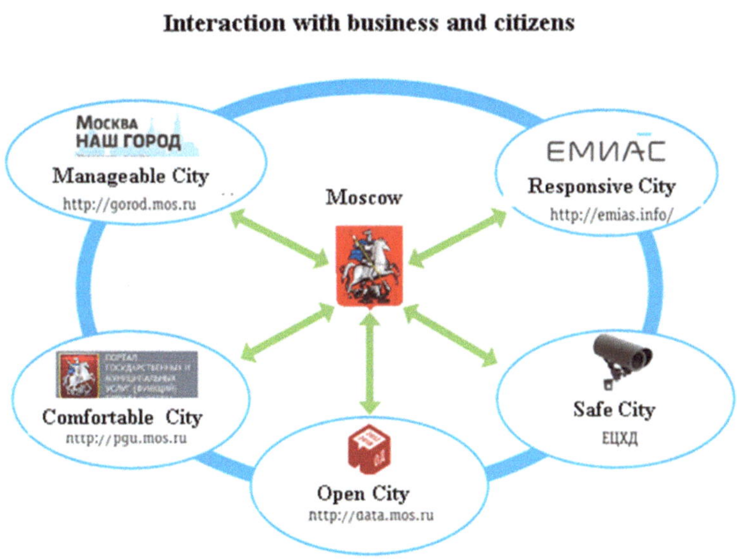

Fig. 1 The Moscow government's "Smart City" concept

There is intense debate in Russian society about what makes a city smart. Sociological surveys in Moscow show that, from the point of view of citizens, smart city development means immediate improvement in various city systems: public transport and road management, public services, housing management, urban architectural design, parking regulation, integration of channels of communication, development of payment tools, and citizens' involvement in city planning.

Pardo and Nam (2011) state that the smart city concept is not only about technology-based urban innovations, but also policy and organizational management. This is not in conflict with the Moscow Government's priority of controlling various city systems by maintaining high-quality data on when and how citizens use transport, health care, education, and utility infrastructure.

Chourabi et al. (2012) also state that there are at least six definitions of the smart city phenomenon. Those closest to what can be considered smart for the Moscow authorities involve technological control of city infrastructure, the connection of social and technological infrastructure, and the use of technologies in various areas such as health care, education, transportation, etc.

Moscow's approach to smart city formation is based on intensive interaction with citizens, while also tending to be what Hollands (2008) defines as technological. The intensive growth of problem-oriented virtual-communication platforms in Moscow confirms the idea expressed by Y. Benkler (2006) that new systems of interaction should be built on the collective production and exchange of information, culture, and knowledge.

Though this chapter focuses on technological solutions implemented by the Moscow authorities, it is important to mention the problem-oriented approach taken by the Moscow Government toward citizens. Problems on a citywide scale (traffic congestion, inadequate quality of utilities and public services, and disputes and conflicts in urban development) tend to be solved both by creating new technological opportunities and by imposing penalties and restrictions on citizens in the context of limited city infrastructure.

3 Method and Research Question

Given the importance of a broad-based, multidimensional view of smart cities, the authors utilized Antovski and Gusev's (2005) approach to mobile platform development in combination with Janssen and Estevez's (2013) lean government concept as a foundation for analyzing Moscow's IMP. Using five principles of mobile platform development, together with the orchestration approach to governance, we formulated the following research question: What are the key factors enabling efficient implementation of platform-based governance in Moscow? The authors conclude the chapter by outlining the key factors involved in the "doing more with less" principle of lean government based on IMP implementation. The authors also foreground challenges which need to be overcome by Moscow city authorities.

The primary research method used in this chapter is the case study method of collecting data. The case study approach is particularly appropriate, since the smart city is a contemporary phenomenon, especially in the Russian context, with considerable real-life contexts and applications (Yin 1984). Data sources include Moscow Government reports, statistics, policy documents, and regulations. Russian media and journalistic sources were used by the authors to better understand the societal context of platform governance development. In addition, since the authors consider smart city projects in Moscow, much information in this chapter was obtained from personal and telephone communication with Moscow city employees, using deep unstructured interviews as a research tool.

4 Smart City Policy in Moscow

The Moscow Government's DIT is seen as a champion of ICT in the city. DIT not only develops ICT infrastructure but also provides the business logic for collaboration between departments and with citizens. ICT implementation is strongly linked to the quality of services provided to Muscovites, their engagement in problem identification, and their involvement through various channels of communication. Due to the high penetration level of the Internet and mobile devices, the Moscow government wants to establish an ecosystem that integrates public services with real-time dialogue between citizens and the authorities (Renda 2010; Internet Society 2010).

The Internet is not the most popular channel of communication between citizens and the Government of Moscow. Landline phones and personal visits to city offices are still popular forms of communication. The official initiative is to stimulate demand for mobile communication and build a mobile platform for interaction with citizens by establishing an open data portal and public services portal as a front office (Kushchu and Kuscu 2003).

Another important role of virtual communication is to provide citizens with an alternative means of obtaining public services. The best example is payment for parking in the city center. In 2012, the Moscow Government began charging for public parking in the city's central zone. In addition to meters, payment can be made from a mobile phone. This was done to improve citizens' negative perception of pay parking by making such payments more convenient (Rossel et al. 2006).

The same policy is pursued for other public payments. Muscovites can pay for utilities and entertainment tickets as well as pay parking fines. The back-office information system not only accepts payments from mobile devices but must also maintain data on payments, payment status, and payment history to ensure the high information quality of mobile service (Minhee et al. 2001). All such data have to be available to users from a mobile device or personal account on the Moscow Public Services Portal 24 h a day, 7 days a week.

The core concept of DIT policy is to involve as many users as possible in mobile interaction, but they must register through the Moscow Public Services Portal. At present, over 1.6 million users have personal accounts on the portal. The more citizens officially registered on the portal, the easier it is to use their data in the interagency information-sharing process. The Moscow Government will be able to predict user needs, keep track of citizens' demand for information and services, and be proactive by sending reminders to citizens about their rights and obligations (fines and tax payments, social support, school enrollment and summer camp offers, etc.)

The Moscow authorities know not only the services and applications that citizens use but also where the interaction takes place (based on exact GPS coordinates) and are thus able to predict real-time transport flows and road use, better control payments, and maintain sufficient "big data" on citizens' behavior.

The sale of depersonalized user data to interested stakeholders is a promising opportunity for the Moscow authorities, as is the possibility of using the mobile platform to provide extra paid services. For example, the "doctor appointment" service is a right and thus free of charge, but reminders to visit a doctor can be charged. Kim et al. (2011) emphasize usefulness and value for money as key determinants of high-demand mobile services. The policy here is to determine the upper limit of payment for mobile services and strike a fair balance between the financial interests of the Moscow Government and the mobile operator.

The same balance must be found in publishing open data. At present, open data is seen by the Moscow Government as an environment for building a mobile application closely linked to the Moscow Electronic Map or Atlas, which provides useful information on skating rinks, sports facilities, theaters, Wi-Fi access points and gas stations with high- and low-quality fuel. The Open Data Portal often provides data by means of an API at the same time that the Moscow Government creates an informative map which is actually competitive with those produced by Russian IT companies. In other words, the Moscow Government is not only an open data supplier in the sense of warming up the market but also a major player in the open data market.

SMS communication from mobile devices is a very important channel of interaction for DIT. DIT has analyzed the interaction scheme as it is now (see Fig. 2).

The scheme shows that the mobile operator currently functions as the front office between the Government and citizens. As a result, the Government cannot guarantee the quality of services or regulate the amount charged by the operator for citizens' access to information systems and resources managed by the Moscow authorities.

The target model of interaction puts the Moscow Government in the front office in communication with citizens (see Fig. 3).

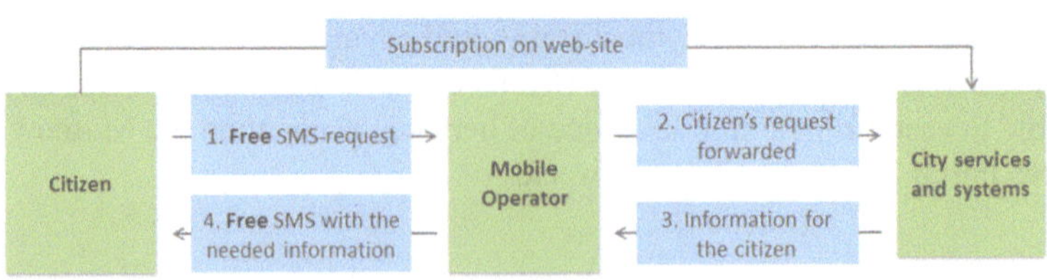

Fig. 2 Current mobile interaction between citizens and the government of Moscow

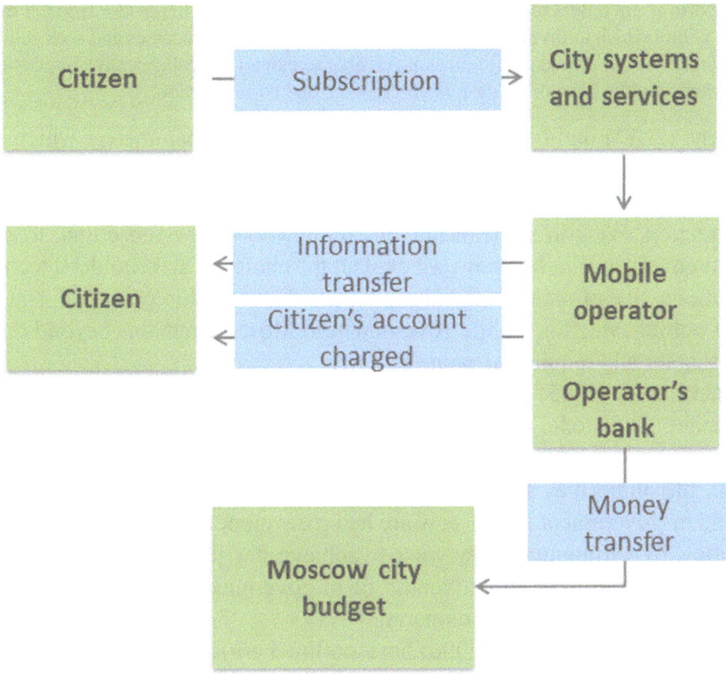

Fig. 3 Target model of mobile interaction between the Moscow Government and citizens

Citizens subscribe to services by registering on the official public services portal. The city information system provides information through the mobile operator's infrastructure. The mobile operator charges a citizen's subscriber account and retains the price of SMS traffic and service provision. The rest of the money goes to the city budget.

This reengineering of the communication process serves to improve the transactional component in public service provision (Shankar et al. 2003). The overall satisfaction level is a function of the accumulated positive mobile transaction services between the citizens and the authorities (Kumar et al. 2007).

The Moscow Government is now able to form a new business model based on extra-value mobile services. Mobile applications and service design companies can now be included in the model. As a result, mobile service providers can form a competitive mobile services market and communicate with citizens via the Moscow Government's IMP.

5 Mobile Government Implementation Models

The role of platform thinking cannot be underestimated in a city's strategic ICT development (Wachhaus 2011; Bailey 2011). Anttiroiko et al. (2013) state that

> a platform is not only a tool for managing information but also, in a wider sense, a framework within which to involve key stakeholders in governance processes and seek solutions to complex social problems. It makes it possible to extend the collaborative dimension of governance in the form of co-design, co-creation, and co-production.

Walravens (2012) outlines very important public value parameters which expand the business matrix to the public matrix of governance. The model is important because it connects technological and managerial factors with the process of public value creation. We utilize the Walravens framework in the sense that, to describe platform governance in Moscow, we explain the choice of stakeholders, technological solutions, use of data, financial models, and partnership potential. The assessment of public value for mobile government in Moscow remains beyond the scope of this chapter and requires separate research.

Borucki et al. (2005) have developed an enhanced response model for mobile government. The model emphasizes high demand for mobile services on the part of citizens as a key impetus for the adoption of mobile governance. In Moscow, by contrast, the authorities stimulate citizens' demand for m-government. E-government and m-government in Moscow are being developed in parallel by the Moscow authorities. At a minimum, they want to achieve the goal of multichannel service delivery to citizens. The effectiveness of m-government and e-government platforms in Moscow is yet to be compared.

Bremer and López Prado (2006) have outlined critical success factors in building an m-government model on the municipal level in Mexico city. Among these are ease of access to services, robustness of the platform, the ability to use different interfaces, the capacity to sustain a high workload (more than 100,000 SMSs at one time), and simple customization of mobile service. All these factors apply to Moscow as well. Mobile services must be simple enough to be understood by a diverse population (in terms of income, level of education, and age). Due to the city's size, the Moscow mobile platform must be robust and able to provide high SMS and MMS capacity simultaneously.

Chourabi et al. (2012) propose a two-level model of factors involved in smart city development. On the first level, the smart city concept means interconnection between policy, technology, and organization. On the second level are factors such as developed IT infrastructure, attention to economic growth due to m-government implementation, the social structure of city communities, the specific nature of governance, and the natural environment. In this chapter, we emphasize technological, economic, and governance factors and describe their influence on the process of building mobile government in Moscow.

The OECD (2011) offers a multifactor mobile value chain model developed by Cable (2011). In this model, value from mobile government can be created only with close collaboration among mobile platform stakeholders. The multistakeholder approach requires government to orchestrate relations between service providers, application developers, content providers, Internet providers, and hardware providers. At the same time, the city government must have a well-developed and detailed legal and policy framework regulating mobile services provision. Moscow has a highly developed IT infrastructure, very competitive hardware and Internet markets as well as strong purchasing power on the part of citizens, and a budget sufficient for mobile service development.

Antovski and Gusev (2005) emphasize that

> a common mobile public services framework must first and foremost incorporate the following five principles: interoperability, security, openness, flexibility, scalability.

A mobile platform thus enables open and transparent partnership between mobile service providers, consumers, and the authorities. Implementation of these five principles in technological, legal, and organizational dimensions is an essential precondition of successful mobile platform adoption by various stakeholders.

In our case study, we will demonstrate how the development and implementation of the Moscow's IMP fit Antovski and Gusev's (2005) five principles. These principles are universal for any government platform.

Interoperability is very important for Moscow, because various Moscow Government departments own information systems developed at specific times for specific purposes. The IMP now has to be interoperable with the Open Data Portal, Public Services Portal, and Citizens Participation Portal.

The security principle has also been incorporated in the IMP because the Moscow Government has to protect citizens' personal data and guarantee their right to view their history of transactions and receive secure and reliable services. Citizens' connections via mobile applications and data transfer are encrypted through suitable Internet protocols. The same security requirements apply to the IMP's service providers and are part of mobile services quality control. Security requirements are reflected in a set of regulations for developers who want to provide services through the IMP.

Openness is another important principle which has already been reflected in the federal Open Government Standard and replicated on the regional level (Open Government Standard 2013). For the IMP, this means clear rules and regulations on joining the platform, open quality control criteria for applications and services, and as a consequence an open decision-making process. Records of all funds received from citizens must be accessible. The technological dimension of IMP architecture and IT solutions must also be open and available.

Flexibility and scalability fit well in the IMP context because the Moscow Government wants as many service providers and application developers as possible as well as citizens to use the platform. In this case, each community or stakeholder group may come up with new initiatives concerning, for example, data formats, software choice, payment processes, etc. Thus, the IMP architecture must be flexible and scalable to adjust to changing needs and requirements on the part of stakeholders, to respond to changing limitations from a legal perspective, and to provide the necessary technological capacity for growing traffic due to the increasing number of IMP users.

To explain mobile platform development in Moscow, we also use the "lean government" concept formulated by Janssen and Estevez (2013). The concept underscores the need to reduce the role of the public sector, involve communities in participation and problem solving, and use platforms as solutions for effective service provision and interaction with citizens. The role of governments in the "lean government" concept shifts to orchestration: collaboration management among many public and private stakeholders. The idea is to do more with less. According to our interviews with Moscow Government employees responsible for IMP management, the role of the Moscow authorities is fully in keeping with the "orchestration" principle. The IMP is seen as a technological and organizational framework which connects application developers, service providers, authorities, and citizens. The Moscow Government has to regulate relations between IMP stakeholders in an open and transparent manner as an arbiter in financial issues as well as in quality control of service provision.

6 The Moscow City Integrated Mobile Platform

The Moscow city IMP is a public information system closely connected with such front office systems as the Moscow Public Services Portal (PSP), through which citizens obtain authorization to use mobile services. Authorization and data encryption protocols form part of the IMP's security principle. The IMP is also connected to payment service providers and mobile service providers.

The IMP was developed on the basis of J2EE technology by a consortium of three leading Russian IT companies. USD 0.5 million was spent over 3 years. The platform's official operator is "Moscow Registry" Public Unitary Enterprise, which operates under the close supervision of the Moscow DIT.

The IMP stores and guarantees the security of each user's personal data and access history. To interact with the Moscow authorities through the IMP, citizens must use the keyboards of their mobile devices. Voice interaction is not supported.

The Moscow Government regards the IMP as a commercial platform and is willing to provide information from public databases through the IMP only under a commercial agreement between the IMP operator and a mobile service provider. This seems fair, since mobile service providers use this data to provide high-value mobile services to citizens. Under the Law on Personal Data, the Moscow Government can provide only depersonalized data that is not a state secret and is not restricted to use within government bodies.

Below we describe the logic involved in regulating stakeholder relations with the Moscow authorities by means of the IMP. These regulations are adopted by a special Moscow Government Regulatory Act. They follow the principles of openness and transparency and take the interests of all collaborating stakeholder groups into account.

For a contract to be approved, the IMP operator must consider a mobile service provider's electronic application within 5 days. If application data are incomplete or inaccurate, the operator may request that the applicant correct the application within 5 days. Otherwise the application is denied.

A communication agent is a company that provides for the exchange of data between citizens and a mobile service provider (whether a company, public authority or public organization). If there is a charge, the service provider transmits information to the payment agent, which must notify the citizen. All participants are obligated to provide full and accurate information about each use of the system by citizens.

To implement scalability and flexibility principles, at least 100,000 users should be able to use the IMP simultaneously, and the system should have at least 4 million active users per month. Data are backed up on a daily basis and crash recovery should not take longer than 48 h.

Interaction between IMP components should be based on an HTTP protocol and secured against third-party penetration. Communication agents should act in accordance with instructions set down in their contract with the IMP operator. In the event of violations, the operator requests that the communication agent eliminate them within 10 days. A communication agent who refuses or fails to eliminate a violation within this time can be disconnected from the IMP.

The IMP operator must connect a service provider to the system within 10 days and inform the communication agent of any charge.

A service provider must store transaction information in the system, pass citizens' requests to the communication agent, and keep information on the service in the system. Each service provider must have a free user-support call service and claim investigation service.

A payment agent must have a special contract with a bank or other credit institution and ensure a reliable exchange of payment information between the credit institution and the IMP. A payment agent must provide verification of payment to any participant entitled to such information.

The IMP operator can charge participants for connection to the system, for IMP transport layer usage and for access to public databases.

IMP functionality is illustrated in Fig. 4.

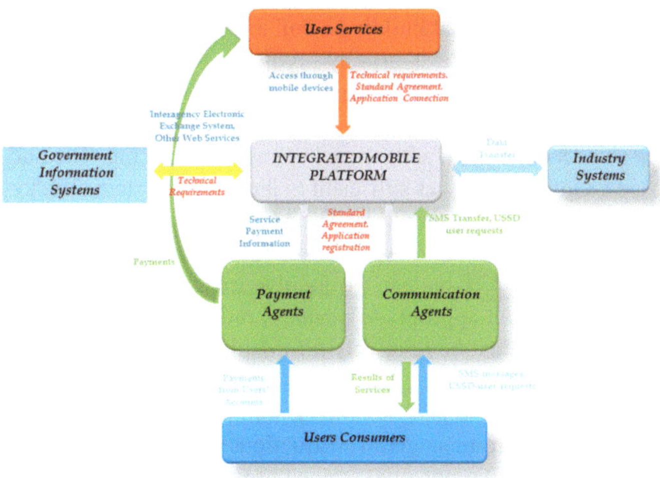

Fig. 4 Interaction scheme for IMP participants

The IMP interaction scheme shows that various information systems, both governmental and commercial, are interconnected. IMP management provides technical requirements for software interfaces, data formats, and protocols, thus enabling the interoperability principle.

At present, 40 SMS-USSD services are implemented through the IMP. Seven mobile applications are available to citizens. The three most popular applications are as follows:

Moscow Transport: Allows drivers to exchange messages about illegal parking based only on a car's official registry number. The application contains an interactive map with public transport stops and metro stations. Users can also check whether a driver was fined by road police and see the whole history of fines.

Mobile Inquiries: This application helps citizens write messages to the mayor of Moscow (130 typical problems classified). Users can track the status of their messages. A decision is made within 8 days.

Moscow Utilities: This application enables citizens to look through their history of utility payments, access their current account, and input meter readings (only for water).

As a result, 6 million SMSs are sent via the IMP on a monthly basis. The number of application downloads each week stands at 10,000. There are 2.2 million users of mass mobile services and 600,000 mobile application users.

The Moscow DIT is now launching the first phase of mobile services: reminders of doctor appointments and notices of upcoming utility payments. In the second phase, mobile services will be expanded to include SMS notifications that children have arrived safely at school, car towing notices, and notices of traffic fines.

7 Discussion of the IMP Case

The IMP case study shows that building a collaborative information system as part of the basic infrastructure of a smart city involves the same spectrum of problems as portals, geographic information systems, and other basic smart government projects do.

The IMP project has a great advantage in terms of financing. The budget utilized by the Moscow Government was sufficient to launch the system. The key challenge is to involve as many users as possible. The Moscow Government aims not only to find money for IMP support and exploitation but also regards the IMP as a source for enlarging the city's ICT budget. To involve more citizens, the Moscow Government has to choose an appropriate business model. The final product—mobile services for citizens—must be relatively affordable, and the service provider must be reliable.

IMP is built on a scalable and flexible IT solution with a high-level capacity to include more users. More managerial instructions still need to be put in place on how to choose and certify a potential service provider through the IMP. The package agreement between DIT and a service provider must regulate financial interests among DIT, the service provider, and any intermediaries.

The service has to be of acceptable value for citizens. One good example is the mobile application "Moms Watching" (www.mamnadzor.ru). This application is integrated with a map and shows all useful places for moms with kids: hospitals, playgrounds, cafes, etc. Moms can exchange opinions and make recommendations on visiting various sites in Moscow with kids.

DIT has the right to reject a service or application provider, but there are no instructions in place on this issue, because the system is new and practical experience of collaboration with service providers has not yet accumulated or been analyzed.

Another problem is maintaining uninterrupted mobile service in the event that a service provider ceases operations. This eventuality should be envisaged in the agreement between users, the IMP, and service providers. The best solution is probably to invite other companies to provide similar services. The Moscow Government is the owner of the platform and applications belong to companies. The Moscow Government must guarantee a fair and competitive application market for service providers to prevent the formation of a monopoly.

A key role is played by the company that is the IMP developer and operator. It is DIT's responsibility to have all necessary documentation on the IMP's development history, structure, and modifications. This has to do with DIT's general standards and requirements with respect to system development, acceptance, and support. The risk of IMP operator change still exists and could cause an interruption in the IMP's operation.

Finally, a serious challenge is posed by demand on the part of citizens. The Moscow Government needs to make access to mobile services as convenient, secure, and easy as possible and at the same time launch a popularization campaign to raise citizens' awareness of the opportunities provided by the IMP. This challenge can be met if many public organizations that are not part of the Moscow Government (public organizations, hospitals, schools, civil registry offices, etc.) are connected to the system and can provide services through the IMP.

8 Future Research

Future research on IMP development should concentrate on stimulating citizens' demand for mobile services. To ensure sustainable growth in demand for mobile services, further research should be done to find key factors influencing citizens' use of mobile applications. These factors can show how citizens make decisions to download and use a mobile application from the IMP. A theoretical approach may embrace a public value measurement framework for mobile applications.

Another research cluster can be built around the measurement of mobile applications' social and economic impact. The theoretical framework may include requirements for monitoring citizens' behavior while interacting with IMP resources, measuring overall satisfaction and payment transaction volume via the IMP. Based on impact measurement research, it will be possible to formulate IMP effectiveness indicators and validate the lean government concept in terms of "doing more with less."

Mobile service marketing is a very promising area for research. The Moscow Government is interested in providing infrastructure and data at a fair price and in keeping with private mobile services provided via the IMP on a level affordable for the majority of citizens.

Finally, the IMP creates extensive opportunities for open government technologies. For example, the mobile application "Active Citizen" was developed by the Moscow authorities to collect citizens' feedback on various problems, topics, and decisions made in the city. Future research is needed to determine the forms of feedback to be gathered via mobile applications. The feedback could be consultative in nature, strengthened by collecting citizens' formal evaluation of various government activities. More research is thus needed to understand how to use the powerful tool of citizens' formal evaluations in making governance decisions and modifying legal and regulatory frameworks.

The influence of m-government based on IMP functionality in Moscow will continue to grow, as will the volume and value of the gathered data. The smartness of government will thus be measured by the capacity to utilize data so as to stimulate city economic activity and increase Moscow's transparency and attractiveness in global competition among cities.

9 Conclusion

The Moscow Government's concept of a smart city means developing technology-based solutions that involve citizens and make the Government of Moscow more accountable. The business model for smart governance requires citizens to pay for high quality or, more precisely, for extra options in addition to public services that are free of charge by law. Integration projects such as the public services portal, citizens' complaints and ideas portal (gorod.mos.ru), open data portal, and IMP set new standards of collaboration not only for citizens but also for government departments.

The IMP case study analyzed in this chapter shows that the Moscow Government's approach to providing mobile services is a typical example of platform-based governance in keeping with the principle of doing more with less (Janssen and Estevez 2013). The IMP's development in Moscow was analyzed according to Antovski and Gusev's (2005) framework for m-government. We conclude that all five enabling principles—interoperability, security, openness, flexibility, and scalability—are incorporated in the platform.

We believe that ongoing stimulation of citizens' demand for mobile services together with increasing simplification of IMP usage by application developers, service providers and payment agents can, in the medium term, reduce transaction expenses, contribute to the Moscow city budget and, most importantly, provide citizens with smart solutions to their everyday problems.

As a result of these solutions, the Moscow authorities are definitely getting to know their citizens better. The Moscow Government will have to be more analytical, using "big data" to identify the demand for services and information as well as new means of collaboration. Bertot and Choi (2013) emphasize that government's use of big data is sustainable only if special attention is given to digital asset management and archiving, privacy, and security policy.

Extensive Internet and mobile penetration and a relatively high level of wealth (Moscow is the leading Russian region in terms of average salary) are prerequisites for starting a smart dialogue with citizens. Gil-Garcia and Pardo (2005) add good IT project management, clear formulation of goals and results, and effective leadership to the critical success factors influencing smart city projects. Seven million Muscovites are smart phone owners and potential mobile application users. The owners of the 18 million mobile phones registered in Moscow are potential SMS users. Right now the formal right and the initiative in formulating issues and providing services belong to the Moscow authorities.

But, even the Moscow Government's current state of technological development enables us to conclude that the next stage of smart governance will involve the expression of citizens' needs and demands in an interactive mode so that the authorities can respond automatically in real time. Such a proactive approach to governance can be achieved very soon on condition that civil society initiatives and proposals continue to gain momentum.

References

Antovski, L., & Gusev, M. (2005). M-Government framework. In Proceedings of Euro mGov 2005: The first European mobile government conference (pp. 36–44). Brighton, UK: Mobile Government Consortium International LLC.

Anttiroiko, A.-V., Valkama, P., & Bailey, S. (2013). Smart cities in a new economy: Building platforms for smart cities. *AI & Society*. doi:10.1007/s00146-013-0464-0.

Bailey, S. J. (2011). The evolving governance of public services in England: Extending competition, choice, co-design and coproduction. In A.-V. Anttiroiko, S. J. Bailey, & P. Valkama (Eds.), *Innovations in public governance*. (pp. 68–88). Amsterdam: IOS Press.

Benkler, Y. (2006). *The wealth of networks*. New Haven: Yale University Press.

Bertot, J. C., & Choi, H. (2013). Big data and e-government: Issues, policies, and recommendations. In Sehl Mellouli, Luis F. Luna-Reyes, & Jing Zhang (Eds.), 14th annual international conference on digital government research (pp. 1–10). http://dx.doi.org/10.1145/2479724.2479730.

Borucki, C., Arat, S., & Kushchu, I. (2005). Mobile government and organizational effectiveness. In Proceedings of Euro mGov 2005: The first European mobile government conference (pp. 56–66). Brighton, UK: Mobile Government Consortium International LLC.

Bremer, A. A., & López Prado, L. A. (2006). Municipal m-services using SMS. In Proceedings of Euro mGov 2006: The second European conference on mobile government (pp. 45–50). Brighton: Mobile Government Consortium International LLC.

Cable, S. (2011). Profiles of the 2010–2012 citizen-engaged communities. Public Technology Institute. Alexandria.

Chourabi, H., Nam, T., Walker, S., Gil-Garcia, J. R., Mellouli, S., Nahon, K., et al. (2012). Understanding smart cities: An integrative framework. In Proceedings of the 2012 45th Hawaii international conference on system sciences (pp. 2289–2297). Washington, DC, USA: IEEE Computer Society. doi:10.1109/HICSS.2012.615.

Gil-García, J. R., & Pardo, T. A. (2005). E-government success factors: Mapping practical tools to theoretical foundations. *Government Information Quarterly, 22*(2), 187–216.
Hollands, R. G. (2008). Will the real smart city please stand up? *City, 12*(3), 303–320.
Internet Society. (2010) The internet ecosystem. www.isoc.org/pubpolpillar/docs/internetmodel.pdf. Accessed 26 April 2014.
Janssen, M., & Estevez, E. (2013). Lean government and platform-based governance—Doing more with less. *Government Information Quarterly, 30,* S1–S8.
Kim, H., Lee, H., & Choi, S. (2011). An exploratory study on the determinants of mobile application purchase. *Journal of Society for E-Business Studies, 16*(4), 173–195.
Kumar, V., Mukerji, B., Butt, I., & Persaud, A. (2007). Factors for successful e-government adoption: A conceptual framework. *Electronic Journal of E-Government, 5*(1), 63–76.
Kushchu, I., & Kuscu, M. H. (2003). From E-Government to M-Government: Facing the inevitable. In Proceedings of European Conference on E-Government (ECEG 2003).
Minhee, C., Jinwoo, K., Hoyoung, K., & Hosung, R. (2001). Information quality for mobile internet services: A theoretical model with empirical validation. *Electronic Markets, 12,* 38–46.
OECD, & ITU. (2011). *M-Government. Mobile technologies for responsive governments and connected societies*. Paris: Organisation for Economic Co-operation and Development. http://www.oecd-ilibrary.org/content/book/9789264118706-en.
Open Government Standard for Federal Authorities. (2013). http://xn–80abeamcuufxbhgound0h9cl.xn–p1ai/events/5508409/. Accessed 1 July 2014.
Pardo, T., & Nam, T. (2011). Smart City as Urban innovation: Focusing on management, policy, and context. Fifth International Conference on Theory and Practice of Electronic Governance (ICEGOV2011).
Renda, A. (2010). Neutrality and diversity in the internet ecosystem. http://dx.doi.org/10.2139/ssrn.1680446. Accessed 26 April 2014.
Rossel, P., Finger, M., & Misuraca, G. (2006). 'Mobile' E-government options: Between technology-driven and user-centric. *Electronic Journal of E-Government, 4,* 79–86.
Sairamesh, J., Lee, A., & Anania, L. (2004). Information cities. *Communications of the ACM, 47*(2), 28–31.
Shankar, V., Smith, A. K., & Rangaswamy, A. (2003). Customer satisfaction and loyalty in online and offline environment. *Journal of Research in Marketing, 20,* 153–175. http://dx.doi.org/10.1016/S0167-8116(03)00016–8.
Sproull, L., & Patterson, J. F. (2004). Making information cities livable. *Communications of the ACM, 47*(2), 33–37.
United Nations E-Government Survey: E-Government for the People. (2012). http://unpan1.un.org/intradoc/groups/public/documents/un/unpan048065.pdf. Accessed 1 June 2014.
Wachhaus, T. A. (2011). Governance as a framework to support informatics. *Innovation Journal, 16*(1), 1–14.
Walravens, N. (2012). Mobile business and the smart city: Developing a business model framework to include public design parameters for mobile city services. *Journal of Theoretical and Applied Electronic Commerce Research, 7*(3), 121–135. doi:10.4067/S0718-18762012000300011.
Washburn, D., Sindhu, U., Balaouras, S., Dines, R. A., Hayes, N. M., & Nelson, L. E. (2010). *Helping CIOs understand 'Smart City' initiatives: Defining the Smart City, its drivers, and the role of the CIO.* Cambridge: Forrester Research http://public.dhe.ibm.com/partnerworld/pub/smb/smarterplanet/forr_help_cios_und_smart_city_initiatives.pdf.
Yin, R. K. (1984). *Case study research: Design and methods*. Newbury Park: Sage.

Evgeny Styrin is a senior research analyst and an associate professor affiliated with National Research University Higher School of Economics. Mr. Styrin provides expertise to Russian public administration bodies. He is an advisor to Russian Open Government Council. His fields of expertise are open government, electronic government, interagency information sharing, open data, and public IT strategy development and evaluation. Mr. Styrin develops and provides trainings, teaching courses for students and civil servants in the area of e-government, open data, strategic IT planning, and civil service. As Russian e-government expert, Dr. Styrin participates in international research projects with World Bank, UNDP, and OECD.

Artem Kostyrko holds a masters degree in sociology. He is the head of Administrative Modeling Department in Institute for Public and Municipal Administration, National Research University Higher School of Economics. He has broad interests in e-government, IT program management, and information society development on national and regional levels of governance in Russia. Mr. Kostyrko is an adviser to Russian Ministry of Natural Resources and Environment. He has vast experience in managing public IT programs such as: Russian E-Government System Project, and Moscow Information City Program. Mr. Kostyrko provides expertise to Russian Federal Ministries in business process reengineering, electronic services development, and monitoring. He actively participates in Moscow city's smart development projects as an expert in IT usage and adoption.

Smart Public Safety: Application of Mobile Electronic System Integration (MOBESE) in Istanbul

Sedat Kula and Ahmet Guler

Abstract This chapter is a case study about the application of mobile electronic system integration (MOBESE) in Istanbul, Turkey. To fight crime in a more effective and efficient way and to manage the traffic problem, the MOBESE system has been put into practice since 2005. From the perspective of smart city and innovation, this chapter explains the process of the MOBESE system and its applications for crime prevention, crime fighting, and traffic management. The chapter first gives brief information about the case and then reviews the smart city research from information science literature and surveillance research from the criminal justice literature. Then, the case is explained to give an idea about the application of the system. Finally, its relevance and contribution to the literature is discussed.

Keywords Smart city · CCTV · MOBESE · Law enforcement · Public safety

List of Abbreviations

CCTV	Closed circuit television
GIS	Geographic information system
GPRS	General packet radio service
HD	High definition
IT	Information technology
MOBESE	Mobile electronic system integration
PDA	Personal digital assistant
TBS	Traffic information system
TNP	Turkish National Police

S. Kula (✉)
Kars Emniyet Mudurlugu, Turkish National Police, 36100 Ortakapi Mah., Kars, Turkey
e-mail: sedatkula@yahoo.com

A. Guler
Lock Haven University, Department of Criminal Justice, USA
e-mail: axg250@lhup.edu.

J. R. Gil-Garcia et al. (eds.), *Smarter as the New Urban Agenda A Comprehensive View of the 21st Century City,* Public Administration and Information Technology 11,
DOI 10.1007/978-3-319-17620-8_13

1 Introduction

Modern communities have faced great transition from industrial society to information society over the last decades. In this context, developments in information and communication technologies have led to rapid and radical changes in every aspect of the government. Knowing that the success of public organizations depends increasingly on how efficiently they utilize the technological advancements in adjusting to contextual changes, e-government applications have begun to play an important role in the modernization of government administration. In this sense, the utilization of mobile applications and technologies has emerged not only in developed but developing countries as well.

Turkey has made considerable efforts in e-government over the past two decades. A number of impressive mobile government applications have been used in the country in recent years. Traffic information system (TBS) is one of the prime examples of country-wide mobile government applications in Turkey, which connects the mobile traffic enforcement units equipped with tablet-PCs to a central information system via general packet radio service (GPRS) to detect driver license information, vehicle registration, and citizen identification (Cilingir and Kushchu 2004).

Another example is mobile electronic system integration (MOBESE), which is one of the leading m-government applications in the law enforcement area. It is a system consisting of the integrated applications of hardware and software productions used for urban security. The aim of this system is to maximize the efficiency and effectiveness of the law enforcement agencies in crime prevention and crime fighting. Detection through its surveillance capability, the prevention of crime, and disorder through deterrence are common goals of MOBESE, like most closed circuit television (CCTV) systems. MOBESE is also an important factor to reduce fear of crime and to foster the image of the police in society.

In this study, the application of MOBESE system in Istanbul, Turkey is explained. Besides technical information, we provide information on the advantages and disadvantages of MOBESE for the society. As a new law enforcement application of smart cities in the world, we discuss its benefits for the society in the fight against crime. The criticisms from human rights are also emphasized.

2 Literature Review

The literature about smart city gains momentum and the research on smart city swiftly expands (Calderoni et al. 2012). Smart city is a fuzzy concept (Tranos and Gertner 2012) and a buzz word (Baron 2012) that defines similar, but not same initiatives to explain the application of information technology (IT) for city management. Moreover, there are alternative definitions to describe IT-related initiatives. These are "digital city" (Schuurman et al. 2012; Lombardi et al. 2012), "wired city" (Baron 2012; Paskaleva 2011), "intelligent city" (Schuurman et al. 2012), "ubiquitous city" (Schuurman et al. 2012), and "smart city" (Gil-Garcia

2012; Nam and Pardo 2011). While digital city, wired city, intelligent city, and ubiquitous city are used to define the technological side of development, smart city is more of a user-centered approach (Schuurman et al. 2012) and requires not only technological but also socioeconomic development (Nam and Pardo 2011). Even though there are different definitions to explain the trend in using IT in city managements, Nam and Pardo (Nam and Pardo 2011) identify three common themes in these working definitions: infrastructure, process, and vision for the future. While technology is a critical infrastructure to realize the smart city initiative, it will be naive not to consider the human side in the equation of the smart city initiatives (Hollands 2008). To advance the smart city idea, human development is crucial for technological enhancement and to foster innovation.

Moreover, using technology as an enabler of smart city and benefiting from it to connect and coordinate different systems and subsystems for a better city management will be crucial to provide the basis for realizing the idea of smarter applications. However, the process of service transformation and fundamental changes in service delivery are crucial to provide the philosophy behind the smart city. This philosophy should envisage smart economy, smart mobility, smart environment, smart people, smart living, and smart governance (Nam and Pardo 2011, www.smart-cities.eu).

According to several researchers (Nam and Pardo 2011; Chourabi et al. 2012; Batagan 2011; Hancke and Hancke Jr. 2012), cities have to be "smart" due to population shift from rural areas to urban areas. The migration trend from rural to urban areas has accelerated and the population of people living in urban areas has passed the population of people living in rural areas in 2010 (www.who.int). People move from rural areas to urban areas to find jobs, to get better education, or to have better quality of life (www.unfpa.org). While migrating from rural to urban areas creates opportunities for individuals to live in better conditions, immense human movement from rural to urban areas brings new problems for the management of cities and state governments. Rapid population changes in cities cause economic, social, and environmental problems such as high urban density of population, inappropriate use of land for housing, traffic congestion, upsurge in crime, and insufficiency of public services (Doytsher et al. 2010). All these issues force city administrations and state governments to find solutions using IT (Nam and Pardo 2011). Due to technical, political, or legitimacy concerns, local, state, or federal governments try to reform their administrations to be more efficient, more effective, and more coordinated, while responding to the abovementioned urban problems (Gil-Garcia 2012).

Moreover, some researchers (Tranos and Gertner 2012; Lombardi et al. 2012; Nam and Pardo 2011; Aoun 2013; Clarke 2013) discuss that smart city as an innovative concept is crucial for cities to be attractive for talented individuals and competitive for economic development. To remain competitive, city administrations need to attract, retain, and generate talented professionals by applying smart solutions to their problems in their cities. According to this idea, if cities become smarter, this image will attract talented young professionals. This will foster economic development and bring positive contribution to the smarter cities.

The quality of a city is determined by several factors, but being safe and peaceful are always at the top of the list (Rogerson 1999). Providing security and safety for all citizens living in cities is the main *raison d'être* for law enforcement organizations. Law enforcement organizations have to provide safety services to the citizens 24 hours a day and 365 days a year. This critical nature of the service requires police organizations to search for effective and efficient ways of policing for city security management. Without being present in all the streets of a city, it is not possible to be sure about the safety and security of a city. To reach this overarching goal, police organizations traditionally have tried to increase their presence in crowded streets and important spots of the city using different patrolling strategies (Braga et al. 2012). Even though the use of motor patrols and foot patrols to cover critical spots of the city helps police organizations to realize their duties, it mostly becomes impossible to cover all the streets and corners of the city due to lack of resources. At this point, police organizations have started to use technologies to provide security to their cities (Byrne and Marx 2011). With technological adaptation, police organizations not only gain the ability to respond to a specific crime that is happening in real time, but also develop proactive strategies to prevent crime by using collected data about current crime trends in their city (Grant and Terry 2008). Most of the city police departments have tried to utilize IT developments with their limited budgets. To benefit from IT inventions in public security field, police departments have tried to adapt themselves by not only buying new gadgets and equipments, but also adjusting its personnel to new ideas that come with new technological developments (Foster 2005).

In today's world, it is widely accepted that policing is all about gathering and using information (Roberg et al. 2009). To provide social control and to maintain order, police departments in developed countries started to benefit from surveillance technology in the 1990s (Kruegle 2007; Goold 2003). CCTV cameras have been widely used to decrease thefts, crimes against personnel and assets, and terrorism (Kruegle 2007). Although CCTV is widely discussed in the literature lately, it is difficult to find a shared definition of CCTV. Simply, we can define CCTV as a system that has a stationary camera, a monitor, and a recorder (Goold 2004). However, current CCTV systems have different features and technical capabilities. New features and applications are added to the system with the innovations in surveillance technology. Thus, there are different surveillance systems in cities. As a comprehensive definition, CCTV refers to electronic monitoring systems, which make use of video cameras, connected by means of a "closed" (or nonbroadcast) circuit, to capture, collect, record, and/or relay visual information about the event-status of a given space over time (Deisman 2003; Gill and Spriggs 2005).

The theory behind using CCTV in crime fighting by the police is succinctly summarized by Armitage (Armitage 2002). These are deterrence, efficient deployment, self-discipline, presence of a capable guardian, and detection. For the deterrence, if there is a camera in a place, a potential offender calculates the cost and benefit of offending crime and decides not to commit or go to other suitable places to offend. Secondly, CCTV system helps the police to deploy their human and other resources wisely by giving enough attention at the right time to a spot when police assistance is needed. CCTV system also brings self-discipline not only to offenders, but also to victims. When victims see the presence of CCTV cameras in place, they become alert against crime. Thus, they try to reduce the risk of being a prey of a criminal by taking personal precautions. For the offenders, CCTV cameras give the sense of being watched all the time and they try to control their behavior to not be caught in misbehavior. According to routine activity theory offered by Cohen and Felson (1979), there should be convergence in space and time of the likely offenders, suitable targets, and the absence of capable guardians. CCTV cameras prevent the convergence of these three elements by providing a capable guardian all the time. The last but not the least, CCTV cameras help the police to detect offenders by recording images of offences taking place. These records play an important role to arrest criminals, give punishment, and incarcerate them. Thus, criminals are caught and cannot perpetuate new crimes and justice is served for the society.

Specifically, 9/11 terrorist attacks have forced governments to invest in surveillance technology to protect their citizens against horrific attacks. To prevent similar attacks in the future, law enforcement agencies have densely started to use surveillance systems, especially CCTV, to monitor human activities at airports, seaports, borders, and crowded city streets (Grant and Terry 2008). This technology provides risk assessment tools such as dataveillance and biometrics for the police to collect, organize, and store information about persons and use these data to identify a person from biometric features (Simon 2005). Dataveillance was coined by Clarke (1999) and he defines it as the systematic use of personal data systems in the investigation or monitoring of the actions or communications of one or more persons. Dataveillance consists of several techniques such as front-end verification, computer matching, profiling, and data trail (Clarke 1999). On the other hand, according to Woodvard (2001), biometrics refers to the use of a person's physical characteristics or personal traits to identify or verify the claimed identity of that individual. This technology includes several biometric identification technologies such as voice/speech recognition, fingerprint scanning, lip movement recognition, retinal scanning, facial recognition software, DNA profiling, and thermal imagery (Grant and Terry 2008).

The research about the benefit of CCTV systems in crime fighting is incomplete, confusing, and inconsistent (Dempsey and Forst 2012). While some researchers (Gill and Spriggs 2005; Carli 2009; Squires 2003; Phillips 1999) argue that CCTV systems are necessary tool for the police to reduce crime, respond to crime in a timely manner, maintain order, and provide social control; others (Deisman 2003; Armitage 2002) claim that there is no significant data that indicates the benefit of CCTV in crime fighting. According to the literature review about the CCTV done by Deisman (2003), the effect of CCTV on crime is variable and unpredictable. The deterrence effects of CCTV also change according to type of a crime, location of a crime, and time of a crime. Similarly, Welsh and Farrington (2008) did an extensive systematic review to assess the effects of closed circuit television surveillance on crime. They concluded that CCTV has a modest but significant desirable effect on reducing crime in car parks, especially vehicle crimes, and is more effective in reducing crime in the UK than other countries.

Moreover, there is limited research about the role of CCTV in preventing ongoing criminal activity and arresting the suspects (Bekkers and Moody 2011). However, there is majority support from the public for deploying CCTV cameras in public places, because it increases feeling of safety among citizens and decreases the fear of crime and victimization. Even though there are some concerns about profiling and the breach of privacy (Greenhalgh 2003), people generally support the usage of security cameras in crime fighting.

Armitage (2002) reviews the studies about the evaluation of CCTV systems in the literature and concludes that there are several methodological issues in the studies due to several reasons. These are inadequate pre- and post-CCTV time periods in which data are collected, no account of seasonal variations is taken, no control areas were used for comparison, little discussion of displacement or diffusion of benefits, unspecified sample size, and lack of independent evaluation. Thus, there should be methodically sound empirical papers in the literature to know to what extent surveillance plays a role in crime prevention and reduction.

However, using security cameras to watch streets, roads, and other public places in the cities raises critical discussion about the "panopticon" in the literature (Goold 2003; Simon 2005). The idea of panopticon was first projected by the utilitarian philosopher, Jemery Benthem, in 1787 by designing a prison which required minimum supervision and provided maximum control over prisoners. According to his design, the prison building is constructed as a circular shaped building in which all cells are located at the periphery of the circle floor by floor, facing towards the guard tower placed in the center. Only a single guard can watch all prison inmates behind the mesh screen of the watch tower, but the prisoners cannot see whether they are being watched or not all the time. All prisoners should assume that they are being observed by the guards and have to show appropriate behavior according to the discipline rules of the prison. If they behave well, they can be rewarded by relocating their cells to better places in the prison or they can be punished if they do not follow the rules as ordered. This design is recognized as a cost effective and highly efficient way of controlling prisons (Kietzmann and Angell 2010). It is argued that there are several prisons all over the world established with the influence of the panopticon idea.

There is a hot debate about using CCTV as a surveillance system in the cities to fight against crime and other disorders. While some scholars argue that camera surveillance is a way to transfer our democratic society to a surveillance society like in a panopticon prison idea, proponents of CCTV say that privacy is a matter of private places such as our homes, not a public place (Weckert 2005). The critical question about the CCTV is "who watches the watchers?" (Franklin 2008) and opponents of CCTV are cautious about how we know that CCTV cameras are not used for profiling, spying, or inquiring into the private lives of innocent people (Mohammed 1999).

3 Research Methodology

This study uses the case study approach to understand how MOBESE, a CCTV system for Istanbul, has been inaugurated as a new law enforcement application of the smart city by the Turkish National Police (TNP), and the results of this project in crime prevention and fighting. Even though this system has been used since 2005, there is no empirical study to assess the effectiveness of the system on crime. We use the case study approach as a research methodology to explore the components of the system and unveil the issues around the system. As stated by Yin (2002), the case study is a distinctive way of empirical inquiry that can be used to investigate a complex phenomenon in real life. Thus, this study will explore the CCTV system to understand its basic components and its applications in crime fighting and will provide a basis for further studies on the MOBESE system.

4 Study Context: Istanbul MOBESE Project

With a population of approximately 76 million as of 2014, Turkey is a unitary structured country governed by a centralized administrative system. Istanbul, which has been the capital of three great empires, namely, the Roman, Byzantine, and Ottoman, is the capital city of Turkey where more than 15 million people reside.

In Turkey, the police and gendarmerie forces are responsible for security. The police generally serves in urban areas whereas the gendarmerie serves in rural areas. These two security forces are centrally structured as General Directorate of Turkish National Police (TNP) and General Command of Gendarmerie. General directors of these organizations perform under the responsibility to the Minister of Interior. The Governor and the District Governor are responsible for the duties performed by the officials in cities and towns, respectively.

The use of CCTV, with the aim of preventing crime and disorder through deterrence for watching crowding town centers, has received a flourish of publicity in Turkey in the first decade of the twenty-first century. Surveillance capability and opportunity provided by the CCTV systems help gain public reassurance by reducing fear of crime. Another important aspect of CCTV is that law enforcement officials are quickly deployed to the incident scene and investigations are done according to the information recorded by the system (Brown 1995).

In Turkey, MOBESE has received pertinence from the public as a top-notch application. Open street surveillance of MOBESE cameras has been widespread in the country in the first decade of the 2000s. MOBESE was first carried out by the intelligence department of police in Diyarbakır City in 2001 with the support of the General Directorate of TNP and then later became widespread in Mersin, Ankara, and Istanbul. In 2008, with the support and declaration of the Ministry of Interior, MOBESE system was established in all the cities of Turkey and in some big towns as well (Çoban 2005).

Especially, the results obtained for crime fighting by using the MOBESE system in Istanbul accelerated the process of widespreading the system across the country. In 2005, the Head of the IT Department of TNP was assigned with the duty to set up MOBESE standards across the country. Standard MOBESE specifications were sent to all the cities in 2007 (Çevik and Filiz 2008). This project was also politically supported. In early 2008, the Minister of Interior declared the government's intention to establish the MOBESE system in all cities of Turkey and in some districts having dense population. The MOBESE system is being used in different provinces of Turkey by law enforcement officers (Cilingir and Kushchu 2004; Istanbul Emniyet Müdürlügü (IEM) 2014).

Like many metropolitan cities, Istanbul is exposing electronic surveillance for its security in the pretext of creating a secure, safe, and an attractive place for tourism, consumption, and entertainment as Lyon (2001) points out. Fear of crime is also an important factor for city life. Therefore, surveillance is considered an inevitable feature of cities, especially metropolitan cities.

Besides the mixed texture of the population, uneven distribution of income among the citizens and social injustice lead to high rates of crime in Istanbul. One of the important reasons for this is the sudden and constant increase in the population of Istanbul in recent years. Most of the people came to Istanbul with the hope of better living conditions. Not finding a job and the fancy lifestyles of others that they see around everyday create stress on them, which in turn, lead them to engage in criminal activities (Özden 2008).

The MOBESE-Istanbul system is considered as an authentic project developed by the police itself in terms of design and process (Çoban 2005; Yıldırım 2006). It became operational in Istanbul Police Headquarter with the monetary support of Istanbul governorship through Istanbul Provincial Special Administration Unit in 2005. The declared objectives of the system at that time were to improve civil services, to facilitate the management function, to regulate the mukhtar services, and to reduce the crime rate (IEM 2014).

Watching the city 24 h a day and 7 days a week, this system aims to prevent crime and increase the self control of the police force patrolling in the street. The cameras are placed at specific locations where the density of population and crime rates are high. Most of the cameras are placed inside the city centers, while some are at the bridges and critical points of the motorways. These cameras record what is monitored. The recordings of the cameras are watched and analyzed by the main command and control center located in the Istanbul Police Headquarter at Vatan Street.

Being insufficient for the population density, bridges and roads negatively affect the living conditions of Istanbul by creating too many traffic problems (Özden 2008; Kurt 2010). The purpose of placing cameras at the bridges and critical points of the motorways is to monitor traffic jams and control the traffic. Through these cameras, tracking the license plates of the cars to detect suspicious and stolen cars is possible (IEM 2014; Kurt 2010)

The districts of Eminonu, Beyoglu, Kadıkoy, Besiktas, and Sisli have the largest number of MOBESE cameras since most of the crimes such as pickpocketing, stealing, and robbery are committed in these places. These places are under the surveillance of MOBESE cameras to prevent crime from being committed or to detect and catch the criminals if any crimes are committed (Özden 2008). Istanbul governorship and Istanbul police department officials assert that, among other precautions and reasons, MOBESE cameras are considered an important reason for the decrease in crime rates in Istanbul.

4.1 *Components of Istanbul MOBESE*

The MOBESE system comprises the following components:

4.1.1 Command and Control Center

The main command and control center has twelve 2×1.5 m long, one 3×2 m long, and one 9×16.5 m long screens. The activities and actions of the police forces in the field are displayed on the satellite maps of Istanbul. The images coming from the cameras located at different places of the city are broadcasted 24 h a day on the screens. Based on the calls coming from citizens through 911 calls (155 for Turkey), police patrols are assigned to the scene of the incidents very quickly to handle the issue. This center also plays an important role in social events, fire, and natural disasters occurred in terms of communication and coordination. District command and control centers provide an opportunity for local police forces to track their vehicles through the vehicle detection system to handle the security issues in their districts (Özden 2008).

The main command and control center has the technical capability to store, retrieve, and print the visual data obtained. The system infrastructure is developed in a way in which it is compatible to all necessary updates and expands. In case of big scale security crises, either a crisis desk can be constituted at the command and control center of MOBESE or the images of the incident scene can be transmitted to crisis management center where the Governor or Chief of police of Istanbul can easily watch the scene and manage the crisis (IEM 2014).

4.1.2 Vehicle Track System

The vehicle track system is developed to track and assign police vehicles on a numeric map of Istanbul compatible with the geographic information system (GIS). This system provides officials to follow-up all vehicles and motor vehicles of the Public Order Unit instantly at 34 determined control centers. The exact addresses of the vehicles and the duration of stop and standby of vehicles can be determined (IEM 2014).

4.1.3 Mobile Vehicle Inquiry System

Mobile vehicle inquiry system is designed in a way in which mobile computers and PDAs provide officials information about vehicles, individuals, or events. It enables officials to make online queries about drivers, vehicles, and personal identifications via GPRS. Text or push talk messaging, sending emergency calls, and forming and printing incident and performance reports are the options available to use (IEM 2014).

4.1.4 Detention Center Control and Development System

The main purpose of the system is to place cameras on all police station detention centers to follow up 24 h a day 7 days a week, to prevent excessive use of force by officials against suspects. Recordings are saved in the police station centers for a specific period of time. Simultaneously, the images of videos can be watched online from the command and control center with the help of specific software developed. It also helps to prevent any suspect from committing suicide. The whole interrogation process of suspects is recorded by the system (IEM 2014).

4.1.5 Zone Imaging System

Several places in Istanbul with high density of population and high rates of crime, have been monitored by 4212 MOBESE cameras at 1206 selected points. Of the 4212 cameras, 160 cameras at the most important places have the ability to record the images in high definition (HD) quality. The recordings of the cameras are watched and analyzed at the main command and control center within the Istanbul police department. For cities such as Istanbul, where social protests predominantly take place, these cameras play an important role in analyzing the seriousness of the events and to assign the required patrol force as soon as possible (IEM 2014).

The camera images are recorded for not more than a week and the images of the recordings are sent in case of the court's request even though they are not accepted as an evidence solely (Özden 2008). Besides the fixed cameras located at different places around the city, portable cameras are also used for security purposes, especially for big events, to transmit the images of the incidents to the main command and control center. Police helicopters are used to capture the images of specific places if needed (IEM 2014).

MOBESE cameras are mostly placed in the street corners to see the maximum range of street views. People walking on the square and the trafficare watched at the same time. Most of the cameras have the capability to capture the view at night and also have pan, tilt, and zoom functions to capture the images as clearly as possible (Özden 2008).

Generally, 6-m long columns, resistant to the hits and crushes that might come from the citizens, are used to fix the cameras. Their resemblance to the city lights make them unnoticeable by the public even though warning signs are placed to inform the public that cameras are in motion. Warning signs such as "This area is monitored by the MOBESE Cameras" inform the individuals that they are being recorded by cameras (Özden 2008).

One of the historic districts of Istanbul, Eminonu has many historical buildings and tourist attractions such as Blue Mosque and Hagia Sophia, one of the most visited museums and prominent monuments in the world in terms of art and the history of architecture. Even though this historical place attracts many visitors, it also draws criminals to commit crimes, especially pickpocketing. To provide security and to eliminate the pickpocketing, MOBESE cameras are mostly placed in this district. Beyoglu, Kadıkoy, Besiktas, and Sisli are some of the other districts where people spend their time. These are the leading districts of Istanbul in the number of MOBESE cameras. Taksim square, in Beyoglu, is generally where people gather toprotest (Özden 2008).

Due to the large number of protests and public disorders, many MOBESE cameras have been installed to control these activities and to catch those who break the law. Beyoglu has many foreign embassies as well. This is also an important factor for the positioning of MOBESE cameras here. Sisli is one of the commerce and business centers of Istanbul with a number of banks and financial institutions. The headquarters of many universal companies and modern financial centers are being built here. Human and money trafficking takes the attention of potential criminals as well. Thus, the number of MOBESE cameras is remarkably high.

4.1.6 License Plate Detection System

Some MOBESE cameras are placed at important transit routes of the city. These cameras are equipped to detect the suspicious and stolen cars by tracking the license plates of the vehicles and reporting them to the central command and control center. In addition to detecting the suspicious and stolen vehicles, this system, with the support of special software, is capable of detecting uninspected vehicles in traffic (IEM 2014).

4.1.7 Mobile Command and Control Center

This center is the mobile version of the main command and control center. A 7.5-m long truck is equipped with the whole system that the main command and control center has. Inside the truck, zone imaging system, wireless communication system, satellite communication system, and mobile automation system are included as modular systems. The meeting room inside the truck serves to discuss and determine how to handle the social event or crisis. Two cameras are placed both in front and the back of the truck for providing security (IEM 2014).

5 Pros and Cons of MOBESE System

Even though people generally support the use of security cameras in crime fighting, there are some concerns about profiling and the breach of privacy (Greenhalgh 2003). One of the strong arguments of proponents of CCTV cameras is public safety and security. Security is the most successful argument in the achievement of widespread existence of CCTV cameras around. On the other hand, the main argument raised by the opponents of CCTV cameras is the foundation of the cameras without the will or informed consent of the citizens.

Security discourse by officials creates a perception in the society that surveillance is necessary and beneficial for crime fighting activities, especially for societies that comprise different ethnic and cultural segments. In addition, uneven distribution of income and social injustice among people (Özden 2008) lead to many social problems in the city of Istanbul. Although warning signs are placed to inform the public that cameras are in motion, these are unnoticeable by the public as they resemble city lights. The idea of being unnoticeable is based on the idea of panopticon of seeing without being seen (Özden 2008).

Coleman (2004) and Lyon (2001) point out that the cameras around us do not solely function as a crime preventing tool by law enforcement officials. Besides, cameras are used as social ordering strategy or social orchestration metaphor tool to adjust and control the behavior of people. They argue that being watched by the cameras constantly create a perception that citizens have to control and adjust their actions. In this respect, media plays an important role to legitimize the widespread functioning of CCTV cameras accepted by the public by stating that cameras are vital to provide a secure and safe society without questioning them (Kurt 2010).

Another argument raised by the opponents of MOBESE is that there are no options left to citizens without surveillance. This situation, being assumed as potential criminals, ruins the authenticity of the public space life of people. Being watched and recorded by cameras result in loss of privacy and self-correction of behaviors of citizens. In addition, always being watched creates a feeling of guilt, even though they are doing nothing wrong (Özden 2008).

To prevent the misuse of the MOBESE system in the command and control center, the employees watching the screens are also watched, and if anyone is detected watching proceedings other than crimes, punishment is given (Özden 2008).

Schwartz (1968) claims that surveillance itself has the potential to create disorder rather than prevention. Under the surveillance of public spaces, unwanted people may choose to isolate themselves from the society and communicate with only those who resemble them, which may result in a new life far from the public space for them.

Great amount of power given to officials using the system is another point of criticism. The question raised by the critics (Franklin 2008; Mohammed 1999) is: Who can guarantee that the recorded images of people cannot be used later as a threat for some reason? An important response by the officials to the criticism that the cameras being used have a potential to intervene in the private life of the citizens is that the software program is designed in such a way that private areas of the buildings are blocked by putting black images. Therefore, private places are never recorded (Özden 2008).

6 Discussion and Conclusion

The security and safety of public has dramatically influenced the design of cities in recent decades. Most of the metropolitan cities are under constant changes due to security issues. They are being designed in a way, in which public safety is promoted through maximum visibility and surveillance. In the name of creating secure, safe, and attractive places for tourism, entertainment, and consumption, Istanbul, like many other metropolis cities (London, New York, etc.), is under surveillance by MOBESE cameras. The diversity of people living in Istanbul, and being an important tourist attraction with its historic and cultural places, Istanbul receives attention not only from people across the world, but from potential criminals as well. To fight crime in a more effective and efficient way and to control the traffic problem, MOBESE system was put into practice in Istanbul. Having 4212 MOBESE cameras at 1206 selected points, Istanbul has been under surveillance since 2005.

As mentioned by several scholars (Nam and Pardo 2011; Chourabi et al. 2012; Batagan 2011; Hancke and Hancke Jr. 2012), population shift from rural areas to urban areas with the hope of finding jobs, getting better education or having better quality of life create many problems that cities have to manage. This situation is also true for the city of Istanbul, the biggest city of Turkey. High density of population in urban areas, inappropriate use of land for housing, traffic congestion, high crime rates, and insufficiency of public services have left a great impact on the city administration of Istanbul to find more efficient and effective solutions to the problems. Among the several problems, providing security is considered a major concern for city administrators of Istanbul to handle, since a citizen's perception of living in cities as being safe and peaceful is the most important determining factor of quality of life (Rogerson 1999). Therefore, the MOBESE Project was put into practice by administrators of the city of Istanbul as an important crime fighting tool with the help of Istanbul governorship.

Police presence, covering critical spots of the city, is one of the traditional strategies of policing to prevent crime. According to the research about hot spots policing, this policing strategy seems to generate positive results in terms of controlling various types of crime (Braga et al. 2012). Considering more than 15 million people are living in Istanbul, it is almost impossible to assign either motor or foot patrols to cover the critical spots of Istanbul. Each district of Istanbul has many important hot spots which take the attention of people. Thus, under the scarce resources, the only way of carrying out this important policing strategy is to benefit from technological blessings. Like many law enforcement organizations that started using technological developments to provide security and maintain order (Byrne and Marx 2011; Kruegle 2007), Istanbul police department started installing MOBESE cameras at the important hot spots of Istanbul where the population density and crime rates are high. Some cameras are also installed at the bridges and critical points of motorways to manage traffic.

The application of MOBESE cameras, like other CCTV cameras, is based on the assumptions of the deterrence theory (Armitage 2002). Presence of a capable guardian, MOBESE cameras, has a potential of deterring crime since the presence of cameras at the important hot spots of the city centers makes a potential criminal calculate the costs and benefits of committing a crime, and the risk of being watched can prevent the potential offenders from committing a crime.

Although some researchers (Gill and Spriggs 2005; Carli 2009) state that the use of CCTV cameras is considered a necessary policing strategy for the police forces to prevent crime, to maintain order, and to provide social control, other researchers

(Deisman 2003; Armitage 2002) argue that there is no significant data available indicating the benefits of CCTV in crime fighting. Similar to the arguments made by Armitage (2002) and Deisman (2003), no empirical research has been conducted about the benefits of MOBESE system in crime fighting, even though the MOBESE system has been in practice since 2005. To conduct a research to measure the efficiency and effectiveness of MOBESE system is inevitable. As stated by Armitage (2002), pre- and posttest studies are not adequately conducted to evaluate to what extent surveillance decrease crime rates considering seasonal variations, control areas for comparison, and displacement effects. Moreover, the perceptions of citizens and officials using the system about the benefits of MOBESE system should be studied to evaluate the views of public and users about the system.

Since the MOBESE system is being used in Istanbul for almost 10 years, this experience of Istanbul could be used by other city officials across the world to make their systems more beneficial from different aspects. One important note for future researchers would be that different cities, using similar systems like MOBESE, might be compared to Istanbul to snapshoot the system capabilities. Comparisons would yield beneficial outcomes for all city managers and security officials.

References

Aoun, C. (2013). The smart city cornerstone: Urban efficiency. *Schneider Electric White Paper.*

Armitage, R. (2002). To CCTV or not to CCTV. *A review of current research into the effectiveness of CCTV systems in reducing crime*. London: Nacro.

Baron, M. (2012). Do we need smart cities for resilience. *Journal of Economics & Management, 10,* 32–46.

Batagan, L. (2011). Smart cities and sustainability models. *Informatica Economica, 15*(3), 1–8.

Bekkers, V., & Moody, R. (2011). Visual events and electronic government: What do pictures mean in digital government for citizen relations? *Government Information Quarterly, 28*(4), 457–465.

Braga, A. A., Papachristos, A. V., & Hureau, D. M. (2012). *Police programs to prevent crime in hot spot areas*. No. 7 of crime prevention research review. Washington, DC: U.S. Department of Justice, Office of Community Oriented Policing Services.

Brown, B. (1995). *CCTV in town centres: Three case studies*. (Crime Detection and Prevention Series Paper No.68). London, UK. Home Office.

Byrne, J., & Marx, G. (2011). Technological innovations in crime prevention and policing. A review of the research on implementation and impact. *Journal of Police Studies, 20*(3), 17–38.

Calderoni, L., Maio, D., & Palmieri, P. (2012). Location-aware mobile services for a smart city: Design, implementation and deployment. *Journal of Theoretical and Applied Electronic Commerce Research, 7*(3), 74–87.

Carli, V. (2009). Assessing CCTV as an effective safety and management tool for crime-solving, prevention and reduction. *International Centre for the Prevention of Crime Report.*

Çevik, H. H., & Filiz, O. (2008). Polis teşkilatında bilgi yönetimi. *Türk İdare Dergisi,* 165–84.

Chourabi, H., Nam, T., Walker, S., Gil-Garcia, J. R., Mellouli, S., Nahon, K., Pardo, T. A., & Scholl, H. J. (2012). Understanding smart cities: An integrative framework. Paper presented at System Sciences (HICSS), 45th Hawaii International Conference on System Sciences.

Cilingir, D., & Kushchu, I. (2004). E-government and m-government: Concurrent leaps by Turkey. Paper presented at Proceedings of the 4th European Conference on e-Government.

Clarke, R. (1999). Introduction to dataveillance and information privacy, and definitions of terms. *Roger Clarke's Dataveillance and Information Privacy Pages*.
Clarke, R. Y. (2013). Smart cities and the internet of everything: The foundation for delivering next-generation citizen services. *Cisco White Paper*.
Çoban, E. (2005). Ülkemizin yeni bilgi ve güvenlik konsepti MOBESE. *Polis Dergisi, 45,* 21–28.
Cohen, L. E., & Felson, M. (1979). Social change and crime rate trends: A routine activity approach. *American Sociological Review, 44,* 588–608.
Coleman, R. (2004). *Reclaiming the streets: Surveillance, social control and the city*. Portland: Willan Publishing.
Deisman, W. (2003). *CCTV: Literature review and bibliography*. Research and Evaluation Branch, Community, Contract and Aboriginal Policing Services Directorate, Royal Canadian Mounted Police.
Dempsey, J. S., & Forst, L. S. (2012). *An introduction to policing*. Clifton Park: Delmar Cengage Learning.
Doytsher, Y., Kelly, P., Khouri, R., McLAREN, R., & Potsiou, C. (2010). Rapid urbanization and mega cities: The need for spatial information management. *Research Study by FIG Commission 3. FIG Publication no* 48.
Foster, R. E. (2005). *Police technology*. Upper Saddle River: Pearson Prentice Hall.
Franklin, S. B. (2008). Watching the watchers: Establishing limits on public video surveillance. *National Association of Criminal Defense Lawyers White Paper*.
Gil-Garcia, J. R. (2012). Towards a smart state? Inter-agency collaboration, information integration, and beyond. *Information Polity, 17*(3), 269–280.
Gill, M., & Spriggs, A. (2005). *Assessing the impact of CCTV*. Home Office Research, Development and Statistics Directorate London.
Goold, B. J. (2003). Public area surveillance and police work: The impact of CCTV on police behaviour and autonomy. *Surveillance & Society, 1*(2), 191–203.
Goold, B. J. (2004). *CCTV and policing: Public area surveillance and police practices in Britain*. Oxford: Oxford University Press.
Grant, H. B., & Terry, K. J. (2008). *Law enforcement in the 21st century*. Boston: Pearson/Allyn and Bacon.
Greenhalgh, S. (2003). *Literature review on issues of privacy and surveillance affecting social behaviour*. Edmonton: Office of the Information and Privacy Commissioner of Alberta.
Hancke, G. P., & Hancke Jr., G. P. (2012). The role of advanced sensing in smart cities. *Sensors, 13*(1), 393–425.
Hollands, R. G. (2008). Will the real smart city please stand up? Intelligent, progressive or entrepreneurial? *City, 12*(3), 303–320.
Istanbul Emniyet Müdürlügü (IEM). (2014). MOBESE(Mobil Elektronik Sistem Entegrasyonu). http://mobese.iem.gov.tr/. Accessed 12 Apr 2014.
Kietzmann, J., & Angell, I. (2010). Panopticon revisited. *Communications of the ACM, 53*(6), 135–138.
Kruegle, H. (2007). *CCTV surveillance analog and digital video practices and technology*. Burlington: Elsevier Butterworth Heinemann.
Kurt, E. (2010). *Surveillance and control in the age of ınformation: A critical analysis of the technology-power relationship*. Unpublished Master Thesis, Middle East Technical University.
Lombardi, P., Giordano, S., Farouh, H., & Yousef, W. (2012). Modelling the smart city performance. *Innovation: The European Journal of Social Science Research, 25*(2), 137–149.
Lyon, D. (2001). *Surveillance society: Monitoring everyday life*. New York: McGraw-Hill International.
Mohammed, E. (1999) An examination of surveillance technology and their implications for privacy and related issues-the philosophical legal perspective. *The Journal of Information, Law and Technology (JILT), 1999*(2), 99-2.
Nam, T., & Pardo, T. A. (2011). *Smart city as urban innovation: Focusing on management, policy, and context*. Paper presented at Proceedings of the 5th International Conference on Theory and Practice of Electronic Governance.

Özden, Ö. (2008). *Istanbul: An urban panopticon*. Unpublished Master Thesis. Middle East Technical University.

Paskaleva, K. A. (2011). The smart city: A nexus for open innovation? *Intelligent Buildings International, 3*(3), 153–171.

Phillips, C. (1999). A review of CCTV evaluations: Crime reduction effects and attitudes towards its use. *Crime Prevention Studies, 10*(1), 123–155.

Roberg, R. R., Novak, K. J., & Cordner, G. W. (2009). *Police & society*. New York: Oxford University Press.

Rogerson, R. J. (1999). Quality of life and city competitiveness. *Urban Studies, 36*(5–6), 969–985.

Schuurman, D., Lieven De Marez, B. B., & Mechant, P. (2012). Smart ideas for smart cities: Investigating crowdsourcing for generating and selecting ideas for ICT innovation in a city context. *Journal of Theoretical and Applied Electronic Commerce Research, 7*(3), 49–62.

Schwartz, B. (1968). The social psychology of privacy. *American Journal of Sociology, 73*, 741–752.

Simon, B. (2005). The return of panopticism: Supervision, subjection and the new surveillance. *Surveillance & Society, 3*(1), 1–20.

Squires, P. (2003). An independent evaluation of the installation of CCTV cameras for crime prevention on the whitehawk estate, brighton.

Tranos, E., & Gertner, D. (2012). Smart networked cities? *Innovation: The European Journal of Social Science Research, 25*(2), 175–190.

Weckert, J. (2005). *Electronic monitoring in the workplace controversies and solutions*. Hershey: Idea Group Pub.

Welsh, B. C., & Farrington, D. P. (2008). Effects of closed circuit television surveillance on crime. *campbell Systematic Reviews*.

Woodward, J. D. (2001). *Biometrics: Facing up to terrorism*. Santa Monica: Rand Cooperation.

Yıldırım, H., Kaplan, V., Çakmak, T., & Üstün, C. C. (2006). *E-devlet başa*. Istanbul: Imyay Yayıncılık.

Yin, R. K. (2002). *Case study research: Design and methods*. Thousand Oaks: Sage.

Sedat Kula is a police superintendent at the Turkish National Police. He holds a PhD in public affairs from the University of Central Florida. His research interests focus on information technologies in law enforcement agencies, police management, and police behavior.

Ahmet Guler is a police superintendent at the Turkish National Police. He did his doctoral study in public administration at the Rockefeller College of Public Affairs and Policy, University at Albany. His research interests are in organizational theory and behavior in criminal justice organizations, information systems and applications in security organizations, and crime prevention strategies.

Building an Intelligent Government, Intelligent City, and Intelligent Citizenry Through ICTs: Smart City Innovations in New Taipei City, Taiwan

Kevin Y. Wang and Chao-Ming Wu

Abstract This chapter presents a case study of the smart city initiatives in New Taipei City (NTPC) in Taiwan. Using the integrative framework proposed by Chourabi et al. (Understanding smart cities: An integrative framework. In 2012 45th Hawaii International Conference (HICSS) on System Science (pp. 2289–2297). IEEE. 2012) as an analytical lens, the authors discuss the characteristics and scope of NTPC's innovations in delivering services and managing resources using ICTs, as well as the policies, human, social, and cultural contexts that shape the adoption and development of these smart technologies. A SWOT (strengths, weaknesses, opportunities, and threats) analysis was conducted in response to the research question, contributing to the identification of lessons learned, from this particular case, which may have broader relevance for other smart city initiatives.

Keywords Smart city · SWOT analysis · New Taipei City · Taiwan

1 Introduction

In recent years, more and more communities around the world have been planning to leverage the power of information communication technologies (ICTs) to make their cities "smarter." Whether this means setting up infrastructure to better manage energy consumption or creating free Wi-Fi zones to bridge the gaps in digital access, millions of dollars are being invested in the research, development, and implementation of these technologies.

K. Y. Wang (✉)
College of Communication, Butler University,
4600 Sunset Avenue, Indianapolis, IN 46208, USA
e-mail: kwang@butler.edu

C.-M. Wu
Research, Development and Evaluation Commission,
New Taipei City Government, No. 161, Sec. 1,
Zhongshan Rd., Banqiao District, New Taipei City, Taiwan
e-mail: mislighter@gmail.com

J. R. Gil-Garcia et al. (eds.), *Smarter as the New Urban Agenda A Comprehensive View of the 21st Century City,* Public Administration and Information Technology 11,
DOI 10.1007/978-3-319-17620-8_14

This push toward smart cities is being met with cautious optimism by citizens. For instance, a new multi-country survey released by IBM (2014) indicated that nearly half of the respondents had positive attitudes toward various innovations designed to make cities smarter, more convenient, and safer.

While ICTs continue to be a fundamental pillar in creating sustainable, livable, and intelligent cities, research on this complex and emerging phenomenon has been noted as limited (Chourabi et al. 2012) and often piecemeal. Therefore, there is a need for a more complete and coherent perspective that informs policymakers and researchers who are pursuing or studying these initiatives.

To that end, this chapter takes a comprehensive view to examine the smart city initiatives of New Taipei City (NTPC) in Taiwan. Recognized by the Intelligent Community Forum (ICF) in 2012 as one of the Smart21 Intelligent Communities (ICF 2014), NTPC has a unique experience in developing and delivering smart city initiatives situated under Taiwan's longstanding electronic government (e-government) and ICT agenda. The authors not only detail the city's innovations in delivering services and managing resources using ICTs, but also discuss the policies, human, social, and cultural contexts that shape the adoption and development of these smart technologies.

2 Literature Review

2.1 Defining a Smart City

Since "smart city" is a relatively new concept, its meaning is also quite fluid, often invoking different conceptualizations in different contexts. Because it is an extension of e-government initiatives, many scholars focus on the role of technology in the day-to-day operation of a city. For instance, Hollands (2008) described smart city as the "utilization of networked infrastructure to improve economic and political efficiency and enable social, cultural, and urban development" (p. 308). Likewise, Washburn et al. (2010) defined smart city as "the use of smart computing technologies to make critical infrastructure components and services of a city—which include city administration, education, health care, public safety, real estate, transportation, and utilities—more intelligent, interconnected, and efficient" (p. 2). This focus on technology is echoed by Caragliu et al. (2011), Toppeta (2010), and Mortensen et al. (2012), all of whom emphasized the use of ICTs to create more sustainable and livable communities through smart management and participatory governance. Although the integration of technology in city planning and administration is indeed critical, one must also recognize that such a perspective alone may not capture the complexity of smart city initiatives.

Moving beyond technological infrastructure, Giffinger et al. (2007) argued that the term "smart city" has been used in the literature to describe piecemeal implementation of technology in various narrow capacities without addressing the holistic attributes of a city. As an alternative, they offered a definition that described a smart city as "a city well performing in a forward-looking way in economy, people, governance, mobility, environment, and living built on the smart combination of endowments and activities of self-decisive, independent, and aware citizens" (p. 11). Such a perspective focuses more on the strategic nature of smart city, as communities look for ways to maximize various resources and activities, including technology, to improve the lives of citizens.

Other definitions of smart city address the human/social capital and human relations aspect as the outcome of ICT use. For example, Coe et al. (2001) considered a smart city to be a community whose economic, social, and political life has been transformed by the increased use of technology. Rios (2008) argued that smart cities should inspire and motivate their inhabitants to create culture and knowledge. Similarly, scholars such as Florida (2002) and Poelhekke (2006) noted that by building a smart community that attracts a concentration of skilled workers, a city could produce spillover benefits that positively affect its entire region.

All in all, these various definitions suggest that "smart city" is a polysemic concept that encapsulates multiple layers of meaning in different perspectives and contexts. Recognizing such complexity, a broader and more comprehensive approach to understand smart city initiatives is offered by Nam and Pardo (2011), who examined various definitions and built a list of common components surrounding the concept of smart cities. Through this process they identified three core components—technology factors (e.g., physical/digital infrastructure and smart computing), institutional factors (e.g., governance, policy, and regulations), and human factors (e.g., human infrastructure and social capital).

In this chapter, the authors adapt such an integrative perspective to examine the smart city experiences in NTPC. By acknowledging smart city as a multidimensional concept, one could better systematically tease apart the factors that inform the success of smart city initiatives.

2.2 Benchmarking Smart City Success

As investments in the development and implementation of smart city initiatives continue around the world, there have also been widespread efforts to assess and benchmark the outcome of these initiatives. Giffinger et al. (2007) proposed a benchmarking framework that includes six indicator areas:

First, *economy* refers to factors surrounding economic competitiveness such as innovation, entrepreneurship, productivity, and the integration/flexibility of labor markets. *People* describe the social and human capital that can be seen through education, diversity, creativity, cosmopolitanism, and engagement in public life. *Governance* emphasizes citizen participation in decision-making processes, delivery of services, and governmental transparency. *Mobility* means the accessibility of ICTs as well as the use of smart technologies to provide innovative and safe transportation systems. *Environment* refers to the management of natural resources to create sustainable environmental conditions. Finally, *living* denotes various facilities and institutions to promote increased quality of life for residents and tourists alike.

Whether used as a ranking mechanism or as an evaluative tool, this framework provides a useful lens to study smart city initiatives and it has been applied by Giffinger et al. (2007) to study 70 medium-sized European cities, and also by the Committee of Digital and Knowledge-Based Cities of United Cities and Local Governments (2012) to examine 28 cities in Europe, Africa, Asia, and Latin America.

Another framework worth noting is Chourabi et al.'s (2012) integrative framework. Building on the conceptual comprehensiveness of a smart city, the authors outlined eight clusters of factors that can be used to study and determine the success factors of smart city initiatives or projects. These factors include issues related to:

- *Management and organization*—Organizational communication and project management as identified by Gil-Garcia and Pardo (2005) play an important role in the development and implementation of smart city initiatives.
- *Technology*—ICT infrastructure, particularly with regards to availability, capacity, institutional willingness and inequality, and culture.
- *Governance*—Implementation of processes with smart city stakeholders to achieve goals and objectives. Focus is placed on collaboration, leadership, transparency, accountability, and service/application integration.
- *Policy context*—Political and institutional components such as laws, regulations, and changes in policy context to support and drive smart city initiatives.
- *People and communities*—Empowering citizens and communities through the use of ICTs to enhance education, quality of life, communication, and active participation.
- *Economy*—Economic outcomes of smart city initiatives such as business/job creation, entrepreneurship, productivity, and regional as well as global economic integration.
- *Built infrastructure*—Physical implementation of ICT infrastructure, including interoperability, security, privacy, maintenance, etc.
- *Natural environment*—Use of smart technology to increase sustainability and manage natural resources effectively.

Taken together, these two frameworks collectively represent a useful set of tools for researchers and practitioners to systematically examine smart city projects and initiatives. The Giffinger et al.'s (2007) framework is more outcome-based and can be used for evaluation or comparison between different smart city projects. In contrast, the Chourabi et al.'s (2012) framework takes a more balanced approach by incorporating various success factors in addition to outcomes and can thus be useful for planning and management as well as assessment.

For the purpose of this chapter, the authors adopted the Chourabi et al.'s (2012) framework to analyze NTPC's smart city success, as it helps to better *explain* the underlying relationships between these factors and smart city initiatives.

2.3 *Why Taiwan? Why New Taipei City (NTPC)?*

For over a decade, Taiwan has invested heavily in its e-government initiatives, with considerable research and development devoted to create technological infrastructure, integrate different levels of services, and encourage citizen usage and participation. Taiwan's ambition to build a world-class e-government system has been led by the Research, Development and Evaluation Commission (RDEC)[1] under the Executive Yuan. In 2002, Taiwan's government portal (http://www.gov.tw) was officially launched, and more than 4400 governmental agencies had established a Web presence (Lau et al. 2008). In the years following the launch, Taiwan consistently received high rankings in the annual global e-government survey conducted by Brown University from 2002 to 2007 (see West 2007, for instance), making it one of the world's leaders in this area. Taiwan also has a legacy of developing smart city projects and initiatives, with several cities being recognized by the ICF among the Smart21 Intelligent Communities (ICF 2014) in the past few years.

Standing in the shadow of Taiwan's capital city of Taipei, NTPC is a former county area that was incorporated in 2010 as the nation's largest and newest municipality. With nearly 3.9 million residents, the city covers roughly 800 square miles and includes urban, rural, and mountainous regions (NTPC 2014). The vast geographic area creates difficulties to deliver government services to all residents and communities. The urban–rural divide was a longstanding problem for NTPC even before it became a municipality (Lin and Lin 2011). Further, with the neighboring capital city of Taipei drawing more resources from the central government and attracting greater business opportunities from home and abroad, many NTPC residents feel a lack of pride and sense of community. According to a 2011 survey, for instance, only 76% of the residents in the sample reported that their family was "happy" living in NTPC, the lowest among all five special municipalities in Taiwan and 27% said they would consider living elsewhere if they had a choice (Chen 2011).

From a city planning and management standpoint, NTPC therefore faces several existing challenges, such as fostering community participation and satisfaction, as well as balancing urban–rural development while delivering quality government services over a large terrain despite budgetary limitations. To address some of these issues and concerns, the city administration, under the leadership of Mayor Eric Chu, turned to the power of ICTs to improve city governance and the quality of life of NTPC residents.

In recent years, NTPC has made tremendous strides in this area and its effort has won several international awards and accolades. In 2013, NTPC received the Cloud Security STAR award from the Cloud Security Alliance and was placed third in the LivCom Awards. Most recently the ICF named NTPC as one of the top seven intelligent communities in 2014 (NTPC 2014). NTPC's unique experience in developing and delivering smart city initiatives, situated under Taiwan's longstanding e-government and ICT agenda, is therefore worthy of further exploration.

[1] The national RDEC was dissolved in January 2014 and superseded by the National Development Council (NDC).

Thus, the present chapter asks the following research question: *What are the characteristics and scope of NTPC's smart city initiatives?*

3 Methodology

With regards to research techniques, this chapter primarily follows the case study method to collect data. Case study is a particularly appropriate approach since smart city is a contemporary phenomenon that involves considerable real-life contexts and applications (Yin 1984). Data sources include government reports, statistics, policy documents, and media and journalistic sources from Taiwan. In addition, since one of the authors works closely with the smart city projects in NTPC, much of the information presented in this chapter also derives from personal communication with other city employees who are involved.

Given the importance of a broad-based, multidimensional approach to examine smart cities, as discussed in the literature review, the authors utilized Chourabi et al.'s (2012) framework to examine the NTPC case. Moving through each of the eight clusters defined in the framework, the analysis will shed light on NTPC's smart city experiences and various initiatives. In addition, a SWOT analysis was conducted in response to the research question, contributing to the identification of lessons learned from this particular case as well as practical and research recommendations.

4 The Case of NTPC

In the following section, the authors discuss the characteristics and scope of NTPC's smart city initiatives by examining the success factors identified by Chourabi et al. (2012). Given NTPC's social, political, and technological context, the arrangement of each of the factors presented here is slightly different from Chourabi et al's initial ordering to indicate the priorities and importance in relation to the NTPC case.

4.1 Policy Context

NTPC's push for smart city initiatives began in 2010 when Mayor Eric Chu took office after the city became the nation's newest municipality. Prior to NTPC, Mayor Chu served as the Magistrate of Taoyuan County between 2001 and 2009 and was known for his effort to bring technology development into city management and governance. For instance, the ICF awarded him its Founders Award and recognized Taoyuan County as one of the Smart21 Intelligent Communities in 2009 (ICF 2014; "Profile: Eric Chu" 2009). With a successful track record in this area, Mayor Chu sought to utilize digital technology to address many of the challenges facing NTPC, as described previously. In other words, NTPC's smart city initiatives were best characterized as a top-down driven effort, led by a leader who embraced ICTs and guided by a set of policies and visions to set objectives and support the implementation processes.

As articulated in various city documents and publications (NTPC 2013; Personal communication 2014), NTPC's smart city initiatives were guided by the core values of **3Os**, **3Ts,** and **3Is**.

The 3Os rubric refers to programs and services that aimed at providing *Open Government*, *One Government*, and *Government On Demand.* In many ways, these visions were built upon Taiwan's national e-government policy, which aims to provide a centralized portal that represents a "one stop" destination for citizens to interact with government officials or request various services, while providing a greater level of transparency through openness and digitization of government information (Lau et al. 2008).

NTPC's official city government website (http://www.ntpc.gov.tw) has attracted more than nearly 10 million visitors (NTPC 2013) and 16 million page views since the city was incorporated in December 2010, and its "e-service" Web portal (see Fig. 1 below) provides nearly 200 government service items and contains more than 800 downloadable forms and information documents. The single service window approach makes government services more efficient, accessible, and convenient for residents.

The 3Ts in NTPCs smart city plan refer to the utilization of *Service Technology, Cloud Technology,* and *Mobile Technology* to facilitate various government services. NTPC embraced the advent of mobile technologies and has developed

Fig. 1 Screenshot of NTPC's "e-service" Web portal

a range of mobile phone apps (see Fig. 2 below), allowing residents to use their mobile devices to complete or request government services on the go. These apps are spread across different service categories and are compatible with iOS, Android, and Windows platforms. As of January 2014, there have been more than 1.1 million total downloads on these 14 apps, updating the notion of government on demand to serve the more mobile lifestyles of contemporary end users. The most popular apps include a travel information app (271,617 downloads), a citizen-petition/suggestion app (122,845 downloads), and a health care information app (119,384 downloads). Further, NTPC has utilized cloud technologies to increase the efficiency of internal operations (discussed further in Sect. 4.4 below) while strengthening the backend information exchange systems for various Web and mobile services.

Finally, by strategically using the 3Os and 3Ts principles, NTPC hopes to achieve the policy objective of 3Is—Intelligent Government, Intelligent City, and Intelligent Citizen. In other words, by becoming a smart city, NTPC effectively utilizes ICTs to facilitate governance and manage city and natural resources, all in the service of an intelligent citizenry that, thereby, enjoys a high quality of life. As Mayor Eric Chu noted, "through the core values of 3O, 3T, and 3I, we wish to transform the public service sector in NTPC from a government-centered philosophy to a user-driven and citizen-based thinking" (Personal communication 2014). To date, NTPC has made considerable progress toward the objective of 3Is, and specific examples of these successful transformations will be discussed in the sections below.

Fig. 2 Screenshot of NTPC's mobile service apps

4.2 Management and Organization

To realize the abovementioned policy objectives, NTPC's RDEC is the office responsible for managing the general smart city planning and development with a budget that accounts for roughly 0.8% of NTPC's nearly $160 million NTD annual budget. The implementation of specific projects and tasks is carried out by respective agencies throughout the city government. As Gil-Garcia and Pardo (2005) noted, managerial and organizational concerns are critical to the success of any IT and e-government initiatives, which is especially true for a large and complex organization like NTPC.

NTPC's smart city effort also had to tackle several institutional obstacles before the effort was embraced by the municipal organization and the employees. For example, the hierarchical and territorial nature of government bureaucracy in Taiwan makes communication between NTPC's 27 departments and 29 district administrative offices problematic. In addition, since government employees in Taiwan have career tenure, they also tend to resist change and innovation.

To overcome these organizational challenges and foster a norm of innovation and service, NTPC instituted a series of mandatory training programs—including face-to-face, online, and hybrid learning platforms—to provide education for its employees, while putting in place a performance-based evaluative and incentive scheme to stimulate participation and adoption. The implementation of these policies, along with encouragement at the leadership and management level, resulted in a transformation of the organizational culture that can be seen in several performance measures reported in 2013.

For instance, the number of caseloads processed by all NTPC agencies grew from 3.1 million in 2012 to 3.3 million in 2013, a 6.7% increase, while the average time for caseload processing decreased by 0.38 day. On-time processing of citizen service requests saw a 7.37% increase to 99.02%, and on-time processing of inter-agency requests improved by 5.49 % to 93.3% (Personal communication 2014). In general, these figures indicate that the smart city initiatives have helped the NTPC government to become a more efficient and effective organization.

4.3 Governance

A related factor to organization and management is the notion of governance, which refers to collaboration between different stakeholders to achieve smart city goals and objectives. As noted by Chourabi et al (2012) and Gil-Garcia and Aldama-Nalda (2013), the presence of leadership is central to this process. In NTPC's push for the adoption of e-government and smart city initiatives, Mayor Eric Chu has championed the cause by encouraging communication, collaboration, participation, and partnership among NTPC's different offices and constituents.

For example, a high-level cabinet meeting involving various executives of NTPC departments has been held regularly since 2011 to discuss ways to improve the quality of online services and to enhance the coordination and communication process among different units. Through these meetings as well as the actual implementation of technological tools that facilitate data exchange and interoperability (discussed in Sect. 4.4 below), various NTPC agencies are able to come together as an organization.

In addition to communicating with its internal constituents, NTPC also regularly consults with local business and community leaders to assess business needs and has recently established a partnership with IBM to further incorporate smart technologies into its police force (IBM 2013; Liao 2013). There are also a number of citizen participation channels (discussed in Sect. 4.5) that encourage civic engagement. Finally, NTPC works closely with the national RDEC, which has been in charge of IT and e-government initiatives in Taiwan. With the formation of Taiwan's NDC in 2014, NTPC is expected to fully collaborate with this new agency to ensure seamless cooperation across different levels of government departments and offices (Personal communication 2014).

These efforts brought together the multiple stakeholders involved in NTPC's city governance through formal and informal channels of communication enabled by technology, thus pursuing the city's smart city policy objective of *Open Government, One Government,* and *Government On Demand.*

4.4 Technology

Technology is the foundation for any smart city project or initiative. Building on the 3Ts principle of service, cloud, and mobile technology, NTPC has successfully developed several innovations in this area to deliver more convenient online/offline services to residents while at the same time increasing the operational efficiency within the organization. With regards to services, some highlights (Personal communication 2014) of NTPC's technological accomplishments include:

- *Cross-district services*—Taking advantage of cloud technologies, residents of NTPC no longer need to return to the district office where their permanent address is officially registered to request or complete different social services. The ability to provide cross-district services saves considerable travel time and energy for residents because of the city's 29 administrative districts covering nearly 800 square miles. As such, this is the most requested service among NTPC's online service offerings (Personal communication 2014).
- *Cross-city services*—In the same vein, NTPC also allows residents whose official household registration is in another city to request or complete government services at any district administrative office. With a large number of NTPC residents originally coming from different parts of Taiwan, such cross-city services also save considerable travel time and energy for residents.
- *Change of address notifications*—When NTPC residents move and change their official household registration address, they can choose to utilize an automatic notification system that alerts 12 agencies (i.e., water or gas companies, health

services, departments of revenue and motor vehicles, etc.) about their change of address, saving residents the trouble of contacting individual offices.
- *Paperless documentation*—Residents of NTPC no longer need to provide paper duplicate copies of documentation for verification purposes when requesting or completing services at any district administrative office. Through digitization and cloud servers, employees can easily look up such information in the database and save time and hassle for residents.

With regards to increasing operational efficiency across NTPC's 27 departments and 29 district offices, some highlights (Personal communication 2014) of NTPC's technological accomplishments include the following:

- *Virtual meetings*—Utilizing video conferencing technologies, NTPC employees are able to attend meetings, trainings, or educational seminars. With 27 departments spanning a wide geographical region, this initiative considerably reduces commuting time for employees and trims operational costs.
- *Virtual storage*—Virtual storage and share drives have been created to allow employees of different units to access information virtually via various devices, including personal computers, tablets, and smart phones. This initiative has reduced storage and printing costs by $10 million NTD.
- *Virtual bookshelf*—A virtual bookshelf was created to allow employees of different units to access electronic books, magazines, and newspapers. By utilizing this virtual space, departments can eliminate redundancies in subscriptions for reading materials and reduce costs.
- *Digital and mobile signature*—NTPC also developed a digital signature system that is connected to different departments and offices, allowing employees to virtually process documents and information. The system is also compatible with mobile devices, further increasing the flexibility, productivity, and efficiency of government information processing.

4.5 People and Communities

One of the focal points of NTPC's smart city initiatives is to use ICTs to empower people and communities. Since 2007, NTPC has developed policies and programs to systematically combat the digital divide and provide digital opportunities to residents, many of whom fall on the short end of the social and economic spectrum.

For instance, the city offers free computer classes for residents in various age brackets and with different skill levels. Reflecting the growing popularity of mobile technologies in recent years, these computer classes now include training sessions on smart phones as well as tablets. To date, more than 9000 sessions have been offered, serving nearly 180,000 residents (Personal communication 2014). Since the city's geography encompasses several mountainous and remote regions, NTPC also reaches out to the residents of these areas by commissioning six mobile learning buses (see Fig. 3 below) that are fully equipped with IT and computers, and sending them out to hold free computer lessons and clinics.

In addition, NTPC has set up computer stations or information kiosks at various district administrative offices for visitors to use. The city also gives away refurbished computers and offers discounted Internet subscription rates for low-income families. All in all, these efforts to bridge the digital divide in NTPC have been quite successful, as the Internet penetration rate has reached 90.70 %, and 92.2 % of the households in NTPC have access to computers (Personal communication 2014).

Beyond providing digital opportunities to its residents, NTPC has also made strides in encouraging citizen participation and engagement through digital channels. For instance, the city hosted an online event in 2012 that invited young residents aged 20–45 to submit ideas to improve the quality of life in NTPC. Through a microsite as well as social media, the event attracted more than 120,000 participants over a period of 50 days. NTPC has also strengthened its social media presence through a Facebook fan page (see Fig. 4 below) and a LINE messenger account. Finally, as mentioned previously, the citizen-petition/feedback app is one of the most popular mobile apps downloaded by NTPC residents.

Overall, NTPC residents are quite satisfied with NTPC's smart city efforts. NTPC has been conducting regular surveys every 6 months since 2011, and as Fig. 5 illustrates, majority of the respondents are happy with the menu of digital services and opportunities, with the satisfaction rate continuing to grow over the years.

Fig. 3 NTPC's mobile learning buses

Fig. 4 NTPC's Facebook fan page

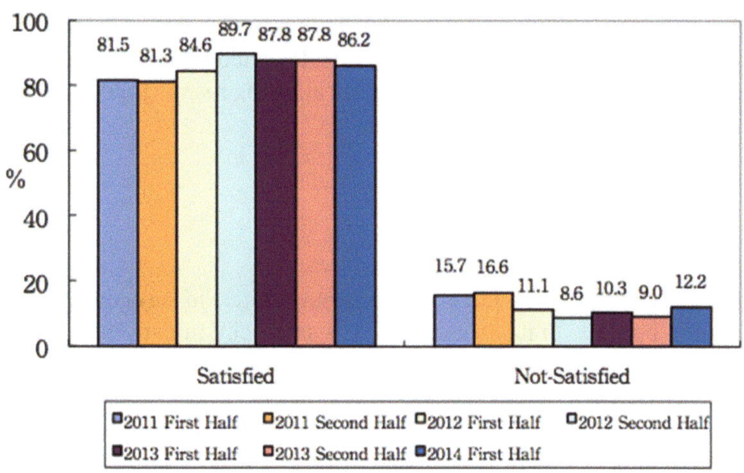

Fig. 5 Citizen satisfaction of NTPC's smart city services

4.6 Built Infrastructure

An important part of NTPC's effort to bridge the digital divide also includes a strong commitment to build and strengthen the city's information technology infrastructure, particularly through the creation of city-wide Wi-Fi zones. NTPC launched the free NewTaipei public wireless Internet service in 2012, with more than 271 indoor and outdoor hotspots spreading across all NTPC office buildings,

Fig. 6 One of NTPC's travel apps available on Google Play

libraries, hospitals, museums, parking structures, and tourist attractions (NTPC 2013). According to NTPC statistics, the Wi-Fi system has served more than 12 million distinct user sessions to date (Personal communication 2014).

To further expand the coverage area of the Wi-Fi system, NTPC collaborated with other government and business partners to allow registered users of the NewTaipei Wi-Fi system to have free roaming Wi-Fi access when they travel outside of NTPC. This is, especially, helpful for the large number of NTPC residents who commute regularly to work in the nearby capital city of Taipei, where they can seamlessly connect to Taipei's TPE-Free (Taipei Free) Wi-Fi network. In total, users have the ability to connect to more than 8000 hotspots throughout Taiwan. Recently, in an effort to expand on the city's international collaboration, NTPC established a partnership with the city of Fukuoka in Japan, allowing roaming access to registered users of both cities' public Wi-Fi systems.

4.7 Economy

In Chourabi et al.'s (2012) framework, "economy" refers to the economic outcomes of smart city policies. In this regard, NTPC's smart city initiatives are yet to yield significant business or job creation that can be directly linked to these efforts. However, as discussed in previous sections, NTPC has made significant progress to deliver many of its services, including services to businesses, using digital and mobile communication technologies. By streamlining information processing and reducing bureaucratic barriers, NTPC has created a business-friendly environment that emphasizes efficiency and productivity—with the hope that it will attract jobs and boost economic growth in the years to come.

One economic aspect for which smart technologies have made a positive contribution to NTPC is tourism. With a rich history and a spectacular natural setting, the city has long depended on tourism as an important part of its economy. To better serve tourists, NTPC has developed several travel-specific mobile apps (see Fig. 6 below) that serve as information guidebooks, all of which have attracted a large number of downloads. In addition, its bus app goes beyond providing a simple route map and timetable to include destination information (i.e., dining and lodging) that might be of interest to tourists. Finally, the city's tourism bureau (http://tour.ntpc.gov.tw) has established an active and strong Web and social media presence to connect with citizens/visitors and market the city's numerous tourism destinations and festivals.

Consequently, NTPC has seen a steady increase in its annual tourism numbers over the years, from 27 million in 2010, to 30 million in 2011, and 33 million in 2012 (National Statistics of R.O.C. 2013). While it is difficult to directly attribute this growth to smart city efforts, one could say that technology has positively contributed to this important area of NTPC's economy.

4.8 Natural Environment

Finally, NTPC has also used ICTs to better manage its natural resources as well as to improve the overall living environment for its residents. This is particularly important, as the city covers a large swath of urban, rural, and mountainous areas, making the management of natural resources complex. NTPC's effort in this area has been concentrated on natural disaster prevention, as it uses various smart technologies to monitor and collect environmental information ranging from weather forecasts to geological data (i.e., river level, earthquake) and radiation levels. This information is fed through an integrated information system for policymakers to make necessary decisions in the event of a crisis. NTPC residents are also able to look up such information in real time through the Disaster Prevention and Rescue Information website (http://www.dsc.ntpc.gov.tw/).

NTPC has also made significant progress in using smart technology within the realm of crime prevention and security. Since 2011, the city has moved toward merging various databases such as its emergency dispatch system, police vehicle positioning systems, a network of 13,000 digital security cameras, geographic information system (GIS), and existing security/crime records to create *one* integrated database to allow more efficient detection, analysis, and prevention of criminal activities. As mentioned earlier, NTPC partnered with IBM in 2013 to further improve its security and crime prevention measures with big data technologies.

5 Discussion and Conclusions

As described in the previous sections, NTPC has invested heavily in its smart city initiatives over the past few years, with various projects touching on nearly every factor discussed in Chourabi et al.'s (2012) evaluative framework. While these efforts have already made considerable progress, they are not without potential challenges and areas of improvement. In the concluding section, the authors briefly discuss the strengths, weaknesses, opportunities, and threats (SWOT) surrounding NTPC's smart city programs.

Perhaps one of the most notable *strengths* of NTPC's smart city initiatives is the clear vision that the leadership embraces to inform the city's policy objectives as well as the subsequent project implementation. The core values of *3O, 3T, and 3I* build on the foundation of Taiwan's long-term e-government strategy present a coherent policy context that brings together disparate NTPC departments and agencies as well as the central government to collaborate on smart city initiatives.

Another area of strength is the culture of innovation and service that has become the strong underpinning of NTPC's smart city initiatives. The transformation away from traditional bureaucratic culture starts at the leadership level with Mayor Eric Chu, who has championed the smart city cause since taking office in 2010 and the effects have trickled down through department heads and middle managers to employees. To further cultivate normative organizational change, NTPC also combines the necessary training and educational programs with incentive schemes, making "change" attainable and less threatening to employees.

With regards to *weaknesses*, it was mentioned earlier that NTPC has yet to reap significant economic benefits from its investments in smart city initiatives. This can be attributed to a number of reasons. First, economic development is a complex process that involves multiple national and global factors and often takes time. Since the majority of NTPC's smart city programs have only been introduced in the last couple of years, it is reasonable to argue that economic influence may not yet have materialized. Second, Taiwan's overall economy is still recovering from the global economic recession, with a weak GDP increase of 1.74 % in 2012 and 1.74 % in 2013 (Chung Hua Institution 2013). This less-than-ideal macroeconomic environment has contributed to the lack of economic impact from the city's various innovation projects. Finally, since NTPC borders the Taiwanese capital city of Taipei, competing against such an economic and talent magnet has long been a challenge, and it may also help to explain the lack of immediate economic results deriving from smart city programs.

That being said, NTPC's progress so far has paved the way for many *opportunities* and future growth areas. For instance, the collaboration with IBM on capitalizing big data technologies to enhance existing smart city initiatives is promising. Such technology could certainly be applied in the areas of crime prevention and security, as discussed previously. Big data also shows promise in the area of citizen participation and engagement. NTPC's growing presence on social media provides increasing opportunities to interact with NTPC residents via these informal channels. Capturing or tapping into these social media conversations using big data analytics thus represents an important opportunity for NTPC as it hopes to take a more active role in facilitating communication between people/communities and the government.

Along the same vein, another growth area for NTPC lies in the technological front. Taiwan introduced high-speed 4G LTE (Long Term Evolution) mobile services in the third quarter of 2014 (Yu and Hwang 2014). As consumers upgrade to newer and faster technologies, opportunities are created for NTPC to take advantage of the mobile platform to better deliver services and interact with constituents in line with the city's smart city policy objectives.

In order for NTPC to continue with its smart city initiatives, it must also find ways to address a number of potential *threats* that may hamper future growth. First, while NTPC has made tremendous progress in changing the organizational culture from territorial bureaucracy to that of innovation and service, work remains to be done. For example, as Lau et al. (2008) noted in their study of Taiwan's e-government programs, local government websites sometimes become unavailable or offline during nonbusiness hours. This is problematic and ironic, considering that 24 × 7 constituent access to government services and information was intended as a major advantage of online information systems. In this research, the authors also observed that a few NTPC agency websites became offline on weekends without displaying messages to site visitors about scheduled maintenance or downtime. Whether this stems from isolated technical difficulties or is a symptom of a larger, more persistent organizational culture is unclear, but it certainly warrants further improvement.

Finally, as discussed earlier, much of NTPC's smart city success can be attributed to strong leadership at the management level. Since leadership roles in any city are subject to change via election or political appointments, NTPC must also find ways to systematically carry out its smart city programs beyond the current city administration so that the established policy objectives and implementations can become meaningful, sustainable, and coherent over the long run. Table 1 provides a summary of the SWOT that stands out in the NTPC case:

Taken together, the SWOT analysis highlights three important lessons that can be learned from NTPC's smart city experience, which may also serve as recommendations for practitioners who wish to pursue smart city initiatives elsewhere. First, it is evident from the NTPC case that political leadership is a key variable that drives smart city adoption. As noted by Gil-Garcia and Aldama-Nalda (2013), leadership support can help foster multiple institutional arrangements and organizational structures to facilitate and deliver smart city objectives—as was true for Mayor Eric Chu of NTPC.

However, political leadership alone is not sufficient to create sustained innovation, as politicians come and go with each election cycle. To that end, the second lesson that can be learned from this case is to establish a clear strategic vision as well as a set of policies and incentives that facilitate long-term organizational change. For NTPC, the core values of 3Os, 3Ts, and 3Is are expected to guide the city's future smart city efforts, while the various training opportunities that the current administration implemented have already seen success in cultivating a more

Table 1 SWOT analysis of NTPC smart city initiatives

Strengths	Weaknesses	Opportunities	Threats
Leadership	Short-term economic benefits	Big data	Continuous service
Collaboration		4G network	Political cycle
Innovation and service culture			

collaborative and citizen-first culture among employees, while making the organization more efficient and effective in its overall operation.

Finally, the third lesson that can be learned from the NTPC case is a balanced focus on technology. While ICTs is at the core of any smart city initiatives, it is important to recognize that technology is a means to an end, not an end in itself. Smart technologies should be utilized to solve problems and to improve people's quality of life. For NTPC, technology proved to be useful in addressing numerous existing challenges such as delivering quality services to all residents, lessening the urban–rural divide, and managing resources that cover a large geographic region. NTPC also reaches out to internal and external stakeholders through multiple channels in order to assess their needs and prioritize emphasis areas. Such a balanced perspective on technology and other social and human factors was critical in NTPC's smart city success.

5.1 Recommendations for Future Research

To summarize, this chapter examined the factors that have shaped the adoption and development of smart technologies in NTPC, Taiwan, using Chourabi et al.'s (2012) framework. As ICTs continue to expand into the public arena, more and more cities around the world are considering ways to reinvent their communities using such technologies. The experience of NTPC in approaching smart city initiatives from multiple angles may serve as an example for policymakers who wish to pursue these endeavors. The present study also opens the door for other future research, some of which can expand on the findings of this study or address related issues.

For instance, with political leadership playing a key role in NTPC's smart city initiatives, it would be interesting to examine in future studies whether and to what extent leadership change may influence NTPC's smart city efforts. It will also be valuable to investigate whether the evidence for organizational and cultural change found in this study is a short-term phenomenon resulting from a top-down policy change or a transformation that can be sustained in the years to come.

In addition, while NTPC has been conducting surveys about the quality of its various e-services, this method is one-dimensional and may not fully capture the nature and extent of user satisfaction and experience. Future studies may employ market research techniques from both qualitative (e.g., interview and focus groups) and quantitative (e.g., surveys or social media metrics) perspectives to obtain a more comprehensive picture of how NTPC residents and communities are using these smart technologies.

Finally, as smart city programs and projects are often extensions of e-government initiatives, future studies can also connect with the existing e-government research agenda in policy, management, or citizen participation to provide a theoretically and empirically grounded view on smart cities.

Acknowledgments The authors would like to thank Ms. Hsin-Yu Huang of the RDEC in NTPC for her assistance in collecting and compiling information.

References

Caragliu, A., Del Bo, C., & Nijkamp, P. (2011). Smart cities in Europe. *Journal of Urban Technology, 18*(2), 65–82.
Chen, Y. (2011, July). Family friendliness among five special municipalities. *Parenting Magazine* (in Mandarin Chinese). http://www.parenting.com.tw/article/article.action?id=5020933. Accessed 11 March 2014.
Chourabi, H., Nam, T., Walker, S., Gil-Garcia, J. R., Mellouli, S., Nahon, K., Pardo, T. A., & Scholl, H. J. (2012, January). Understanding smart cities: An integrative framework. In 2012 45th Hawaii International Conference on System Science (HICSS) (pp. 2289–2297). IEEE.
Chung Hua Institution for Economic Research. (2013). Taiwan economic forecasts 2014 (in Mandarin Chinese). http://www.cier.edu.tw/ct.asp?xItem=12375&CtNode=105&mp=1. Accessed 11 March 2014.
Coe, A., Paquet, G., & Roy, J. (2001). E-governance and smart communities a social learning challenge. *Social Science Computer Review, 19*(1), 80–93. Accessed 11 March 2014.
Committee of Digital and Knowledge-based Cities of UCLG. (2012). Smart city study: International study on the situation of ICT, innovation and knowledge in cities. http://www.cities-localgovernments.org/committees/cdc/. Accessed 11 March 2014.
Florida, R. L. (2002). *The rise of the creative class: And how it's transforming work, leisure, community and everyday life*. New York: Basic books.
Giffinger, R., Fertner, C., Kramar, H., Kalasek, R., Pichler-Milanovic, N., & Jeijers, E. (2007). *Smart cities: Ranking of European medium-sized cities.* Vienna: Centre of Regional Science (SRF), Vienna University of Technology.
Gil-García, J. R., & Aldama-Nalda, A. (2013). Making a city smarter through information integration: Angel network and the role of political leadership. In System Sciences (HICSS), 2013 46th Hawaii International Conference on (pp. 1724–1733). IEEE.
Gil-García, J. R., & Pardo, T. A. (2005). E-government success factors: Mapping practical tools to theoretical foundations. *Government Information Quarterly, 22* (2), 187–216.
Hollands, R. G. (2008). Will the real smart city please stand up? Intelligent, progressive or entrepreneurial? *City, 12*(3), 303–320.
IBM. (2013, April). New Taipei City and IBM joins force to build a smart city. *IBM Blue View* (in Mandarin Chinese). http://www-07.ibm.com/tw/blueview/2013apr/8.html. Accessed 11 March 2014.
IBM. (2014). The vote is in: Citizens support 'smart cities' with driverless cars, public service drones and surroundings that sense activities. http://newsroom.intel.com/community/intel_newsroom/blog/2014/02/10/the-vote-is-in-citizens-support-smart-cities-with-driverless-cars-public-service-drones-and-surroundings-that-sense-activities. Accessed 11 March 2014.
Intelligent Community Forum. (2014). The Smart21 communities. https://www.intelligentcommunity.org/index.php?submenu=Awards&src=gendocs&ref=Smart21&category=Events&link=Smart21. Accessed 11 March 2014.
Lau, T. Y., Wang, K. Y., & Atkin, D. (2008). Public service in the information age: A study of e-government in Taiwan. *Media Asia, 35*(3), 190–197.
Liao, Y. (2013, March 27). New Taipei City and IBM initiates smart city challenge. *DigiTimes* (in Mandarin Chinese). http://www.digitimes.com.tw/tw/things/shwnws.asp?cnlid=15&cat=10&cat1=20&id=0000327323_RZS4W7U01CYLWB49FMR9K. Accessed 11 March 2014.
Lin, C., & Lin, P. (2011, January). Interview with New Taipei City Mayor. *Global Views Monthly* (in Mandarin Chinese). http://www.gvm.com.tw/Boardcontent_17225_1.html. Accessed 11 March 2014.
Mortensen, J., Rohde, F., Kristiansen, K., Kanstrup-Clusen, M., & Lubanski, M. (2012). Danish smart cities: Sustainable living in an urban world. Copenhagen Cleantech Cluster. http://www.cphcleantech.com/home/publications/reports/smart-city-report-2012. Accessed 11 March 2014.
Nam, T., & Pardo, T. A. (2011, June). *Conceptualizing smart city with dimensions of technology, people, and institutions*. In Proceedings of the 12th annual international digital government research conference: Digital government innovation in challenging times (pp. 282–291). ACM.

National Statistics of R.O.C. (2013). New Taipei City tourism statistics 2012 (in Mandarin Chinese). http://www.stat.gov.tw. Accessed 11 March 2014.
NTPC. (2013). *Nine lessons in building smart cities in New Taipei City* (in Mandarin Chinese). New Taipei City: Research, Development and Evaluation Commission. Accessed 11 March 2014.
NTPC. (2014). About New Taipei City. http://foreigner.ntpc.gov.tw/_file/2968/SG/44813/D.html. Accessed 11 March 2014.
Poelhekke, S. (2006). Do amenities and diversity encourage city growth? A link through skilled labor. EUI Economics Working Papers # ECO2006/10.
Profile: Appointment as vice premier will put Taoyuan County's Eric Chu to the test. (2009, September 9). *Taipei Times*. http://www.taipeitimes.com/News/taiwan/archives/2009/09/09/2003453131/1.
Rios, P. (2008). Creating "the smart city." http://archive.udmercy.edu:8080/handle/10429/393. Accessed 11 March 2014.
Toppeta, D. (2010). *The Smart City vision: How innovation and ICT can build smart, "livable", sustainable cities*. The Innovation Knowledge Foundation.
Washburn, D., Sindhu, U., Balaouras, S., Dines, R. A., Hayes, N. M., & Nelson, L. E. (2010). *Helping CIOs understand "smart city" initiatives: Defining the smart city, its drivers, and the role of the CIO*. Cambridge: Forrester Research, Inc.
West, D. (2007). *Global e-government, 2007*. Brown University. http://www.brown.edu/academics/taubman-center/research-and-initiatives/e-government. Accessed 11 March 2014.
Yin, R. K. (1984). *Case study research: Design and methods*. Newbury Park: Sage.
Yu, C., & Hwang, A. (2014, February 17). Taiwan market: Chunghwa Telecom aims to kick off 4G operation in July. *DigiTimes*. http://www.digitimes.com/news/a20140217PD212.html. Accessed 11 March 2014.

Kevin Y. Wang received a PhD degree in mass communication from the University of Minnesota; an M.C. degree in digital media and a BA degree in communications from the University of Washington. He is an assistant professor at the Butler University's College of Communication in Indianapolis, USA. Dr. Wang teaches public relations and advertising in the Strategic Communication Department and conducts research on the social and political implication of new communication technologies. He has authored several journal articles and book chapters on e-government, online consultation, online public diplomacy, political campaigns on the Web, and virtual communities. He also regularly presents his work at national and international conferences.

Chao-Ming Wu holds a PhD degree in information management from National Central University, Taiwan. He is the chairperson of Research, Development and Evaluation Commission of New Taipei City Government. Dr. Wu was temporarily transferred to serve in New Taipei City Government from Chung Yuan Christian University to manage New Taipei City Government's policy research and developments, performance control, and evaluations. Dr. Wu also directed several information services projects such as New Taipei City Government's Smart City, Government Cloud, Efficient Government, E-Governance, and Municipal Service Procedure Improvement.

Social Media Experiences at County Level: The Case of the State of Mexico

Rodrigo Sandoval-Almazán and Juan Carlos Núñez Armas

Abstract Social media started to be used in public administration in the 1990s. After the Web 2.0 tools impacted government websites, the new trend of Facebook, Twitter, YouTube, and Instagram were implemented in home pages. However, very little is known regarding the consequences of these new technologies and their relationship with citizens and public officials. The focus of this chapter is to understand the link between social media and smart cities. Since citizens use information technologies to communicate, collaborate, and interact, these same activities can be used in smart cities. In order to understand this linkage, we developed a questionnaire based on two different but complementary frameworks related to social media and smart cities and interviewed the chief information officers (CIO)'s of eight cities in the State of Mexico. Our findings reveal that social media tools are immersed into the smart cities' practices. As a consequence, they are bringing changes in the departments, systems, and relationships among internal agencies. If social media tools are going to be used, then important issues such as inclusion, empowerment, and information quality need to be addressed.

Keywords Smart · Cities · Social media · Twitter · Facebook · Mixed research

R. Sandoval-Almazán (✉)
Universidad Autónoma del Estado de México, Pino Suares Sur 607, Col. Americas, Toluca, 50130 MEX, Mexico
e-mail: rsandovala@uaemex.mx

J. C. Núñez Armas
Universidad Autónoma del Estado de México, Guillermo Menes Servin 107, Residencial, Colón, Toluca, 50120 MEX, Mexico
e-mail: jnuneza2012@gmail.com

J. R. Gil-Garcia et al. (eds.), *Smarter as the New Urban Agenda A Comprehensive View of the 21st Century City,* Public Administration and Information Technology 11,
DOI 10.1007/978-3-319-17620-8_15

1 Introduction

Web 2.0 technologies have recently been implemented in public administration (Chun et al. 2012) and their impact on citizen services and government website developments just started to be analyzed (Sandoval-Almazán et al. 2011). In general, research about the use of technology in local governments started long time ago (Bertot 1998; O'Toole Jr et al. 1998), when expectations of public officials about the efficiency and the use of new information technologies were implemented in government services and processes (Deb 1999). The Japanese government took a lead on this perspective, introducing technology on local government with nonstop initiatives and several innovations for local government digital information according to Sekiguchi and Andersen (1999). Later on, Brown (2001) researched a major police department in the USA analyzing costs, benefits, and productivity, becoming one of the first attempts to research local governments. After this, several scholars continued doing research on technology and government, for example, the challenges of Y2K (Ho and Smith 2001), the government process (Becker et al. 2003; Holden et al. 2003; Hoogwout 2002), and politics and democracy on local governments (Lodge 2003; Parvez 2003).

Another trend is related to information and services. Under this perspective, the case of the Korean Fire Service using the GIS technology and information systems to improve its system is one of the examples used for disaster management (Joo et al. 2004). Following research in this area became a new trend for e-government research (Scholl 2006). Another use of technology in local governments is related to the survey of different adoptions or appropriation of technology (Griffin and Halpin 2005). Other two different surveys about local government chief information officers (CIO) represent the impact of this trend. The first one analyzes cities of more than 50,000 people and reveals that public administrations became revitalized with the use of technology (West and Berman 2001).

Further research has been made on local governments to introduce e-government practices through technology (Kim and Bretschneider 2004). Most of them are related to services (Asgarkhani 2005) other to the implementation of the mobile government (Griffin et al. 2006; Holland and Cahill 2006), but very few are related to the e-administration or the back end of the government process (Attour-Oueslati et al. 2007; Koussouris et al. 2007). Most recently, some government municipal portals have changed their perspective, thus becoming a central agent that handles information, promotes collaboration, and shares data (Sandoval-Almazán and Gil-Garcia 2008). This new kind of government portal is related to a citizen-centric perspective but in a personalized way.

Finally, the use of these technologies in an extraordinary number of local governments, focusing on social media tools such as blogs, Twitter and Facebook platforms, YouTube, and Instagram has reached the vision of the government CIO's and is introduced as a common practice in government portals (Shah and Lim 2011; Bonsón et al. 2012). Most of the Mexican portals have introduced these kinds of elements and started using them on a regular basis (Sandoval-Almazán et al. 2011). However, very little research has been made regarding the effects—advantages or disadvantages—of social media in smart cities.

Social media is the gasoline for smart cities' interaction between citizens and government. Nowadays citizens—that can be called smart citizens—are using smartphones and Web 2.0 tools, such as Twitter and Facebook, to communicate, to do business, and to look for information. Cities in the USA and Spain are also using these tools (Mainka et al. 2014; Mossberger et al. 2013; Pérez et al. 2013). A smart city cannot work without citizens' input; the need for feedback from citizens and administrators has to be permanent and is a part of this change (Zavattaro and Sementelli 2014).

Social media is an intrinsic part of smart cities. Oliveira and Welch (2013) mention this relationship more clearly: "Social media technologies have begun to enter the governmental workplace as tools to accomplish improved public service engagement" (p. 397). The main purpose of social media in smart cities is to provoke citizens' engagement for their actions, policies, and changes. On the other hand, social media has become the new marketing strategy for cities around the world (Huerta-Carvajal and Luna-Reyes 2009; Zhou and Wang 2014).

This overview of different perspectives of the use of social media technologies in local government leads us to the question: How is social media linked to smart cities? How is this link working in cities? What are the smart city functions or tasks that are related to social media? The emergent use of technology in cities is a starting point for innovation and the most common place for citizens to take advantage of this kind of technologies and improve their government relationship (Linders 2012; Wohlers and Bernier 2012).

We will divide this chapter into five sections to describe our research. This introductory section that presents the research topic and its context; the second section describes a literature review about social media tools on smart cities and previous research on municipal level using social media, case studies, and theoretical frameworks related to it; the third section presents the methodology, interview description, questionnaire, and variables. The fourth section describes our findings and discussion. Finally, the fifth section presents our future research on urban agenda and recommendations for the twenty-first century city related to social media.

2 Social Media and Municipalities

The purpose of this literature review is to understand the different theoretical approaches and ideas related to social media and municipalities, which is our first section. It is also important to summarize both research models that are being followed on this research, the Mergel (2013a) and the Chourabi et al. (2012) in the second section of this review.

The use of technology for municipal diffusion of government activities is not new in the scientific literature. The research of Weare et al. (1999), studying 454 websites in California, is one of the first studies with this perspective, in which cities are conceived as data providers through the use of Internet. Later on, the e-MuniS project introduced new ideas for sharing information and best practices for the implementation of online services (Dobrev et al. 2002). Research on the use of computer technology and information systems in the cities started in Poland, when the analysis of 15 polish cities revealed the important trends for the federal government (Pawlowska 2002). More research on city level was done in Japan (Thomson 2002a, b) and European cities (Lodge 2003; Meer and Winden 2003).

Survey studies regarding the use of the Internet technology in cities was also held by Holden et al. (2003) in the USA and by Bochicchio et al. (2004)—who found some interesting needs in Italian cities—for front office procedures, and normative and bureaucratic procedures. A Romanian study of 165 cities analyzing and understanding "intelligent citizen" and the urban perspective are other contributions in this field (Stoica and Ilas 2009). Finally, Agostino's research (2013) with over 119 Italian municipalities is related to the engagement using social media tools. More research on cities and the use of technology to implement services and to promote the online government has been done along time (Ho 2002; Hoogwout 2002; Jacumeit 2002). Most of them refer to the idea of one-stop online services and the diffusion of government activities.

The case of Corpus Christi, Texas, presents a different research approach based on the idea of integrating changes and citizen interaction to enable the online government transformation researching over e-services for citizens and their diffusion (Jorgensen and Cable 2002). Later on, another study of the Los Angeles county related technology with services (Freeman 2004). Later on, in the UK, a change in the concept introduced the e-citizen into the CRM formula, different from the traditional perspective of citizen as a client (Richter et al. 2004).

The Ontario case, using data to predict and increase citizen contact, is a complementary perspective of the use of technology that started supporting the relationship between citizens and governments (Reddick 2005).

This research path led us into the social media use in municipal public administration. An important study that links this novel technology with citizens and government is the Italian IRIS online platform in which citizens can report the urban maintenance problems and expect an answer from the government. This adaptation of local bureaucracy using technology was implemented in Venice (Alfano 2011). Another contribution was provided by England in which social media was used to understand and to reveal government actions (De Saulles 2011). Ideas to understand the problem of transparency and social media openness in municipalities come from the research of Swedish municipalities, where authors stated the difficulties to align such purposes—transparency—with the process and fast updation of social media (Klang and Nolin 2011).

Another contribution in order to understand the use of social media, transparency, and accountability is proposed by Bonsón et al. (2012), whose research introduces the idea of an e-participation and social media. Gil-Garcia and Aldama-Nalda (2013) also contributed with the case in Mexico City with the integration of the Angel Network Technology Program. A new generation of the use of technology in governments is the mobile technology. The launch of the Arlington City App was a success before the Super Bowl and helped the government to promote more services (Raths 2011). This conception leads to understand the city as a platform for interaction more than just services (Walravens and Ballon 2011; Walravens 2013).

Once we have addressed several perspectives on the literature—technology and government, diffusion and municipalities, social media, and municipalities—we did not find any specialized research on the impact of social media at this level of government. The purpose of this chapter is to contribute with empirical evidence in this field. Models for this research consider the smart cities' initiative and the social media tools. For the smart cities, Chourabi et al. (2012) were selected because they clearly state the technological components which we want to study. On the other hand, the Mergel (2013) model is the most recent contribution to the social media in public administration. This model describes three main components—participation, collaboration, and transparency—and two inherent or implicit components—empowerment and inclusion.

Participation 2.0 refers to the use of Internet and social media technologies to engage citizens with governments (p. 147) to inform, consult, collaborate, and include.

Collaboration 2.0 is considered as a new form of information sharing, despite the bureaucratic rules and hierarchy problems (p. 181).

Transparency 2.0 refers to "the use of social technologies to increase transparency and accountability in governments" (p. 211). We translated and considered this component as a query component for our research on social media rather than the whole dimension on transparency that in Mexico has a wider implication.

Empowerment: According to Mergel "Participation 2.0 technologies can also be used to empower citizens, that is to place decision-making authority in their hands…" (p. 151).

Inclusion: Mergel's definition "… participation 2.0 technologies to engage the public in information and to give citizens more influence over decision making" (p. 150).

Chourabi et al.'s (2012) model called smart city initiatives framework was developed by researchers from four countries—Canada, the USA, Mexico, and China. It has eight components in an integrative framework: (1) management and organization; (2) technology; (3) governance; (4) policy context; (5) people and communities; (6) economy; (7) built infrastructure; (8) natural environment.

Each component has its own items which were developed and tested by this research team. We used these model components to analyze our results from the interview sessions.

3 Method

The main purpose of this research is to explore the use of social media in Mexican cities. For such reason, we conducted several interviews with CIO's in the Mexican state of Mexico—located in central Mexico—very close to Mexico City. This section describes the research methods, samples, and variables.

Table 1 Research models for the interview instrument

Smart cities' framework (Chourabi et al. 2012)		Q	Social media in public administration (Mergel 2013)		Q
Component	Description		Component	Description	
Management and organization	Organizational factors	1,6,7,10,11	Information	Provide general information	1–6,18,
Technology	IT skills	8,18	Query	Ask questions	19,20
Governance	Laws, administrative rules, collaboration	4,5,17	Collaboration	Collaborate with government officials	4,5,7,12 13,21,23
Policy	Policy context	2,3	Inclusion	Online participation for public affairs	14–16
People and communities	Digital divide, participation	9,12–16,19	Empowerment	Promote leadership	22
Economy	Business values				
Built infrastructure	ICT infrastructure, security, costs				
Natural environment	Sustainability and resources				

Exploratory research is usually conceived as a way to understand new research paths or novel fields. We choose the qualitative tool of nonstructured, direct interviews in order to explore ideas, confirm research paths, and establish new ideas (Crano and Marilynn 2002; Flick 2002). Research on the use of social media in smart cities is emerging—as described in the previous section. Accordingly we conceived this trend as an exploratory one.

In order to analyze this particular trend, we chose two theoretical frameworks to analyze our interviews. The first one is the smart cities' framework developed by the Center for Technology from the Government (CTG; Chourabi et al. 2012); second, we took the social media framework for public administration developed by Mergel (2013a). We developed an instrument with 15 open questions considering both models and contacted 11 municipalities in the State of Mexico and scheduled interviews of 30–40 min. At the end, we only got eight interviews that were recorded and transcribed for analysis purposes. Table 1 presents the two model variables and the number of question.

3.1 Sample

Some literature states that there is no difference between small cities and big cities for the use of social media. We chose populated areas near Mexico City for our convenience and because they differ about income, education level, and computer access. Our sample was determined by the different municipalities available for interviews during October and November 2013. Finally, we got eight complete interviews with CIOs or responsible ones of the technology department in the following municipalities: Atizapán de Zaragoza, Coacalco, Ecatepec, Huixquilucan, Metepec, Naucalpan, Nezahualcóyotl, and Tlalnepantla.

In order to analyze the interviews, we followed the traditional process for the qualitative research. The first step was the transcription of the interviews by a research assistant; the second step was to categorize and analyze the content using categories from both theoretical frameworks; the third step was to locate each category on the general analysis and summarize it. The last step was to analyze each category and to link it to the smart cities' research.

4 Findings and Discussion

Our findings are organized following the two research models. In the first section, we presented findings related to Chourabi et al. (2012) smart cities' framework containing eight components. The next section presents findings related to Mergel's components (2013), in order to analyze the social media. In both cases, we present evidence from the interviews, interpretations, and extracts accordingly to each case.

4.1 Management and Organization

This component explains the way a county gets organized and distributes and decides about the social media. The findings reveal two main trends: the first one is the most common—six municipalities have a community manager that centralizes all the activities from the different social media platforms. The manager is usually a part of the speaker or communication agency of the municipality. The second trend is that each internal agency has its own Twitter and Facebook account.

A second level of organization used by municipalities—at least two of them, Naucalpan and Coacalco—is that they have distributed responsibilities in each internal agency. For example, Naucalpan has 23 community managers for each one of their inside agencies: public services, light and power. They share this information and coordinate it with the main community manager, who is usually responsible for the municipal account. This can lead to a problem of multiple and conflicting goals among different agencies that share the tasks of social media.

Most of the social media strategies and policies are aligned to two accounts: the personal account—city major—and the institutional account. On the other hand, municipalities of Nezahualcóyotl, Metepec, and Huixquilucan social media accounts are linked to the speaker and the media department. So, it is probable that the media strategies are linked to these accounts; however we did not find any evidence to support this.

Monitoring or following other social media actors, such as neighbor communities, federal or local government officers, and mass media accounts, they are made without any systematic or strategic approach in any case.

4.2 Technology

The large majority of the interviewed cities are using common social media tools such as: YouTube, Twitter, and Facebook platforms to interact with their citizens. The focus of this component is IT skills. Some of the people related to social media accounts in interviewed municipalities are webmasters, community managers, or people with communication skills in traditional media. In the cases of Huixquilucan, Ecatepec, Metepec, Nezahualcóyotl, and Huixquilucan, they do not have a technological profile. They have people under their command to report and help in this kind of tasks. None of them receive extra training for social media or government networking.

The social media tool that has spread the most and is mostly used, according to all the interviewed ones, is Facebook. Accordingly to the Naulcapan's analysis, the Facebook platform is used by low- and middle-income population segments and Twitter is used in higher income segments. Coalcaco mentioned that "Twitter is more political focused and more people know how to use Facebook."

4.3 Governance

Most of the social media accounts started in January 2013 with the new period of majors—every period lasts 3 years with no immediate reelection. The municipality of Ecatepec is the exception, since it continues with the same party and the social media administration. This case has the largest number of followers and friends from the sample, because it started their network presence in 2009.

Municipal terms of 3 years make the administration of social media accounts difficult, as referred by one interviewee: "every administration change is difficult because of getting social media passwords and codes from the previous administration" (Naucalpan interview).

This component is related to collaboration and administrative rules. The cases of Naulcalpan, Metepec, Nezahualcóyotl, and Coacalco divide the use of social media accounts into different agencies. The case of Naulcapan has 23 links with internal agencies to share and collaborate. Some rules or content decisions, time for posting, and internal campaigns are held in regular meetings once a week with the main CIO or the speaker and the mass media department.

Naucalpan, Tlalnepantla and Nezahualcoyotl use tools to measure social media impact, such as Hoot Suite and Social Bro (see Table 2). Naulcalpan has determined that an average of 250 friends on Facebook that give a "like" could multiply the viral effects of the information and give a possible number to measure the impact.

Table 2 Social media tools used by Mexican municipalities on the sample

Municipality	Social media tool for metrics or statistics	
Naucalpan	www.hootsuite.com	www.socialbro.com
Nezahualcóyotl	www.nic.mx, www.akky.mx	www.bitly.com
Tlalnepantla	www.klout.com	www.tweetstats.com

4.4 Policy

Another important finding is that the city major has an important role to promote and impulse the social media tools for their municipality in at least three municipalities—Naucalpan, Valle de Mexico, and Metepec. Usually, the major's account has the same number of followers and friends as their municipal account.

A defined policy or main strategy to rule the use of the social media tools was determined by Naulcapan. It has defined a mission statement for the social media area "attend citizens to avoid bureaucracy in procedures and petitions." This means to reduce the traditional time and the space path for any transactions to a direct petition in the social media platforms. A second example is Tlalnepantla: "provide information and become a bidirectional link with citizens." Nezahualcóyotl understands the use of social media as "a more direct approach to citizens."

Tlalnepantla is the only one to mention a concrete goal of reaching 25,000 followers on Twitter. The rest of the municipalities are in the process of planning general goals such as increasing the number of followers, information of facts and government actions, and introducing government information to 100 % of the social media platforms.

Regarding the social media policies and rules for content sharing, strategy, or tactics, most of them are determined by the people in charge of these departments. There is no evidence to support that city councils or city governments interfere or establish certain rules for content sharing on social media.

An interesting case of policy relation with other agencies occurred in Nezahualcóyotl that reached 60,000 queries or like setups and alert for the federal ministry of security—Homeland Security Agency. We asked the reason for such peak in the social media platforms. The answer was the promotion of a rock concert by the municipality. Same kinds of peaks occurred in Naucalpan with the festival "Luminaria" which is held every year.

4.5 People and Communities

User interaction of social media in municipalities has been heterogeneous and with multiple actions. For example, Atizapán just sends information and provides very few answers. Metepec answers immediately through the platforms. Ecatepec just answers during office hours; Ecatepec, Huixquilucan, and Nezahualcóyotl answer on Twitter and Facebook using chats. Tlalnepantla requires filling out an online format to have an answer.

Table 3 Technological infrastructure of municipalities. (Source: INEGI 2010 Census)

	Inhabitants	Households	House with computer	%	House with Internet	%	Cell phone
ESTADO	15,175,862	3,687,193	1,162,156	32	811,030	22	2,467,712
Atizapán de Zaragoza	489,937	127,487	59,335	47	48,491	38	96,629
Coacalco	278,064	74,048	38,610	52	28,479	38	60,511
Ecatepec	1,656,207	419,207	145,044	35	101,562	24	285,489
Huixquilucan	242,167	59,526	28,719	48	24,339	41	45,797
Metepec	214,162	53,521	28,858	54	23,446	44	42,204
Naucalpan	833,779	212,677	86,669	41	68,079	32	156,081
Nezahual-cóyotl	1,110,565	280,401	98,818	35	72,382	26	188,559
Tlalnepantla	664,225	171,673	77,018	45	59,705	35	127,151

Naucalpan has a different perspective; this municipality fosters citizen relationship through conversations. They ask: How does bureaucracy get along with you? How are you? Was the time for solving your doubts or transactions short? Also in this case, they have published "House Rules" on a Facebook wall that describes: content rules, schedule for answer petitions, language rules, business rules and ethics, and policy rules for social media interactions.

4.6 Built Infrastructure

All the interviewed municipalities have Internet access in their cities. Metepec has 54% of their households with a computer at home. Coacalco presents a similar percentage with 52% according to the last census of 2010 (INEGI 2010). Global infrastructure is described in Table 3. All of these municipalities are above the State of Mexico's average Internet access—22%. Most of them have a regular average of income and a large population directly linked to the Mexico City metropolitan area.

4.7 Economy and Natural Environment

We analyzed these two variables of smart cities' model because they have an indirect approach to the social media interaction model of this study. On the economy perspective, very few cases, like Naucalpan and Coacalco support commerce or any other economic activity using their social media tools. The rest of the interviewee mentioned that they do not promote advertising or any other similar activity rather than their own events.

The natural environment component is different. Most interviewees referred promoting environmental activities such as water use and disposal or trash collections along the city; but especially two municipalities—Naucalpan and Coacalco—mentioned the program of "adoption of dogs" as a great success using social media.

The program "adoption of dogs" has become an alternative instead of sacrificing street dogs. The program has a great number of people interacting with Twitter or Facebook platforms in order to adopt an animal and also for getting more information about the adoption process. The rest of the programs are directly linked to public services claims—garbage collection, street cleaning, water supply, etc.—and for most of the citizens it is the only way to promote their own interest or solve their problems.

The smart cities' framework components (Chourabi et al. 2012) allow us to understand the social media tools interaction with the smart city perspective. The components of management and organization describe that municipalities create new departments or organize their communication departments to allow interaction with social media. The technology component describes the prior use of Facebook as a main platform and the lack of training and specialized personnel for addressing such new tools. On the governance component, very few municipalities have a strategy or goals to determine their actions through the social media and are regulated only by the personnel in charge of this action.

The policy context describes that social media tools are linked to the city council or city government and most of them depend on the city major knowledge of social media and the support of this tool. Finally, most of the interactions with people and communities are different according to their own strategy and just one municipality has clear rules and objectives for citizen interactions.

The combination of all these components represents a different perspective of the impact of social media tools on smart cities' actions. Next section will focus on the social media impact and the relation with citizens.

4.8 Mergel (2013a), Model of Social Media Interaction

The four components proposed by Mergel (2013a) to understand the impact of social media in public offices were applied on federal and local governments. No prior research is known in municipal areas. However, we believe these tools have enough potential to be applied at this government level with important results. Our findings are as follows:

4.8.1 Information

The main activity of all interviewed municipalities is to share information, promote government activities, and create a public image of the local government. Only Naucalpan shares a bidirectional perspective, but the rest are more focused in providing one-way information to citizens.

The target "public" was analyzed in the Nezahualcóyotl municipality that has targeted young people—less than 35 years. Tlalnepantla mentioned that they have a map of public segments in order to plan content, schedule tweets, and promote their messages in an effective way.

Another case is the use of Twitter and Facebook accounts to provide information related to the water supply program and mainly when the Cutzamala System informs about water shortcuts, because the system is going to be repaired. They use the technology—among other mass media—to alert citizens about this issue in order to take important measures to prevent water shortages in their houses. This municipality also provides an online community manager to solve doubts or questions related to the public services using the social media accounts in order to reduce time and produce effectiveness.

4.8.2 Query or Information Requests

The interviewed municipalities mentioned that most of the information requests, claims, and messages received are concerning three main areas: security, traffic in the city, and information about municipal cultural events. The topic of security is based on crime, felonies, or robbery reports. In the topic of traffic, most complaints are about delays, traffic lines, or car crashes.

4.8.3 Collaboration

The collaboration perspective could be the most notorious one in the use of social media. However, from the municipal practice, there are very few examples. Most of the interviewees refer to this lack of participation to the passive behavior of their users, but it could also be considered as a result of the lack of interest in public affairs. For example, the Atizapán case uses the social media tools to promote canine vaccine for street dogs or abandoned pets with great success; another example is to ask for more patrol control in certain areas or to request certain services.

The Metepec case is quite different where the government requested help for social disasters like Guerrero and Oaxaca with great success. A similar call for help was for the assistance for the annual festival Quimera, where they received 57 proposals for help in different instances.

4.8.4 Inclusion

The "good morning" program is an original idea from the Naucalpan municipality. It uses different ways—jokes, photos, phrases—to say good morning to their followers. This idea has been selected by citizens as one of the most important ones and they feel as part of their city just because of these greeting messages.

Two more cases are related to the inclusion of citizen in public policy through the use of social media platforms. The first one is the program: "Sunday on a bike." The idea of closing downtown streets to regular traffic on Sundays and allowing riding bikes came from a Twitter account in the Tlalnepantla municipality. The number of tweets and discussions over this topic created the program.

A second example is the online public hearing. Usually this kind on hearings are held by the city major 1 day a week, but the Naucalpan municipality created the online version as a result of petitions from Facebook and Twitter.

4.8.5 Empowerment

Interviewees answered that there was no support to the question regarding user participation or discussion in public policies to become government actions. A closer case was Tlalnepantla with the participative budget action, which invites citizens to an online training to promote such practice. However, very less is directly related to empower the citizens using social media.

5 Conclusions and Future Research

Smart cities and social media are linked through the information needs of citizens, bureaucracy, and internal process. In order to answer our research question: How is social media linked to smart cities? We made an exploratory research using two theoretical frameworks that complement each other. One framework was only related to social media (Mergel 2013b); the second framework focused on smart cities (Chourabi et al. 2012). Our findings present evidence that there is a link between smart cities and social media in at least four different aspects: (1) New agencies and systems were created in bureaucratic process to support the social media strategies (management and organization). (2) Technology improvements to collect needs and answers for the social media platforms were required in municipalities. (3) The use of social media to promote environmental practices or support government decisions was constant among the interviewees (natural environment). (4) Some cities, such as Naucalpan and Tlalnepantla, tried to create an online community through social media and engage people in the activities and interactions with the government (people and communities).

An important contribution of this chapter is to state that social media could be linked with smart cities' practices or implementation strategies. On the other hand, our findings present some weaknesses in smart cities which could be considered to be improved using these tools, for example, the lack of clear strategies to implement Twitter and Facebook into their systems, the lack of citizen empowerment practices, and more inclusion of citizens in government discussion policies. All of this could be held through social media and promote collaboration with a clearer media strategy.

This qualitative evidence supports the idea that smart cities can benefit from the use of social media. Furthermore, citizens use social media and smart cities can take advantage of this. Nevertheless, this research has its own limitations like the number of cases—eight interviews—and the location, only in the State of Mexico; but the findings are consistent in the whole sample.

Future research opportunities on this path are numerous. A first one is to confirm the use of social media tools in different municipalities, considering size, culture, indigenous context, and location (north and south). Another research could be to analyze quantitative data from the use of social media and smart cities' best practices such as the implementation of video cameras around the city and to request citizens to monitor cameras as a crowdsourcing practice.

A third path of the research is the organizational transformation that produces the input from social media. How is the city government able to process tweets, posts, likes, or videos? How cost effective could the use of social media in a smart city be? How many people can be concentrated on social media practices?

More research can be done on content analysis of the social media tools, metrics for smart cities' practices and social media feedback, etc., but all of these depend on the evolution of smart cities' progress, their own transformation and needs. Social media can help showing the changes and the gradual transformation if bureaucracy is smart enough to capture the needs, views and ideas from the people who use social media.

References

Agostino, D. (2013). Using social media to engage citizens: A study of Italian municipalities. *Public Relations Review, 39,* 232–234. doi:10.1016/j.pubrev.2013.02.009.

Alfano, G. (2011). Adapting bureaucracy to the Internet. The case of Venice Local Government. *Information Polity, 16,* 5–22.

Asgarkhani, M. (2005). The effectiveness of e-service in local government: A case study. *Electronic Journal of E-Government, 3,* 157–166.

Attour-Oueslati, A., Dufresne, D., & Longhi, C. (2007). *The development of the local e-administration: Empirical evidences from the French case*. Berlin: Springer.

Becker, J., Algermissen, L., & Niehaves, B. (2003). Processes in e-government focus: A procedure model for process oriented reorganisation in public administrations on the local level. In R. Traunmüller (Ed.), *Electronic government: Second international conference, EGOV 2003, Prague, Czech Republic, September 2003. Proceedings, lecture notes in computer science* (pp. 147–150). Berlin: Springer.

Bertot, J. C. (1998). Challenges and issues for public managers in the digital era. *Public Manager, 27,* 27.

Bochicchio, M. A., Celentano, M. G., & Longo, A. (2004). Innovation needs in the e-government scenario: A survey. In R. Traunmüller (Ed.), *Electronic government, lecture notes in computer science* (pp. 347–354). Berlin: Springer.

Bonsón, E., Torres, L., Royo, S., & Flores, F. (2012). Local e-government 2.0: Social media and corporate transparency in municipalities. *Government Information Quarterly, 29,* 123–132.

Brown, M. M. (2001). The benefits and costs of information technology innovations: An empirical assessment of a local government agency. *Pubic Performance & Management Review, 24,* 351–366.

Chourabi, H., Nam, T., Walker, S., Gil-Garcia, J. R., Mellouli, S., Nahon, K., Pardo, T. A., & Scholl, H. J. (2012). *Understanding smart cities: An integrative framework.* Proceedings of the 45th Hawaii International Conference on System Sciences, HICSS '12. IEEE Computer Society, Washington, DC, USA, pp. 2289–2297. doi:10.1109/HICSS.2012.615.

Chun, S. A., Luna-Reyes, L. F., & Sandoval-Almazán, R. (2012). Collaborative e-government. *Transforming Goverment: People, Process and Policy, 6,* 5–12. doi:10.1108/17506161211214868.

Crano, W., & Marilynn, B. (2002). *Principles and methods of social research*. New Jersey: Lawrence Erlbaum.

De Saulles, M. (2011). *Social media and local government in England: Who is doing what?* Presented at the 11th European Conference on eGovernment (ECEG 2011), Academic Conference International Ltd., 2011, pp. 187–192.

Deb, G. K. (1999). Electronic governance—A vehicle for the new world order. *Electronics Information & Planning, 27,* 29–33.

Dobrev, B., Stoewer, M., Makris, L., & Getsova, E. (2002). *E-MuniS—Electronic municipal information services—Best practice transfer and improvement project: Project approach and intermediary results*. Berlin: Springer.

Flick, U. (2002). *An introduction to qualitative research*. London: Sage.

Freeman, R. J. (2004). The Los Angeles county experience: A field study of information technology issues challenging local government. *Journal of E-Government, 1,* 93–101.

Gil-Garcia, J. R., Aldama-Nalda, A., (2013). Making a city smarter through information integration: Angel network and the role of political leadership. 46th Hawaii International Conference on System Science (HICSS 2013). pp. 1724–1733.

Griffin, D., & Halpin, E. (2005). An exploratory evaluation of UK local e-government from an accountability perspective. *The Electronic Journal of E-Government, 3,* 13–28.

Griffin, D., Trevorrow, P., & Halpin, E. (2006). Using SMS texting to encourage democratic participation by youth citizens: A case study of a project in an English local authority. *The Electronic Journal of E-Government, 4,* 63–70.

Ho, A. T. K. (2002). Reinventing local governments and the e-government initiative. *Public Administration Review, 62,* 434–444.

Ho, A. T.-K., & Smith, J. F.(2001). Information technology planning and the Y2K problem in local governments. *American Review of Public Administration, 31,* 158–180.

Holden, S. H., Norris, D. F., & Fletcher, P. D. (2003). Electronic government at the local level: Progress to date and future issues. *Public Performance and Management Review, 26,* 325–344.

Holland, C. P., & Cahill, M. (2006). 3G technology in local government: Case examples of business process change and strategic innovation. Mobile government consortium international LLC.

Hoogwout, M. (2002). *Organizing for online service delivery: The effects of network technology on the organization of transactional service delivery in Dutch local government*. Berlin: Springer.

Huerta-Carvajal, M. I., & Luna-Reyes, L. F. (2009). City boosterism through Internet marketing: An institutional perspective. In M. Gascó-Hernández & T. Torres-Coronas (Eds.), *Information communication technologies and city marketing: Digital opportunities for cities around the world* (pp. 130–151). Hershey: IGI Global.

INEGI. (2010). Censo de Población y Vivienda 2010 (Population and Households Census). http://www.inegi.org.mx/sistemas/olap/proyectos/bd/consulta.asp?p=17118&c=27769&s=est#.

Jacumeit, V. (2002). *Administration 2000—Networking municipal front and back offices for one-stop government*. Berlin: Springer.

Joo, I.-H., Kim, K.-S., & Kim, M.-S. (2004). *Fire service in Korea: Advanced emergency 119 system based on GIS technology*. Berlin: Springer.

Jorgensen, D. J., & Cable, S. (2002). Facing the challenges of e-goverment: A case study of the city of Corpus Christi, Texas. *S. A. M. Advanced Management Journal, 67,* 15–21.

Kim, H. J., & Bretschneider, S. (2004). *Local government information technology capacity: An exploratory theory*. Proceedings of the 37th Annual Hawaii International Conference on System Sciences (HICSS'04)—Track 5. Computer Societry Press, Island of Hawaii (Big Island), p. 50121b (1–10).

Klang, M., & Nolin, J. (2011). Disciplining social media: An analysis of social media policies in 26 Swedish municipalities. *First Monday, 16,* 1–21.

Koussouris, S., Charalabidis, Y., Gionis, G., Tsitsanis, T., & Psarras, J. (2007). *Building a local administration services portal for citizens and businesses: Service composition, architecture and back-office interoperability issues*. Berlin: Springer.

Linders, D. (2012). From e-government to we-government: Defining a typology for citizen co-production in the age of social media. *Government Information Quarterly (Selections from the 12th Annual International Conference on Digital Government Research (dg.o2011)), 29,* 446–454. doi:http://dx.doi.org/10.1016/j.giq.2012.06.003.
Lodge, J. (2003). Towards an e-commonwealth? A tool for peace and democracy? *Round Table,* (372), 609–622.
Mainka, A., Hartmann, S., Stock, W. G., & Peters, I. (2014). *Government and social media: A case study of 31 informational world cities.* 47th Hawaii International Conference on System Science (HICSS 2014). p. 10.
Meer, A. van der, & Winden, W. van (2003). E-governance in cities: A comparison of urban information and communication technology policies. *Regional Studies, 37,* 407–419.
Mergel, I. (2013a). *Social media in the public sector: A guide to participation, collaboration and transparency in the networked world* (1st ed.). Australia: Jossey-Bass.
Mergel, I. (2013b). Social media adoption and resulting tactics in the U.S. Federal Government. *Government Information Quarterly, 30,* 123–130. doi:10.1016/j.giq.2012.12.004.
Mossberger, K., Wu, Y., & Crawford, J. (2013). Connecting citizens and local governments? Social media and interactivity in major U.S. cities. *Government Information Quarterly, 30,* 351–358. doi:10.1016/j.giq.2013.05.016.
O'Toole Jr., L. J., Brown, M. M., & Brudney, J. L. (1998). Implementing information technology in government: An empirical assessment of the role of local partnerships. *Journal of Public Administration, 8,* 499–525.
Oliveira, G. H. M., & Welch, E. W. (2013). Social media use in local government: Linkage of technology, task, and organizational context. *Government Information Quarterly, 30,* 397–405. doi:10.1016/j.giq.2013.05.019.
Parvez, Z. (2003). Local democracy shaping e-democracy. In R. Traunmüller (Ed.), *Electronic government: Second international conference, EGOV 2003, Prague, Czech Republic, September 2003. Proceedings, lecture notes in computer science* (pp. 63–68). Berlin: Springer.
Pawlowska, A. (2002). Computing in Polish local administration—New technology, old experience. *Information Polity: The International Journal of Government Democracy in the Information Age, 7,* 49–64.
Pérez, M. D. C. C., De Rosario, A. H., Sáez Martín, A., Fernández, A., & Martínez-Monje, P. (2013). *The Spanish smart cities and the social media.* 7th International Conference on Methodologies, Technologies and Tools Enabling E-Government (MeTTeG 2013). pp. 11–20.
Raths, D. (2011). Cities add citizen engagement mobile apps to their portfolios. *KM World, 20,* 10–11.
Reddick, C. G. (2005). Citizen-initiated contacts with government: Comparing phones and websites. *Journal of E-Government, 2,* 27–53.
Richter, P., Cornford, J., & McLoughlin, I. (2004). The e-citizen as talk, as text and as technology: CRM and e-government. *Electronic Journal of E-Government, 2,* 207–218.
Sandoval-Almazan, R., & Gil-Garcia, R. (2008). E-government portals in Mexico. In A.-V. Anttiroiko (Ed.), *Electronic government.* Hershey: IGI Global.
Sandoval-Almazan, R., Mendoza-Colín, J., Gil-Garcia, J. R., Luna-Reyes, L. F., & Luna-Reyes, D. E. (2011). *Local government websites: Cases of innovation and best practices in Mexico.* Proceedings of the 12th Annual International Digital Government Research Conference: Digital Government Innovation in Challenging Times. ACM, College Park, Maryland, pp. 349–350. doi:10.1145/2037556.2037620.
Scholl, H. J. (2006). *Is e-government research a flash in the pan or here for the long shot?* Berlin: Springer.
Sekiguchi, Y., & Andersen, K. V. (1999). Information systems in Japanese government[a]. *Information Infrastructure and Policy, 6,* 109.
Shah, B. P., & Lim, N. (2011). *Using social media to increase e-government adoption in developing countries.* Proceedings of the 5th International Conference on Theory and Practice of Electronic Governance, ICEGOV '11. ACM, Tallinn, Estonia, pp. 205–213.

Stoica, V., & Ilas, A. (2009). Romanian urban e-government. Digital services and digital democracy in 165 cities. *Electronic Journal of E-Government, 7,* 171–182.
Thomson, C. S. (2002a). Enlisting on-line residents: Expanding the boundaries of e-government in a Japanese rural township. *Government Information Quarterly, 19,* 173–188.
Thomson, C. S. (2002b). Recruiting cyber townspeople: Local government and the Internet in a rural Japanese township. *Technology in Society, 24,* 349–360.
Walravens, N. (2013). The city as a service platform: A typology of city platform roles in mobile service provision. Proceedings of 19th Americas conference on Information System (AMCIS).
Walravens, N., & Ballon, P. (2011). *The city as a platform: Exploring the potential role(s) of the city in mobile service provision through a mobile service platform typology.* Presented at Tenth International Conference on Mobile Business (ICMB), pp. 60–67. doi:10.1109/ICMB.2011.22.
Weare, C., Musso, J. A., & Hale, M. L. (1999). Electronic democracy and the diffusion of municipal web pages in California. *Administration & Society, 31,* 3–27.
West, J. P., & Berman, E. M. (2001). The impact of revitalized management practices on the adoption of information technology: A national survey of local governments. *Public Performance & Management Review, 24,* 233–253.
Wohlers, T. E., & Bernier, L. (2012). Innovative city hall: A comparative case study of policy adoption in the U.S. and France. *Journal of Information Technology & Politics, 9,* 415–432.
Zavattaro, S. M., & Sementelli, A. J. (2014). A critical examination of social media adoption in government: Introducing omnipresence. *Government Information Quarterly.* doi:10.1016/j.giq.2013.10.007.
Zhou, L., & Wang, T. (2014). Social media: A new vehicle for city marketing in China. *Cities, 37,* 27–32. doi:10.1016/j.cities.2013.11.006.

Rodrigo Sandoval-Almazán is a professor with the School of Political and Social Sciences of the Autonomous University of the State of Mexico (UAEM) in Toluca, México. He has been a professor at the Graduate School of Public Administration (EGAP) and Business Administration (EGADE) of the Institute of Technology and Superior Studies of Monterrey (ITESM). He is a member of the National Researchers System Level 2. He has authored or coauthored more than 30 research papers. In 2013, he won the second Latin American Award for Public Administration for the book, the *Long Walk of Open Government: Theory, Metrics and Future.* Former editor of the academic journal RECAI (Journal of Studies on Accounting, Management and Informatics). Dr. Sandoval Almazan is a member of the Digital Government Society Board and NovaGob. His research interests include e-government, information technology organizations, social networks in government, and digital divide.

Juan Carlos Núñez Armas is a civil engineer, was alderman to the City of Toluca from 1991 to 1993, Member of Parliament of the State of Mexico from 1997 to 2000 and Mayor of Toluca that year to 2003; he was elected federal deputy for the XXVI Federal Electoral District of the State of Mexico to the LIX Legislature for the period ending in 2006.

Part III
Smart Cities and Citizen Participation

The Role of Citizen Participation in Municipal Smart City Projects: Lessons Learned from Norway

Lasse Berntzen and Marius Rohde Johannessen

Abstract In this chapter, we examine the role of citizen participation in the development of smart cities through literature and example cases from Norwegian cities. We present an overview of technologies used for participation, including their strengths and weaknesses, discuss how different types of projects should be handled differently in the decision-making process and present recommendations for how practitioners can set up citizen participation projects in Smart City initiatives. We present three different categories of participation: citizen competence and experience, data collection through citizens' use of technology and participation as democratic value. Further, we discuss how these categories can be understood in terms of, who sets the agenda and who makes the final decisions in order to frame the project internally in the municipality and externally so that citizens participating know what the outcome of the project will be. Finally, we offer suggestions for technologies that could be used to collect citizen input in each of the three categories of participation.

Keywords Smart cities · Citizen participation · Technology choice · Multichannel communication · Proprietary systems · Social media

1 Introduction

Smart city is a label or statement to indicate that the city is actively pursuing use of modern technology to increase the quality of life in urban space, both by improving the environmental conditions and delivering better services to its citizens. To create the best conditions for urban population, a strategy is needed. Dameri and Coccia (2013) claim such strategy is used to implement actions, projects and programmes aiming at different goals, such as:

L. Berntzen (✉) · M. R. Johannessen
Department of Business and Management, Buskerud and Vestfold University College, Postboks 235, Drammen, 3603, Kongsberg, Norway
e-mail: lasse.berntzen@hbv.no

M. R. Johannessen
e-mail: marius.johannessen@hbv.no

J. R. Gil-Garcia et al. (eds.), *Smarter as the New Urban Agenda A Comprehensive View of the 21st Century City,* Public Administration and Information Technology 11,
DOI 10.1007/978-3-319-17620-8_16

- To improve environmental quality in urban space, reducing CO_2 emissions, traffic and waste;
- To optimize energy consumption by building efficiency and renewable energy production;
- To increase quality of life, delivering better public and private services, such as local public transport, health services and so on.

Citizen participation may be an important factor in implementing such actions, projects and programmes. Such participation may fall within the following three categories:

Citizens' Competence and Experience First, citizens have a lot of experience that can aid development of better plans, solutions and services. Some citizens may even have important competence that the city does not possess. By listening to the citizens, potential problems can be addressed early and thereby the risk of failure can be reduced.

Citizens as Data Collectors Second, citizens can help collecting environmental and other data by using smartphones and other technologies. One example is "FixMyStreet.com", a web application to report problems with roads and other infrastructure. Another example is the Green Watch project (Ratti and Townsend 2011). The project distributed 200 smart devices to citizens of Paris. The devices sensed ozone and noise levels as the citizens lived their normal lives and the results were shared through a mapping engine. The project showed how a grassroots-sensing network could reduce monitoring costs dramatically and at the same time engage citizens in environmental monitoring and regulation.

Citizens as Democratic Participants Finally, active participation enhances democracy, especially on the local level. Participation is not only about taking part in decision-making processes but also about building sustainable local communities where citizens care for each other.

The objective of this chapter is to present lessons learned and suggestions for citizen participation in "smart city" projects. We argue that citizen participation is important to achieve better solutions and services, and to promote democratic involvement. Further, we argue that effective participation projects on a municipal level should involve a combination of proprietary software and social media, as well as a combination of online and offline communication channels. Thus, our research question for this chapter is "How can municipalities improve Smart City initiatives through citizen participation?"

The remainder of the chapter is structured as follows: First, we present our research approach for the chapter. Second, we present a brief review of literature related to "smart cities", citizen participation and technologies used in citizen participation projects. Although not a comprehensive literature review, these sections still present a good overview of existing research in the field. Third, we present cases from Norway, which serve as examples of how government may utilize technology to enhance participation, and how citizen participation may influence decision-making process. We then discuss these findings in relation to the categories presented above and the models of democracy presented in Sect. 3.1. Finally, we present lessons learned and suggestions for practitioners.

1.1 Research Approach

Three interpretive case studies form the empirical basis of the chapter. These three cases were selected as they present interesting findings related to participation on the municipal level and cover a variety of actors and activities. The data in the three cases consist of interviews, online content from various social media and proprietary platforms, as well as municipal documents concerning each individual case. Individual findings from the three cases were discussed over three meetings between the authors and compiled into a list of citizen participation issues. The topic for this chapter is derived from those discussions.

The findings presented in this chapter were derived through meetings after the literature review was completed. Each of the authors presented findings from his case(s) relevant to the topic "citizen participation in Smart City initiatives" and the categories used in the discussion section.

The literature review consisted of a keyword search (combinations of the key words "smart city", citizen, participation, eParticipation, social media and technology) in Google Scholar and the EBSCOhost database. Both authors were responsible for this, and results from our individual reviews were combined and used in the chapter.

2 Smart Cities

Dameri and Cocchia (2013) investigated the use of the terms "smart city" and "digital city" over a period of 20 years. Their study was based on 705 selected papers containing the "smart city/smart cities" or "digital city/cities" label. They made the following observation regarding the rather wide usage of the term "smart city": "It regards both sustainable technologies, able to reduce pollution and energy consumption, and communication technologies, based on the large use of smartphones or other smart devices. Moreover, also information and communications technology (ICT) could be at the basis of sustainable urban strategies such as smart software used to support a better local public transport planning. The use of the smart label to address sustainable cities is driven by EU programmes, but the smart city idea overcomes this definition to collect under this urban strategy heterogeneous technologies and policies. Moreover, the smart city concept is not entirely based on technology: also energy savings through more aware behaviour, or larger urban green areas, are sometimes considered smart actions".

Chourabi et al. (2012), observed that there was still no clear and consistent understanding of the phenomena "smart city" among practitioners and academia, and proposed a framework based on an extensive array of literature from various fields, such as eGovernment, information science, urban studies and public administration. People and communities were chosen as one of the eight major success factors of "smart cities", and the authors claimed that addressing the topic of people and communities as part of smart cities was critical, and had been neglected traditionally on the expense of understanding more technological and policy aspects of smart cities.

They also argued: "Projects of smart cities have an impact on the quality of life of citizens and aim to foster more informed, educated, and participatory citizens. Additionally, smart cities initiatives allow members of the city to participate in the governance and management of the city and become active users".

Giffinger et al. (2007) compared 70 European medium-sized cities based on a set of indicators and characteristics of "smart cities". They used "smart governance (participation)" as one of the six key characteristics of smart cities. This finding is confirmed by other studies (Coe et al. 2001; Yang and Pandey 2011) and is essential for the argument presented here that citizen participation should be an integrated part of Smart City initiatives. The next section provides a more detailed description of this argument.

3 Citizen Participation

Most democratic ideals involve some sort of interaction between citizens and government, and communication between citizens and the politicians elected to rule is considered a necessity in the democratic process (Casteel 2010; Putnam 2000).

3.1 Models of Democracy—How are Citizens Participating?

Governments and politicians recognize the value of participation (Bryson et al. 2013), but there is a need to define what the outcome of participation should be. Several models and frameworks have been made to address this question. Held (2006), in his extensive work on historic and present democratic systems, identifies no less than ten different models of democracy. While these models differ from each other, they all agree on a set of common democratic principles: free elections, freedom of speech, inclusive citizenship, freedom to form and become members of organizations and the rule of law. Apart from that, they vary in their normative approach to democracy (Strömbäck 2005).

Päivärinta and Sæbø (2006) presented four models of democracy, separating participation into two dimensions: agenda setting (citizens or government set the agenda) and decision-making (citizens or government have the final decision). While other models present a distinct normative approach, their model presents a synthesis of various normative approaches and discusses the implications of each one. They stress the importance of clarifying the model of democracy being followed at the beginning of a citizen participation project and that this should be communicated to participants in order to avoid confusion about the outcome of the discussion.

3.2 Participation in Smart Cities

While participation is important for democracy as a whole, it can be equally important in the development of smart cities. Participation and collaboration between government, citizens and organizations is seen as essential in the development of smart communities (Coe et al. 2001). Some of the activities (parks and recreation, planning, and community development) typically involved in smart city projects can benefit greatly from citizen participation (Yang and Pandey 2011), and a recent study found a clear correlation between cities' adoption and implementation of sustainability policies and public participation in policy formulation (Portney and Berry 2010). ICT is seen as an essential component in this process (Coe et al. 2001).

A prerequisite for effective participation is that citizens are actually heard and their opinions evaluated. A recommendation (OECD 2010) to governments on how to do consultations specifies that: "A comprehensive summary of comments, responses and ideas received must be compiled. A summary of comments, opinions and views expressed in hearings must also be brought up at the decision-making stage. Proposals for decisions must also detail viewpoints that did not result in changes and reasons for not including them".

The techniques for involvement vary and can include focus groups and interviews (with experts and users), usability, functionality, and accessibility testing, encouraging real-time comments and suggestions, log file and transaction log analysis, providing interactive help screens or telephone assistance and developing and adhering to measures and standards of service quality (Jaeger and Bertot 2010).

Following up on this, we examine some of the technologies typically used for participation in Sect. 5, and discuss the importance of a systematic approach to participation in the discussion section.

4 Description of Example Cases

The authors have been involved in several cases illustrating the opportunities and problems related to the use of participation technology. In this section, we present three different cases that are later used to shed light on the use of technology for participation in relation to smart city development.

The first two cases target the use of technology for planning purposes. The last case shows how a city utilizes social media to get more responsive towards its citizens.

4.1 City planning in Arendal, Norway

After several unsuccessful attempts at deciding what to do with a large area close to the city centre of Arendal, Norway, the city decided, in 2010, to start from scratch and run a new and inclusive process. The process involved both online and offline activities and included a series of workshops for invited participants, as well as two surveys distributed to the general population. Expectations were high, but the activist groups who fought against development were disappointed with the process as they felt the city council had not really listened to the input they provided.

Part of the problem in this case was that participation was not equally distributed. The activist groups and their supporters were far more active in the discussions than those who were in favour of development, and this led the city council to seek advice elsewhere and not pay too much attention to the workshops. Feedback from politicians suggested that a more balanced list of participants would have made the process more interesting and valuable.

Interview respondents from the municipal administration reported that they were concerned that the municipality had not set aside resources to discuss the case in social media. This meant that the only information available in social media was from the activists opposing development, which led to social media becoming a purely activist channel of communication where there was no room for rational discussion on the development project. As the activist's Facebook group had several thousand participants who received a one-sided argument, the omission of social media from the municipality's inclusive process tool box can be seen as a strong lesson that inclusive processes should include a strong presence in social media, where all sides are invited to debate the issue at hand.

4.2 *Digital Planning Dialog*

The Digital Planning Dialog (Berntzen and Trollvik 2007) is a web-based system to facilitate electronic communication between stakeholders in zone planning processes. The system integrates a geographic information system with a document handling system, and presents both maps and all relevant documents to its users. The presentation module of the geographic information system was extended to include a separate panel to access documents. Another panel shows a timeline of the planning process, including periods where stakeholders may submit comments. Stakeholders can submit comments by clicking an icon, the comments are stored inside the document handling system and will be available for all other stakeholders.

Planning processes are complex, and the web-based system makes such processes more accessible for the general population. The use of electronic maps makes it easier for citizens to grasp the spatial aspects of the plan, and the timeline makes it easier to understand the temporal aspects.

The "Digital Planning Dialog" project was initiated by 12 municipalities in the County of Vestfold, and is used as a "showcase" in the Norwegian eGovernment Programme "Digitizing Public Sector Services" (Norwegian Ministries 2012).

4.3 Use of Social Media in Sarpsborg

Most municipalities use social media as an information channel. The city of Sarpsborg is using Facebook for active interaction and dialogue with its citizens. The city has chosen to place responsibility for social media with the municipal service center as one of the several communication channels. The Facebook page is quite popular with more than 10,000 likes. (The city has a population of approximately 52,000.) The city provides regular news updates and answers requests and questions from citizens with short response time.

The authors are monitoring municipal use of social media, and use the city of Sarpsborg as a showcase. In most municipalities the communication is one-way. Many municipalities provide just a replica of the news feed on their own home page. We will discuss reasons for Sarpsborg's success later as part of the lessons learned section.

5 Technologies for Participation

While the value of citizen participation is recognized both in academia and government, recent political trends show that political engagement is decreasing. Across the Western world, fewer people are members of a political party (Van Biezen et al. 2012) or vote in elections (Gray and Caul 2000). Partly as a consequence of this, the past decade has seen a number of technologically driven participation projects.

Citizen participation activities have been identified as online voting, online debates, decision-making, activism, consultation, campaigning and petitioning (Medaglia 2012). The first three are briefly mentioned below, as they are the most relevant for cities wanting to increase participation.

The technologies used for these activities are mostly general purpose systems adapted to the political context (Panopoulou et al. 2010) as well as different social media applications. In this section, we present some of the technologies that have been used for citizen engagement.

5.1 Proprietary Platforms

By proprietary we refer to those technologies that are hosted and controlled by government. Examples include various voting systems, including systems that allow citizens to receive information and discuss the issue at hand as well as voting on it (Salazar et al. 2008). There are also several examples of voting advice applications which aid the user in his/her choice of political party or other issues being voted on (Ladner and Pianzola 2010). Other studies focus on security in voting applications (Ramilli and Prandini 2010).

For debating and consultation, there are numerous examples of existing and proposed systems as well as guidelines on how to design effective and efficient systems (Rose and Sæbø 2010). Studies show that participation may lead to increased inclusion of citizens in the agenda setting stage (Hudson-Smith et al. 2005) and increased inclusion of grassroots movements (Reed 2005). However, online participation has not been found to increase participation overall as most participants are already politically active (Taewoo 2010).

As for online decision-making, findings reflect those of online voting. There is disagreement between those who see online participation as a means of reinvigorating democratic decision-making (Norris 2000), and those who fear that online decision-making will only serve to further empower the already powerful (Sivesind et al. 2002). Planning processes is one area where online decision-making is said to be suitable and examples of technologies which could be applied for this purpose include system for collaborative writing (Lourenço and Costa 2007), geographic information systems (Loukis et al. 2010), participatory budgeting (Kim 2008), urban planning tools which allow weighting of votes (Geldermann and Ludwig 2007), and systems for implementing citizen feedback in the decision-making process (Lourenço and Costa 2006).

One particular case of technology use in planning processes is the "Digital Planning Dialog" discussed in Sect. 4.2.

5.2 *Social Media*

Many eParticipation projects using proprietary technologies have encountered problems (Sæbø et al. 2009; Kolsaker and Kelly 2008). Thus, many researchers point towards social media as a platform for participation (Rainie et al. 2012), as there are indications that social media use increases participation (Effing et al. 2011).

Social media has been used for various participation purposes. Politicians and political parties have established their social media presence to reach citizens during elections (Effing et al. 2011). They report that the objective is to establish two-way communication with citizens, but that this can be challenging in an otherwise busy schedule (Johannessen 2010). Examples of concrete projects include the Irish city of Galway; social media was an essential part of a project to lessen traffic congestion (Porwol et al. 2012). In the USA, social media has been used in low-income neighbourhoods to get young people into after-school programmes and activities (Al-Kodmany et al. 2012). In South Korea, government blogs, where citizens could comment and discuss policies, were found to increase trust between politicians and citizens (Park and Cho 2009). In Italy, social media was applied in a project to generate entrepreneurial ideas in underdeveloped economic areas (Bianchi and Cottica 2010). In Norwegian urban planning, social media has been used by activist groups fighting to preserve valuable areas for the public (Johannessen 2012).

There are also examples of social media use in various smart city-related contexts. One study collects data from a location-based social network to map the movement of citizens across neighbourhoods, showing that the city's neighbourhood division does not reflect the actual movement of citizens (Cranshaw et al. 2012), which could have implications in large cities on how public transport and information should be handled. Schaffers et al. (2011) and Komninos et al. (2011) show how social media, especially crowdsourcing and co-creation, plays a part in smart city innovation. Examples can be seen in Thessaloniki's smart city project, where a bicycle sharing website, aggregation of citizen inputs and municipal blogs are essential elements in the smart city strategy (Komninos and Tsarchopoulos 2012). Similarly, co-production and citizen inputs are important in Manchester's smart city strategy (Carter 2012).

Table 1 presents a summary of the opportunities and challenges related to the technologies presented. For proprietary platforms, the ability to create customized applications stands out, but in practice, this opportunity is seldom used. Therefore, a challenge is the non-adaption to specific context.

Many of the systems that have been developed have only been tested in a single case, so there is a need for commercialization and more widespread promotion of

Table 1 Summary of technology properties

	Opportunities	Challenges
	Possibility for tailoring to purpose, such as voting and discussion systems	Often general purpose systems are used in specific contexts. Not always the best match for the purpose
Proprietary platforms		
	Systems have been developed for several areas: collaborative writing, geographic information systems, participatory budgeting and urban planning tools for implementation of citizen feedback in the decision-making process	Some of these systems remain at a conceptual level and are not tested beyond a single case or project
	Voting advice applications helpful for decision-making	Security issues can be a challenge, especially for voting systems
	Can facilitate inclusion of citizens and grassroots movements	Mostly engages those who already are politically active
Social media	Can be more inclusive than proprietary systems, attracts regular users	Politicians have limited time and resources to follow up their social media presence
	Numerous examples showing different ways of using social media	Few studies go beyond describing a single case. Need for reviews and guidelines
	Can utilize crowdsourcing, co-creation through collaborative systems and location-based data from smartphones etc.	Requires at least some proprietary software for analysis
	Aggregation of data from blogosphere and other social media can provide valuable information	Information overload, while still not covering every viewpoint, danger that analysts can overlook the fact that these are not necessarily representing general public opinion

these type of systems. Also, several cases show that these types of systems often fail to attract a sufficient number of users.

For social media, the main opportunity lies in the user base and the many different applications of social media that can be used for a variety of purposes. The challenges are related to gathering and analysing data, where proprietary systems are often needed. The field is still fairly new, so there are few studies presenting guidelines for social media application in various settings.

5.3 Case Experiences

Based on the empirical findings from our cases, we have identified a number of strengths and weaknesses in both proprietary platforms and social media.

Social media's strengths include their reach and number of users. Most of the informants in our cases—politicians, citizens and other actors are members of at least one social media platform. Further, they report to be regular users of social media, accessing the sites several times a week. With proprietary platforms, use of the service is far more sporadic. Finally, social media provides the government with a means of targeting specific groups they want input from, either through word of mouth (or rather keyboard), through various groups or through targeted advertising.

Social media's weaknesses include a lack of control over the platform. In social media, government has to ensure that things are done with what is there (Johannessen and Munkvold 2012), and what is there may not always be the best choice for a participatory process. Informants also raise some concern about privacy issues, as the municipality does not own the data posted to social media and the municipality has limited control over what is written. For example, many politicians are concerned about citizens writing inappropriate statements invading the privacy of government employees and the time it takes to moderate and remove such comments. Some argue that this makes social media better suited for information than interaction in a political participation process. Control of content is another issue, where one municipality, using a free solution for webcasting of council meetings, found pornographic advertising in the sidebar of the webcast.

Proprietary platforms, on the other hand, offer far more control over who participates, what they contribute, etc., and also provide functionality tailored to the individual issue. The Digital Planning Dialog tool used by Norwegian municipalities offers functionality that would be impossible to copy using available social media tools. The problems with proprietary systems are that they can be expensive to create and maintain, and that even well-tailored systems often fail to attract a sufficient number of users. Here, our findings are in line with those of earlier research in digital participation. In the next section, we discuss how these findings can be applied when setting up a participation platform for smart city initiatives.

Following the OECD (2010) recommendations for involvement could in this case have produced very different results.

6 Offline and Online Participation

The previous section shows examples of participatory projects using both proprietary systems and social media. There are also examples of hybrid systems where social media is included as a feature in government websites in order to increase interaction between citizens and government (Dolson and Young 2012), as well as proposals for a hybrid system with a central hub connecting to social media platforms via their APIs (Charalabidis and Loukis 2011). One of the main lessons learned from the successful presidential campaign of Barack Obama was that this type of hybrid approach was valuable. Social media was used as spokes in a hub, where the hub was the Obama campaign website, where people could volunteer or sign up as contributors to the campaign (Lutz 2009).

Phang and Kankanhalli (2008) point out that successful participation requires clear objectives and careful selection of systems, tools and techniques for fulfilling the objectives. While it can be tempting to apply whichever system is at hand or just set up another Facebook group, successful participation require that technologies and expected outcomes match and that the citizen groups being involved agree on this match (Johannessen et al. 2012).

7 Discussion and Recommendations for Municipalities

So far, the term "smart city" has focused on the use of technology to improve quality of life in urban areas.

We argue that being a "smart city" also includes a dimension of being more attentive to the needs and wants of its citizens. Participation is the key to achieve better solutions, services and democratic involvement. The argument is not only primarily based on our case studies but is also supported by literature. The citizens may contribute positively to city development, but this requires careful planning from the local government.

Participation can be achieved in different ways, with or without the use of technology. Focus groups, surveys, polls, dialogues and town hall meetings may be significant measures to collect citizen opinions. But technology brings new possibilities for citizen involvement. Discussion forums and social media facilitate citizen input independent of time and space. Some citizens may be reluctant to join a focus group or join a town hall meeting due to time and space constraints. Technology may help remove such constraints and provide enhanced opportunities for citizens to involve themselves in the decision-making processes. In such cases, it is essential to communicate beforehand what the input from participation will be used for.

In the introduction of this chapter, we mentioned three categories which outline major reasons for citizen involvement: citizen competence and experience, which can aid in developing better plans, solutions and services; using citizens to collect data through the sensors in their smartphones or through other technological means; and third that participation should be valued by itself, as an important part of democratic thinking and a way of building and sustaining local communities.

A challenge when implementing citizen participation is how to operationalize the role of citizens in a particular project. In our interviews with various government officials, this issue has often been raised. Government employees, in particular, are concerned about having a clear division between their role as facilitator and executor and the decisions made by politicians, in order to maintain the division between bureaucracy and elected politicians. Therefore, we suggest that participation projects should be clear about this and clarify early on in the project which model of democracy (Päivärinta and Sæbø 2006) each individual project is following. Doing this and being explicit about how citizen input will be used in the project, can help alleviate some of the critique from citizens in projects such as the municipal urban planning case. In Table 2 we show how the three categories of citizen involvement can fit in with these models of democracy. This is, of course, only one interpretation, based on our interviews with Norwegian officials. In other contexts, the result may well be very different.

While this first step aids in clarifying the context and formal placement of input from citizen participation, it still remains to set up the project. The actual set-up will

Table 2 Categories of participation and appropriate democracy models

Category	Citizen competence and experience	Collecting data through citizens' technology use	Participation as democratic value
Agenda setting responsibility	Government should normally set the agenda and call for citizen input on specific projects, outlining which types of input they are most interested in	Government should normally set the agenda, but in cases where NGOs such as environmental organizations collect data in order to provide input on issues such as pollution and transport, agenda setting could be the responsibility of these groups	This category includes open fora such as town square meetings and social media groups set up so that politicians can understand citizen concerns, with or without a question guiding the input. Thus, agenda setting rests with citizens as the agenda depends on citizen input
Decision-making responsibility	In order to avoid placing too much emphasis on one source of information, the government administration should treat input from citizens as one source of input in the hearing process, and forward this to politicians	The final decision remains with the elected politicians, but in cases where expert knowledge is needed, such as transport patterns and pollution, expert opinion should influence the decision	As most western democracies are representative, decision-making power should remain with politicians. However, in some cases municipalities may want to hold a referendum and follow public opinion

of course depend on the complexity of the project, but in general, we would like to offer the following recommendations, based on experiences from our cases and from recommendations from literature.

Projects aimed at gathering citizen experience and knowledge should consider a hybrid approach which combines social media and offline activities. Systems such as loomio.org or even proprietary decision-support systems can be useful in cases where consensus is required. The urban planning case clearly shows the danger of not using social media in cases where there are actors with a strong interest. These actors can easily set up their own social media campaigns in order to further their own agenda, but if the municipality shows initiative and acts first, it can be easier to get feedback from more than one group of like-minded people.

The city of Sarpsborg has a successful social media presence. The city facilitates interaction. The good thing is that answers to questions are easily shared among the citizens. One problem is when citizens publish personal information related to a case that violates rules on privacy. From time to time, the city responds that Facebook is not the right channel and asks the citizen to interact through other channels. This shows the dilemma of social media. Citizens are used to shareing information, but the municipality needs to follow the rules of law.

Projects aimed at collecting data through citizens' use of smartphones likewise, should consider a hybrid approach, but here, proprietary systems are probably more necessary in order to collect and do meaningful analyses of data. In these projects, social media could act as a channel for recruiting participants.

For projects where participation is seen mostly as a democratic value, technology use should include a mix of discussion forums, social media accounts and offline activities such as town hall meetings. These projects are aimed at gathering information about people's concerns and it is important to reach out to as many citizens as possible. Projects that are targeted towards a specific policy area could perhaps also consider the use of loomio.org or similar systems.

Finally, we suggest that for proper handling of citizen participation projects, it is important to apply guidelines and frameworks aimed at supporting participation and involvement. The guidelines developed by the OECD provide valuable information on how to compile the results of citizen participation campaigns into a proper format for the formal hearing process.

8 Conclusion

This chapter has discussed participation in the context of "smart cities". The rationale for participation can be found in models of democracy, but a more pragmatic approach looks at citizens as resources that can help local government to implement better plans, services and processes. Different technologies are available, with advantages and disadvantages. Proprietary technology ensures control, while social media attract regular visitors. We suggest a mixed approach, where social media are used to mobilize citizens, and then linked to proprietary technology for citizen involvement.

References

Al-Kodmany, K., Betancur, J., & Vidyarthi, S. (2012). E-civic engagement and the youth: New frontiers and challenges for urban planning. *International Journal of E-Planning Research (IJEPR), 1*(3), 87–104. doi:10.4018/Ijepr.2012070105.

Berntzen, L., & Trollvik, J. A. (2007). *Digital planning dialog*. Proceedings, 3rd international conference on E-Government, Academic Conferences, 2007, pp. 61–68.

Bianchi, T., & Cottica, A. (2010). Harnessing the unexpected: A public administration interacts with creatives on the web. *European Journal of Epractice, 9,* 82–90.

Bryson, J. M., Quick, K. S., Slotterback, C. S., & Crosby, B. C. (2013). Designing public participation processes. *Public Administration Review, 73,* 23–34.

Carter, D. (2012). Urban regeneration, digital development strategies and the knowledge economy: Manchester case study. *Journal of the Knowledge Economy, 4*(2), 169–189.

Casteel, P. D. (2010). *Habermas & communicative actions* (pp. 1–6). New York: Great Neck Publishing.

Charalabidis, Y., & Loukis, E. (2011). *Transforming government agencies' approach to eparticipation through efficient exploitation of social media*. Paper presented at the European Conference on Information Systems (ECIS), Helsinki, Finland. http://Aisel.Aisnet.Org/Ecis2011/84.

Chourabi, H., Nam, T., Walker, S., Gil-Garcia, J. R., Mellouli, S., Nahon, K., Pardo, T. A., Scholl, H. J. (2012). Understanding smart city initiatives: An integrative and comprehensive theoretical framework. In Proceedings of the 45th Hawaii International Conference on System Sciences, pp. 2289–2297.

Coe, A., Paquet, G., & Roy, J. (2001). E-governance and smart communities: A social learning challenge. *Social Science Computer Review, 19,* 80–93.

Cranshaw, J., Schwartz, Z., Hong, J. I., & Sadeh, N. (2012). *The livehoods project: Utilizing social media to understand the dynamics of a city*. Paper presented at the proceedings of the sixth International AAAI Conference on Weblogs and Social Media.

Dameri, R. P., & Cocchia, A. (2013). *Smart city and digital city: Twenty years of terminology evolution*. Proceedings Itais 2013. http://Www.Cersi.It/Itais2013/Pdf/119.Pdf. Accessed 2 Jan 2014.

Dolson, J., & Young, R. (2012). Explaining variation in the e-government features of municipal websites: An analysis of e-content, e-participation, and social media features in Canadian municipal websites. *Canadian Journal of Urban Research, 21*(2), 1–24.

Effing, R., van Hillegersberg, J., & Huibers, T. (2011). Social media and political participation: are Facebook, Twitter and YouTube democratizing our political systems? In E. Tambouris, A. Macintosh, & H. de Bruijn (Eds.), *Electronic participation* (Vol. 6847, pp. 25–35). Berlin: Springer.

Geldermann, J., & Ludwig, J. (2007). Some thoughts on weighting in participatory decision making and e-democracy. *International Journal of Technology, Policy and Management, 7*(2), 178–189.

Giffinger, R., Fertner, C., Kramar, H., Kalasek, R., Pichler-Milanovic, N., & Meijers, E. (2007). *"Smart cities—ranking of european medium-sized cities". Smart cities*. Vienna: Centre of Regional Science.

Gray, M., & Caul, M. (2000). Declining voter turnout in advanced industrial democracies, 1950 to 1997. *Comparative Political Studies, 33,* 1091–1122.

Held, D. (2006). *Models of democracy*. Cambridge: Polity Press.

Hudson-Smith, A., Evans, S., & Batty, M. (2005). Building the virtual city: Public participation through e-democracy. *Knowledge, Technology & Policy, 18*(1), 62–85.

Jaeger, P. T., & Bertot, J. C. (2010). Designing, implementing and evaluating user-centered and citizen-centered egovernment. In C. G. Reddick (Ed.), *Citizens and e-government—evaluating policy and management* (pp. 1–19). Hershey: IGI Global.

Johannessen, M. (2010). Genres of participation in social networking systems: A study of the 2009 norwegian parliamentary election. In E. Tambouris, A. Macintosh, & O. Glassey (Eds.), *Electronic participation* (Vol. 6229, pp. 104–114). Berlin: Springer.

Johannessen, M. R. (2012). *Genres of communication in activist eparticipation: A comparison of new and old media.* Paper presented at the proceedings of the 6th International Conference on Theory and Practice of Electronic Governance, Albany, New York.

Johannessen, M. R., & Munkvold, B. E. (2012). *Defining the IT artefact in social media for eparticipation: An ensemble view*. European Conference on Information Systems (ECIS), 2012 Barcelona.

Johannessen, M. R., Flak, L., & Sæbø, Ø. (2012). Choosing the right medium for municipal eparticipation based on stakeholder expectations. In E. Tambouris, A. Macintosh, & Ø. Sæbø (Eds.), *Electronic participation* (Vol. 7444, pp. 25–36). Berlin: Springer.

Kim, J. (2008). A model and case for supporting participatory public decision making in e-democracy. *Group Decision and Negotiation, 17*(3), 179–193.

Kolsaker, A., & Kelly, L. L. (2008). Citizens' attitudes towards e-government and e-governance: A UK study. *International Journal of Public Sector Management, 21*(7), 723–738.

Komninos, N., & Tsarchopoulos, P. (2012). Toward intelligent Thessaloniki: From an agglomeration of apps to smart districts. *Journal of the Knowledge Economy, 4*(2), 149–168.

Komninos, N., Schaffers, H., & Pallot, M. (2011). Developing a policy roadmap for smart cities and the future Internet paper presented at the e-challenges.

Ladner, A., & Pianzola, J. (2010). Do voting advice applications have an effect on electoral participation and voter turnout? Evidence from the 2007 Swiss Federal Elections.

Loukis, E., Xenakis, A., Peters, R., & Charalabidis, Y. (2010). Using GIS tools to support e-participation—A systematic evaluation. In E. Tambouris, A. Macintosh, & O. Glassey (Eds.), *Electronic participation* (Vol. 6229, pp. 197–210). Berlin: Springer.

Lourenço, R. P., & Costa, J. P. (2006). *Discursive e-democracy support.* Paper presented at the Hawaii International Conference on System Sciences, Hawaii.

Lourenço, R. P., & Costa, J. P. (2007). Incorporating citizens' views in local policy decision making processes. *Decision Support Systems, 43,* 1499–1511.

Lutz, M. (2009). *The social pulpit: Barack Obama's social media toolkit.* Chicago: Edelman.

Medaglia, R. (2012). E-participation research: Moving characterization forward (2006–2011). *Government Information Quarterly, 29,* 346–360.

Norris, P. (2000). *A virtuous circle: Political communications in postindustrial societies.* Cambridge: Cambridge University Press.

Norwegian Ministries. (2012). Digitizing public sector services. http://Www.Regjeringen.No/Upload/Fad/Kampanje/Dan/Regjeringensdigitaliseringsprogram/Digit_Prg_Eng.Pdf. Accessed 12 Dec 2013.

OECD. (2010). Better regulations for Europe: Finland. Http://Www.Oecd.Org/Gov/Regulatory-Policy/45054502.Pdf. Accessed 11 Dec 2013.

Päivärinta, T., & Sæbø, Ø. (2006). Models of E-democracy. *Communications of the Association for Information Systems, 17*(4).

Panopoulou, E., Tambouris, E., & Tarabanis, K. (2010). eParticipation initiatives in Europe: Learning from practitioners. In E. Tambouris, A. Macintosh, & O. Glassey (Eds.), *Electronic participation* (Vol. 6229, pp. 115–126). Berlin: Springer.

Park, J., & Cho, K. (2009). *Declining relational trust between government and publics, and potential prospects of social media in the government public relations.* Paper presented at the Egpa Conference 2009 The Public Service: Service Delivery In The Information Age, St. Julian's, Malta.

Phang, C. W., & Kankanhalli, A. (2008). A framework of ICT exploitation for e-participation initiatives. *Communications of the ACM, 51,* 128–132.

Portney, K. E., & Berry, J. M. (2010). Participation and the pursuit of sustainability in U.S. *Cities. Urban Affairs Review, 46,* 119–139.

Porwol, L., O'donoghue, P., Breslin, J., Mulligan, B., & Coughlan, C. (2012). *eParticipation and transport management: A practical approach.* Paper presented at the proceedings of the 13th Annual International Conference on Digital Government Research, College Park, Maryland.

Putnam, R. D. (2000). *Bowling alone: The collapse and revival of American community*. New York: Simon & Schuster.

Rainie, L., Smith, A., Schlozman, K. L., Brady, H., & Verba, S. (2012). Social media and political engagement Pew Internet & American Life Project. http://www.pewinternet.org/files/old-media//Files/Reports/2012/PIP_SocialMediaAndPoliticalEngagement_PDF.pdf. Accessed 11 Dec 2013.

Ramilli, M., & Prandini, M. (2010). An integrated application of security testing methodologies to e-voting systems. In E. Tambouris, A. Macintosh, & O. Glassey (Eds.), *Electronic Participation* (Vol. 6229, pp. 225–236). Berlin: Springer.

Ratti, C, & Townsend, A. (2011). The social nexus. *Scientific American, 305*(3), 42–48.

Reed, T. V. (2005). *The art of protest: Culture and activism from the civil rights movement to the streets of Seattle*. Minneapolis: Minnesota University Press.

Rose, J., & Sæbø, Ø. (2010). Designing deliberation systems. *The Information Society, 26,* 228–240.

Salazar, J. L., Piles, J. J., Ruíz, J., & Moreno-Jiménez, J. M. (2008). E-cognocracy and its voting process. *Computer Standards and Interfaces, 30,* 124–131.

Sæbø, Ø., Rose, J., & Nyvang, T. (2009). The role of social networking services in e-participation. In A. Macintosh & E. Tambouris (Eds.), *Electronic participation* (Vol. 5694, pp. 46–55). Berlin: Springer.

Schaffers, H., Komninos, N., Pallot, M., Trousse, B., Nilsson, M., & Oliveira, A. (2011). Smart cities and the future Internet: Towards cooperation frameworks for open innovation. In J. Domingue, A. Galis, A. Gavras, T. Zahariadis, D. Lambert, F. Cleary, P. Daras, S. Krco, H. Müller, M.-S. Li, H. Schaffers, V. Lotz, F. Alvarez, B. Stiller, S. Karnouskos, S. Avessta, & M. Nilsson (Eds.), *The future Internet* (Vol. 6656, pp. 431–446). Berlin: Springer.

Sivesind, K. H., Lorentzen, H., Selle, P., & Wollebæk, D. (2002). *The voluntary sector in Norway: Composition, changes and causes*. Oslo: Institute for Social Research, University of Oslo.

Strömbäck, J. (2005). In search of a standard: Four models of democracy and their normative implications for journalism. *Journalism Studies, 6,* 331–345.

Taewoo, N. (2010). Whither digital equality?: An empirical study of the democratic divide. Paper presented at the 43rd Hawaii International Conference on System Sciences (HICSS).

Van Biezen, I., Mair, P., & Poguntke, T. (2012). Going, going,… gone? The decline of party membership in contemporary Europe. *European Journal of Political Research, 51,* 24–56.

Yang, K., & Pandey, S. K. (2011). Further dissecting the black box of citizen participation: When does citizen involvement lead to good outcomes? *Public Administration Review, 71,* 880–892.

Lasse Berntzen is a professor of information systems at Buskerud and Vestfold University College, Norway. Since 2003, his research has focused on the Information Society and Information and Communication Technology (ICT) in public sector, with emphasis on eGovernment and eParticipation. He has been involved in several research projects on national and international level, author and coauthor of several conference papers and articles, and programme committee member and reviewer for some of the most important conferences within his field of research. He has also written several textbooks. He is a fellow of IARIA and member of ACM, IEEE and AIS

Marius Rohde Johannessen is an associate professor of information systems at Buskerud and Vestfold University College, Norway. He earned his Ph.D. from the University of Agder in 2013 on the topic "social media use in eDemocracy". His research interests include digital communication, eGovernment, eParticipation and activism.

eParticipation in Smart Cities of Developing Countries: Research-Based Practical Recommendations

Paulo Roberto de Mello Miranda, Maria Alexandra Viegas Cortez da Cunha and José Maria Pugas Filho

Abstract Cities are facing major challenges and the expansion of citizen participation has been considered a valuable strategy. Citizen participation experiences supported by information and communication technologies (ICT), the so-called eParticipation, have taken hold in cities around the world. In this chapter, we discuss some of the enabling and inhibiting elements for the creation of virtual spaces for participation by governments in a developing country of Latin America—Brazil. A list of enabling and inhibiting elements grouped in five dimensions was produced from a survey with Brazilian experts, including public officials, academic researchers, and consultants involved with the implementation of eDemocracy projects. To contribute to smart city initiatives in developing countries, interviews were made with the chief information officers (CIOs) of four major cities in Brazil—Rio de Janeiro, Porto Alegre, Curitiba, and Campinas. We found out that the enablers are concentrated in the sociocultural and technological dimensions, while the inhibitors are mainly in the political and governance dimensions. The enabling elements are, for the most part, external to the sphere of action of city information and communication technologies (ICT) managers. As far as inhibiting elements are concerned, even though they pose broad challenges that may be perceived to be beyond the reach of the leaders of eParticipation initiatives, there is more scope for managerial action.

Keywords eParticipation · Smart cities · Developing countries · Enabling elements · Inhibiting elements

P. R. de Mello Miranda (✉) · M. A. Viegas Cortez da Cunha
Rua Luiz Manzochi 98, casa 7, Curitiba/PR 80710-590, Brazil
e-mail: prmmiranda@gmail.com

M. A. Viegas Cortez da Cunha
e-mail: mariaalexandra.cunha@gmail.com

J. M. Pugas Filho
Rua Alferes Angelo Sampaio 1100, apt. 902, Curitiba/PR, Brazil
e-mail: jmpugas@gmail.com

J. R. Gil-Garcia et al. (eds.), *Smarter as the New Urban Agenda A Comprehensive View of the 21st Century City,* Public Administration and Information Technology 11,
DOI 10.1007/978-3-319-17620-8_17

1 Introduction

Due to population growth, cities are facing major challenges and the urgency to solve them brings pressure to bear on public managers to seek newer and smarter ways to deal with multiple and complex issues. One strategy that has been adopted by local governments is the expansion of citizen participation in the search for improvements in urban life. Citizen participation experiences supported by information and communication technologies (ICT), the so-called eParticipation, have taken hold in cities around the world. Although eParticipation initiatives could be taken by actors from various sectors of society, in this chapter we discuss some of the facilitating and limiting elements for the creation of such spaces, by governments, in an emerging economy—Brazil.

The concept of a smart city itself is still emerging, and the work of defining and conceptualizing it is in progress. One way to conceptualize a smart city is as an icon of a sustainable and livable city (Chourabi et al. 2012). It is a well-known fact that ICT infrastructure and applications are prerequisites for it to work. However, without real involvement of public institutions, the private sector, voluntary organizations, schools and citizens, as well as their willingness to collaborate and cooperate, there can be no smart community (Lindskog 2004).

There is a growing interest among academic researchers and government officials in new forms of relationship between governments and citizens, especially in the sphere of citizen participation in public decision-making processes (Sæbø et al. 2008). Among practitioners, eParticipation has also been a focus of attention, increasing the practical relevance of related investigations (Bélanger and Carter 2012). Since the active involvement of all community sectors in decision-making is essential (Nam and Pardo 2011), the discussion of eParticipation in smart cities becomes relevant.

Brazil is a developing country and a young democracy; the current constitution was promulgated in 1988 after a military dictatorship that lasted more than 20 years and eParticipation experiences are recent in the country. In this chapter, we seek to understand the facilitating and inhibiting elements of the establishment of eParticipation spaces by the government in the country's context, particularly in its cities. We then advance some research-based practical recommendations for eParticipation in smart cities in emerging economies through the Brazilian perspective. Even though ICT initiatives in local governments are intrinsically linked to its organizational culture, priorities, and strategic vision (Odendaal 2003), the critical examination of the concepts of smart cities and the results of the empirical research aim to contribute with insights for city managers in Brazil on how to face the challenges of building smarter cities. Eventually, this chapter provides insights for city managers in other emerging economies. The chapter is organized so that, after this introduction, a brief conceptualization of smart cities and eParticipation is presented. This is followed by the methodological aspects that guided the empirical investigation. Then we present the data and discuss the results, with the goal of presenting some practical recommendations for smart city managers in Brazil, which are emphasized in the concluding section.

2 Smart Cities and eParticipation

Although the term "smart city" is widely used among the public in general and also in academic circles, a clear and consistent understanding of the concept is still lacking (Nam and Pardo 2011; Chourabi et al. 2012). There is a wide range of definitions, from a "sustainable and livable city" to "a large organic system connecting various subsystems and components, such as public services and urban facilities, as well as systems for data capture and analysis, and sensors and personal devices, geared toward improving sustainability and quality of life in the cities" (Chourabi et al. 2012; Kanter and Litow 2009). The second concept emphasizes the fundamental role of communities (Kanter and Litow 2009). The components of a smart city are institutional, technological, and human (Nam and Pardo 2011). In Europe, cities have developed collaborative digital environments to foster local prosperity and competitiveness, through electronic participation, integrated electronic services and knowledge networks and partnerships (Paskaleva 2009), which highlight the importance of participation and/or collaboration mechanisms. Smart cities should provide collaboration environments supported by ICT, where citizens, companies, and researchers contribute their own contents, or build new applications to *mash-up* with databases available from governments to face challenges such as sustaining an innovation economy, job creation and stability, optimizing energy and water usage, and creating safer cities (Schaffers et al. 2011). In the smart city, investments in human and social capital and infrastructure foster sustainable economic growth and quality of life, with a balanced management of natural resources through participative governance (Caragliu et al. 2009).

Technology creates conditions to increase the desirable possibilities of participative governance (Holzer 2005). Electronic participation, or eParticipation, is usually known as a subfield of electronic democracy (Macintosh 2004) even though there is some discrepancy between these two concepts (Susha and Grönlund 2012). eParticipation refers to the use of new technologies, particularly the Internet, with the implication that technology has the ability to change or transform citizen involvement in deliberation or decision-making processes (Sæbø et al. 2008). eParticipation connects with opportunities for consultation and dialogue between government and citizens using a series of ICT tools, including e-voting—the use of ICT to support the democratic process of voting (Macintosh 2004). However, eParticipation is much more than just voting (Rose et al. 2007). It includes the extension and transformation of participation in societal, democratic, and consultative processes mediated by ICT (Sæbø et al. 2008) and involves three spheres of governance—political, civil, and administrative (Grönlund and Horan 2005). The focal point of eParticipation is to increase citizens' abilities to participate in digital governance, in the processes of public service provision at various stages in the production chain—planning, decision-making, implementation, and evaluation (Grönlund 2001; Sæbø et al. 2008). It is reasonable to assume that it can produce a significant impact on public policies (Åström et al. 2012).

The development of new applications for the Internet, mobile phones, and particularly social media creates demands for higher capacity and speed in access channels. The different levels of government acknowledge these needs and have developed national and regional broadband projects. Brazil is lagging in broadband supply (ITU 2013) and if that is not solved, Internet usage to support economic and social development may be compromised, since digital exclusion deepens socio-economic exclusion (Macadar and Reinhard 2002). The use of ICT in developing country cities, although desirable as a driver of quality of life, could exacerbate inequalities and widen the digital divide (Odeendal 2003). In the political arena, the full exercise of citizenship presupposes the ability to know rights and obligations, limits for action and power, and such ability is critical when relationships are mediated by technologies. More than being able to read and write, digital literacy implies the ability to access, to codify and decode meanings, to think critically about how and why they were produced, and the competence to translate knowledge into consequential actions (Coleman and Rowe 2005). It enables people to use ICT to develop and distribute contents that are meaningful to their own community; "it is able to provide participation tools for the citizenry, not only as receptors, but also allow them to engage in dialogue for policy building" (Macadar and Reinhard 2002).

eGovernment may embody an opportunity to change the nature of government–society relations, from a hierarchical command-and-control model to an interactive collaboration among governments, citizens, companies, public servants, and other government spheres (Esteves and Joseph 2008). In Brazil, the use of electronic tools in government has targeted primarily the improvement of administrative efficiency as well as public services (Cunha and Miranda 2013). Even so, ICT and particularly the Internet have been used for the implementation of a relative variety of participative processes in Brazilian public administration such as public hearings and consultations, cooperation of civil society representative entities, participation of users in public service provision, participative budgeting, public meetings, surveys, use of focus groups, and citizen counsels (Cunha et al. 2013). Various participation strategies are found in government Internet portals (Holzer and Manoharan 2007), but the expansion of electronic participation requires a deeper reflexion about the democratic implications, the goals of transparency and citizen participation, beyond the focus on service provision, which characterizes most eGovernment initiatives in Brazil (Frey 2004; Cunha et al. 2005; Carrizales 2008).

The extant literature refers to diverse elements that facilitate or inhibit democratic electronic participation, although the terminology varies. The discussions bring up political, ideological, social, and cultural motivations, in addition to technological issues, on the implementation of eParticipation spaces (Agre 2002; Cunha et al. 2005; Ruediger 2002). Esteves and Joseph (2008) propose a framework for the evaluation of electronic government initiatives, discussing strategic, technological, economic, operational, and service dimensions. Other authors focus on laying out success factors for eGovernment projects (Gil-García and Pardo 2005) and smart cities (Chourabi et al. 2012). Another framework, Strategy, Technology, Organization, People, and Environment (STOPE) is also used for the evaluation of different issues related to ICT, including eBusiness and eGovernment planning (Bakry 2004). It is structured into five dimensions: strategy, technology, organization, people, and environment. In a "Social Development Note" published by the World Bank (Thindwa et al. 2003), there are four dimensions of external factors that may assist civil society participation in promoting its interests: the legal and regulatory framework, the political and governance context, sociocultural characteristics, and economic conditions.

On the basis of these frameworks, as well as an understanding that the construction of eParticipation spaces occur within the scope of eGovernment, we establish five dimensions to classify the facilitating and inhibiting elements of eParticipation spaces: (1) political and governance-related, (2) sociocultural, (3) economic, (4) technological, and (5) legal and regulatory. We built a table, grouping in these dimensions and elements taken from 41 texts during the literature review. We use them to guide the first stage of empirical research and to discuss the results.

3 Method

This research focuses on initiatives that use ICT to create eParticipation spaces for citizens and society at large, in government decision-making processes. In general, eParticipation spaces are recent initiatives and many of these in Brazil have an experimental nature. They have raised interest among public agents, researchers, technological and/or consulting companies, and nongovernmental organizations, because of their importance and potential for the creation of new forms of relations between government and society. Those groups of actors were included in the empirical investigation, which occurred in two stages. The first one, in 2010, gave rise to a list of facilitating and inhibiting elements that guided the interviews conducted with IT officers of smart cities during the second stage in 2014.

The first stage comprised a literature review and interviews with five experts in the field, with open script. Two sets of elements were obtained, 1 with 27 facilitators and another with 30 inhibitors. Then, applying a web survey, we sent the list to 155 Brazilian experts in the 3 government levels, including 23 respondents at the municipal level, 23 at the state, and 26 at the federal level. The survey also comprised 26 consultants from the private sector and nongovernmental organizations and 41 academic researchers, all involved either with the implementation of eParticipation projects or with scientific research about them. After two rounds, from January to May 2010, 55 among them expressed their degree of agreement with the elements listed and attributed a level of importance. Data analysis was performed with Excel, Sphinx, and Statistical Package for the Social Sciences (SPSS) tools, and the result was a list of elements in the following dimensions: political and governance, sociocultural, economic, technological, and legal–regulatory.

For the second stage, in early 2014, the objective was to investigate those facilitating and inhibiting elements for the construction of eParticipation spaces in smart cities. The interviews focused on the same dimensions investigated in the previous survey with Brazilian experts, bringing them into the context of the urban setting. The list of facilitating and inhibiting elements was used to develop a semi-structured interview script that was applied to municipal government chief information officers (CIOs) in four large cities located in the Southern and Southeastern regions of Brazil: Rio de Janeiro, Porto Alegre, Curitiba, and Campinas. The cities were chosen among the top ranked on Centre for Research and Development (CPQD)—Brazilian Index of Digital Cities 2012[1]: Curitiba (1st), Rio de Janeiro (2nd), Campinas (5th), and Porto Alegre (8th). Three of them, Curitiba, Porto Alegre, and Rio de Janeiro, have also, jointly or separately, figured from 2010 to 2013 on the ranking "Smart21 Communities" published annually by the Intelligent Community Forum (ICF)[2]. The population of Rio de Janeiro is approximately 6.4 million inhabitants, and the populations of Curitiba, Porto Alegre, and Campinas are 1.8, 1.5 and 1.1 million respectively[3]. In economic terms, the GDP of Rio de Janeiro, Curitiba, Porto Alegre e Campinas is US$133.4, 37.0, 29.0, and 25.8 billion[4]. Their respective Human Development Indexes (HDIs) are 0.799, 0.823, 0.805, and 0.805[5].

The interviews were conducted in February 2014 and lasted on an average for 55 min. They were recorded and transcribed. After the analysis, the facilitating and inhibiting elements were organized into five analytical categories: political and governance, sociocultural, economic, technological and legal–regulatory.

4 Analysis and Discussion of Results

The results were obtained in two stages. In the first one, we built a list of facilitating and inhibiting elements for the creation of eParticipation spaces. In the second stage, we looked into eParticipation in Brazilian cities.

4.1 Facilitating and Inhibiting Elements for eParticipation

In the first phase of the research we identified a set of elements that may influence the implementation of eParticipation spaces by governments—facilitating it or inhibiting it. We gathered opinions from actors involved in electronic democracy

[1] Available at http://www.wirelessmundi.inf.br/component/content/article/51-edicoes/edicao-n-9/904-ranking-cidades-digitais.

[2] Available at http://www.intelligentcommunity.org/index.php?ubmenu=Awards&src=gendocs&ref=Smart21&category=Events&link=Smart21.

[3] Population estimate 2013, Instituto Brasileiro de Geografia e Estatística—IBGE http://www.cidades.ibge.gov.br/xtras/home.php.

[4] Average exchange rate July 2011 and GDP base year 2011, Instituto Brasileiro de Geografia e Estatística—IBGE, available at http://www.cidades.ibge.gov.br/xtras/home.php.

[5] Municipal HDI base year 2010, Instituto Brasileiro de Geografia e Estatística—IBGE, available at http://www.cidades.ibge.gov.br/xtras/home.php.

Table 1 Facilitating elements of eParticipation spaces

Rank	Facilitating elements
1	Growth in adoption of computers and the Internet by Brazilians in recent years
2	Choice of an appropriate technology for easy, intuitive, and friendly use
3	Use of social media and other virtual tools
4	Growth in ICT education reduces barriers to their use
5	Exponential growth of resources for discussion, sharing, and collective learning in social media
6	Ubiquitous expansion of mobile telephony
7	Positive experiences of creation and use of these spaces in Brazil
8	Growth of "digital native" population
9	A combination of political will with a well-prepared bureaucracy to create eParticipation spaces
10	Choice of topics that attract public interest for the initial implementation

processes in Brazil, from academia, federal, state and municipal governments, and other actors such as non-governmental organization (NGO) members and consultants, who deemed such elements as important.

For Brazilian specialists, the main facilitating elements for the implementation of eParticipation spaces are related to the growth in use of technologies, expanding education in informatics, and the fact that the new generations are more prepared to use the tools (1st, 3rd, 4th). Other elements are the choice of an appropriate technology, easy and friendly to use (2nd), and the exponential growth of resources for discussion, sharing and collective learning in social networks (5th). Table 1 lists the ten elements that were ranked as most important.

There are differences between the results of this study and other experiences observed elsewhere. The political and governance dimension indicates distinct elements in relation to the issues that are reported internationally. Here, for example, there is no concern about the fact that virtual spaces are complementary to traditional participation spaces, but rather a tool to create and expand participation possibilities (Williamson 2006; New Zealand 2004; Coleman 2005). Another difference is related to legal and regulatory aspects, which are seen in the international literature as well as in practices of national and supranational governing bodies to create favorable laws and norms for the development of eDemocracy (Madrid 2004; Damodaran et al. 2005; Criado 2008; United Kingdom 2009a, b). In our case, concerns with laws or norms to support eParticipation mechanisms are absent.

In the sociocultural dimension, as it would be expected, the research reveals facilitating elements related to the affinity of young people with technology and also to capacity building for the use of technology, with an emphasis on digital inclusion. However, no importance was given to the existence of organized society networks as a facilitator of eParticipation spaces to be created by the government, even though it is present in the literature (Guidi 2001; Busatto and Vargas 2004; Dahlberg 2001). On the other hand, "The growing adoption of computers and the Internet in recent years," was not found in the literature review, yet it appears as the main facilitator for Brazilian specialists.

Table 2 Inhibiting elements

Rank	Inhibiting element
1	Fragmentation of ICT management in governments
2	ICT is seen as a tool rather than inductor of change in public administration
3	ICT is not part of government strategy
4	Lack of maturity of governance models
5	Lack of culture of transparency in governments
6	Government officers and public managers do not ascribe importance to eParticipation
7	Interface or user environment nonintuitive, hard to use
8	Broadband is still incipient in Brazilian households
9	Lack of interest among a large part of public managers to use ICT to expand citizen participation
10	Digital exclusion

Among the inhibiting elements, we should point out the emphasis given by Brazilian specialists to the political and governance dimension as an inhibiting element of virtual participation spaces (Table 2), even though the majority of them are internal to the government. Among ten elements, six belong to this dimension, including the top three. These elements fall basically into two types: those related to strategy and governance, and those that arise from political culture. Several authors identify the predominance of an instrumental view of technology, pointing out that the main focus of Internet use by governments has been the increase in efficiency and the provision of public services (Cunha and Miranda 2013; Frey 2004; United Nations 2010; Damodaran et al. 2005). The research confirmed a lack of understanding among governments and managers about the role that technology may play to foster democratic participation and economic development (Motta 1990; Ainsworth et al. 2005; Damodaran et al. 2005; Schaffers et al. 2011).

The first six elements listed as the most important are connected with politics and/or ICT governance. In the economic dimension, the lack of broadband infrastructure was expected as a inhibiting element, as it happens, but it is remarkable that barriers of high costs of computers and Internet access in Brazil are not among the ten most important elements. Among the social aspects, digital exclusion is an inhibiting element, yet other problems that are significant in developed-country democracies, such as the decline in political engagement, do not appear among the concerns of Brazilian specialists, or the lack of capabilities by government workers as well as citizens, which appeared but did not reach the top ten list of most important elements.

In the phase of data analysis, we inquired whether the groups presented internal homogeneity and whether the answers were significantly different among the groups. For that purpose, we used tests that are suitable for small samples and an absence of normal distribution of the data (Kruskal-Wallis). Some differences were found. There is a difference on the importance given to the absence of technology in government strategy, which takes first place for respondents in government areas and other organizations, but is not even listed for academic researchers among the ten most important elements.

4.2 *Facilitators and Inhibitors of eParticipation in Smart Cities of Brazil*

In the second stage of the study, we delved into eParticipation spaces in smart cities of Brazil. The objective was to verify whether there are distinctive elements that facilitate or inhibit the creation of eParticipation spaces in those cities.

The application of the five dimensions of analysis enabled us to observe an emphasis in elements of the political and governance, sociocultural and technological dimensions, while there were few mentions to the legal and regulatory and economic dimensions. The facilitating elements are clustered in the sociocultural and technological dimensions, whereas the inhibiting elements are concentrated in the political and governance dimension. The facilitating elements are, for the most part, external to the scope of action of municipal governments. The inhibiting elements are more related to elements inside the government and therefore more amenable to governance interventions.

The main facilitating elements in the political dimension include positive eParticipation experiences, trust by the citizenry that public participation will be taken into account in municipal decisions, and explicit involvement of the main officers of the municipal administration. To illustrate the categorization applied to the interviews, the following excerpt is one of those selected for the element "trust by the citizenry:"

> What leads citizens to participate is the belief that they may influence government decisions, warn the government, complain about the government. (Interviewee 1).

In the sociocultural dimension, the adoption of social media as a tool for interactions among individuals was highlighted. They generate a flow of social dialogues that are relevant for municipal management and take place independently from government and sometimes in spite of it, as observed in:

> The June demonstrations[6] were strongly based upon mobilization through social media. (Interviewee 2).

In the technological dimension, the most cited facilitating elements were the popularization of mobile devices such as smartphones and tablets, the diffusion of mobile applications developed by the private sector, and the consolidation of Internet and social media usage by a large part of the population, as observed in:

> The smartphone is a device that is increasingly diffused, it has a great penetration in all social classes. (Interviewee 2).

Lastly, the legal and regulatory dimension appeared less as enabler of eParticipation, but even so, it is deemed important for the assimilation of the participation process. The Law of Access to Information was mentioned[7].

---

[6] In June 2013, several popular demonstrations occurred in the major cities, and they were the largest since the re-democratization period of 1982–1985. The movement was called out spontaneously through social media, without a formal leadership.

[7] Law 12.527, sanctioned by the President in 18 November, 2011, with the purpose of regulating the constitutional right of citizen's access to public information. The rules of the law are applicable to the three powers of the Union, States, Federal District and Municipalities.

> Traditional political culture is not, and never has been, very inclined to transparency and control by society. It is a culture that rejects that kind of thing. Regardless, the transparency portals are here, and the LAI creates new dynamics. (Interviewee 2).

With regard to inhibiting elements, there was consensus about the political and governance dimension as the main consideration in the development of eParticipation spaces, markedly due to the lack of involvement by the municipal government, the lack of maturity of governance models and their inability to institutionalize participation processes in governments, the absence of impacts of virtual decisions in the real world, and the fragmentation or lack of coordination of ICT management in governments, as found in:

> The application of ICT had a merely instrumental character. It was only in early 2013 that government strategy incorporated the intensive use of ICT to create new electronic channels for citizen participation in the government's decision-making processes and public policy formulation, to improve administration mechanisms and to offer new and better public services. (Interviewee 4).

The lack of a match between political will and a well-prepared bureaucracy, given the dearth of skilled personnel in the public sector, is an inhibiting element as well, as observed in:

> The bureaucracy is averse to power sharing. (Interviewee 4).

Transparency is mentioned as a fundamental component in the context of eParticipation. Informational disorganization, the absence of adequate processes of information analysis and the increase in demand hinder transparency and make the government opaque. However, several Brazilian cities have open data policies. An example of categorization is:

> Our bottleneck today regarding Open Data, in addition to publicizing them, is the ability to provide data. Much of them are hidden in our bureaucratic processes. (Interviewee 1).

The technological inhibitors are mostly linked to a deficient connectivity infrastructure and the high costs of telephony and data connections for the Brazilian population, as observed in:

> We cannot do eParticipation without a connectivity infrastructure… The lack of reach by (Internet) providers is a strong inhibitor. (Interviewee 2).

The social dimension highlights the concern with the inclusion in eParticipation of those who are socially and digitally excluded, as found in:

> The cost of a 3G connection still is too high in Brazil. (Interviewee 1).

In the economic dimensions, the amount of investments that are needed in the cities, together with ICT governance and the national telecommunications infrastructure, have a strong impact on eParticipation projects, as seen in:

> Over the years, public administrations have not undertaken the necessary investments to build smart cities. The barriers range from the low quality and lack of integration of information systems and databases (including problems with lack of standards, inconsistency of information, etc.), to weaknesses in the processing environment (data center, information security, etc.), and the lack of a reliable telecommunications infrastructure with high capacity and speed. Any project in this area requires a major investment effort that most cities are unable to tackle. (Interviewee 4).

4.3 Discussion of Results and Recommendations

Notably, the facilitating elements of eParticipation projects are related to environmental variables, particularly in the political and governance dimension, whereas the inhibitors are related to challenges that are internal to the government. Even though the most emphasized dimension for inhibitors is also politics and governance, the strategies to overcome them are internal, dealing with variables of the internal environment. Hence, it seems appropriate to discuss eParticipation spaces not only within the realm of initiatives to develop smarter cities but also within the field of eGovernment.

We found that the role of ICTs in support of citizens' participation and public involvement is a critical theme on urban governance. Collaboration environments and participative governance, supported by ICT, where people can participate in the construction of better life conditions in the city, is a key element of a smart city (Kanter and Litow 2009; Paskaleva 2009), and is also an enabler of a sustainable economic growth and quality of life in a smart city (Caragliu et al. 2009).

Thus smart city initiatives, eParticipation, and eGovernment are interrelated areas, and there is an extensive, consolidated literature on eGovernment. While smart city initiatives differ from general eGovernment initiatives, the two kinds of projects have much in common. Both are government initiatives and both are characterized by intensive use of ICT (Chourabi et al. 2012). In general, the construction of eParticipation spaces in cities is a part of eGovernment efforts.

Despite the advances of ICT in Brazil, particularly in the poorest areas of the country, situations of exclusion are still present, due to digital illiteracy or the lack of technological support. The digital divide phenomenon is linked to social exclusion (Norris 2001). When such situations are identified, concrete actions must be taken in the Smart City project to ensure that those citizens will not be excluded from eParticipation. However, the research indicates that communications infrastructure is the greatest problem in cities. It is known that communications infrastructure and systems are prerequisites to make a smart city, even though they are not sufficient (Chourabi 2012). Yet, communication still is a problem in many regions of Brazil. Efforts should be concentrated to ensure connections with adequate speed for smart city services, at appropriate prices for a developing economy, and able to provide quality support for mobile, virtual, and ubiquitous technologies. Brazil is among the world's most expensive countries with regard to broadband and mobile telephony, and other developing countries face similar situations (ITU 2013). In many cities, eParticipation projects are hindered by precarious connections and high prices. In spite of some federal government initiatives, solutions appear to be far away. The issue of infrastructure is not solved in large Brazilian cities, even those listed as smart cities. Managers of ICT and eParticipation projects must position themselves to participate and exert influence on national, subnational and local actions related to broadband infrastructure, such as the National Plan by the federal government in the Brazilian case. Given the scope of smart city projects, population size and gross domestic product (GDP) of the cities studied, the managers are highly influential stakeholders. The eParticipation projects in specific communities that are digitally excluded may involve actions to mitigate the problem. Yet, actions to make infrastructure available in the city should be the main goal.

Several smart city studies address the technological infrastructure from a systems point of view, particularly their accessibility and availability (Nam and Pardo 2011). Integration, interoperability, compatibility among systems, software, and applications are important issues regarding IT governance in smart cities (Chourabi et al. 2012). The lack of integration of information systems and databases is an obstacle for effective eParticipation and it stems from the absence of ICT governance. Actions to establish the bases for the integration of information systems and databases have positive long-term effects that go beyond the boundaries of eParticipation projects. The establishment of basic processes of ICT governance is a fundamental commitment to the future. Also within the governance field, improvement and adjustment of internal processes that provide consequential responses to electronic participation have a potential to enhance trust in eParticipation projects by citizens and foster their involvement.

A great part of the population already has incorporated the use of mobile devices and applications in their daily lives and such tools embody a great potential for interactions between citizenry and government. If mobile technology was seen a few years ago as an emerging opportunity for initiatives to expand participation (Brücher and Baumberger 2003), one should take into account that by 2014, adoption in developing countries has reached a large scale[8]. Therefore, the people already have electronic devices in their hands that can be used to interact with the government. Similarly, social media can be no longer ignored by governments; they must be incorporated into eParticipation projects.

Transparency is closely connected to eDemocracy and eParticipation projects and it is a way to foster trust among the population (Cunha and Miranda 2013). The development of transparency practices, such as open data, has impacts on the legal and regulatory areas, and it can go beyond that. Public access to data sets made available in open format by government increases transparency and creates new possibilities for interaction and participation. Open data can be associated with actions to support the development of applications, particularly mobile apps to promote the use of that information.

The two phases of the study revealed that public managers, in general, have limited awareness of the importance of incorporating ICT to administration strategies. This is not only observed in developing countriesbut also indicates challenges that are common to smart cities and eGovernment alike (Chourabi 2012). The strategies to deal with such challenges encompass the identification of important stakeholders, the involvement of people, good communication, and planning, among others (Gil-Garcia and Pardo 2005). Leaders of smart city projects must enhance the perception of public managers about the relevant role that ICT can play in citizen participation, providing ICT support to include participation in ongoing initiatives related to key

[8] Data from the National Telecommunication Agency—Agencia Nacional de Telecomunicações (Anatel)—shows that in Brazil the number of mobile phones in January 2014 was 272.4 million or 136.99 phones per 100 inhabitants (available from http://sistemas.anatel.gov.br/SMP/Administracao/Consulta/AcessosPrePosUF/telaConsulta.asp).

sectors, such as public health, education, budgeting, etc. The incorporation of eParticipation tools and processes into strategic projects of public administration holds a great potential to increase the involvement of the targeted communities, expanding the reach of municipal projects. In addition, bureaucracy readiness is another managerial/organizational problem.

Funding is a critical problem for the feasibility of eParticipation. In this area, we envision two points upon which managers may focus: First, to delineate a strategic vision of eParticipation that moves beyond the administrative modernization arguments that have supported eGovernment projects. Instead, eParticipation should be included in a broad smart city strategy that integrates the different areas of administration and is driven by a citizen-centric view. Such strategic vision creates more favorable conditions to justify the necessary investments. The second point concerns the funding sources themselves. It is important to seek partnerships that help to make the project viable within the scope of the smarter city. The partnerships may involve other cities with similar projects and also federal or state governments, which generally have funding schemes for innovation in public management. In smart city projects, one option is the constitution of public–private partnerships, in which private entities fund part of the costs in exchange for permission to exploit commercially part of the services. There is also the possibility of partnerships with international organizations. The European Union, some of its individual members such as Germany, France, and Spain, as well as other countries like the USA and Japan have programs that may fund those projects.

5 Conclusion

Based on the literature and on smart city projects around the world, eParticipation is an important topic of urban agenda. This work was undertaken with the goal of understanding what are the facilitators and inhibitors for the provision of eParticipation spaces in Brazilian smart cities, so that we could advance some recommendations for public managers. For that purpose, we investigated the elements that facilitate or inhibit eParticipation in general, and in the second phase we directed the focus toward smart cities.

We found out that political and governance, sociocultural, and technological aspects are emphasized. The enablers are concentrated in the sociocultural and technological dimensions, while the inhibitors are mainly in the political and governance dimension. The facilitating elements are for the most part, external to the sphere of action of city managers, but their effects may be magnified if undertaken with the support of the municipality. As far as inhibiting elements are concerned, even though they pose broad challenges that may be perceived to be beyond the reach of the leaders of eParticipation initiatives, there is more scope for managerial action.

The main facilitating elements in the political dimension include positive eParticipation experiences, trust by the citizenry, and explicit involvement of municipal executives. In the sociocultural dimension, the adoption of social media was highlighted. In the technological dimension, the most cited were the popularization of mobile devices, the diffusion of mobile apps, and the consolidation of Internet and social media usage by the population. The legal and regulatory dimension was deemed important for the assimilation of the participation process.

Regarding inhibiting elements, political and governance dimension is the main consideration—the lack of involvement by the municipal government, the lack of maturity of governance models, the absence of impacts of virtual decisions in the real world, the weak coordination of ICT management in governments, and the lack of a match between political will and a well-prepared bureaucracy. Information management problems hinder transparency and make the government opaque. Deficient connectivity infrastructure and high communications costs are technological inhibitors. In the social dimension, the concerns are with social and digital exclusions. In economic dimension, investments in the cities, in ICT governance and in national telecommunications infrastructure, have strong impact on eParticipation projects.

It is necessary to emphasize that Smart Cities projects in developing countries may require special attention to facilitators and inhibitors influenced by political, economic, social, and technological characteristics, that may be very specific, such as the limitations of broadband infrastructure.

Our recommendations for smart city and/or ICT managers are in regard to:

- Position ICT managers as important stakeholders in national, subnational and local policies regarding broadband infrastructure
- Delineate a strategic vision of eParticipation within the scope of smart city and citizen-centric actions, beyond the administrative modernization arguments that are recurrent in eGovernment in the country
- Expand ICT governance actions, such as the integration and redefinition of processes as a way to increase trust among the population
- Use mobile technologies and social media
- Develop transparency initiatives, such as Open Data, associated with eParticipation projects
- Define eParticipation projects in the scope of smart city strategy to create favorable condition to justify investments and to take advantage of more innovative forms of funding
- Support initiatives external to the administration, incorporating them into the smart city. This will extend the reach of the project with little additional investment

This research was developed in Brazilian smart cities. As a contribution, this chapter provides a list of facilitators and inhibitors for the construction of public spaces for participation in smart cities in Brazil, which may eventually be useful in other developing countries. There are more general aspects, social, political, and other closer to managerial activities. A limitation of this study is related to their small number, and even the question whether they are in fact smart. However, these cities are moving toward becoming smarter. A better understanding of the challenges they face and a reflection about the positions they assume may also provide a useful contribution for other cities in developing countries.

Public officials may not be sufficiently aware of the possibilities allowed by eParticipation, or be reluctant to allow greater participation, in general. This possibility exists and has additional and partly different managerial implications. This is an interesting topic for future research. Another topic that deserves attention is the use of ICT by social actors to influence government decisions and policies, not necessarily in official virtual spaces.

References

Agre, P. (2002). Real-time politics: The internet and the political process. *The Information Society, 18*(5), 311–331.

Ainsworth, S., Hardy, C., & Harley, B. (2005). Online consultation: E-democracy and e-resistance in the case of the development gateway. *Management Communication Quarterly, 19*(1), 120–145.

Åström, J., Karlsson, M., Linde, J., & Pirannejad, A. (2012). Understanding the rise of e-participation in non-democracies: Domestic and international factors. *Government Information Quarterly, 29,* 142–150. doi:10.1016/j.giq.2011.09.008.

Bakry, S. H. (2004). Development of e-government: A STOPE view. *International Journal of Network Management, 14*(5), 339–350.

Bélanger, F., & Carter L. (2012). Internet voting and political participation: An empirical comparison of technological and political factors. *ACM SIGMIS Database, 43*(3), 26–46.

Busatto, C., & Vargas, P. (2004). *Governança Solidária Local—Fundamentos políticos da mudança em Porto Alegre*. Proceedings of the International Colloquium—The multi-stakeholder aproach in information and communication policy. Veneza, Itália.

Caragliu, A., Del Bo, C., & Nijkamp, P. (2009). *Smart cities in Europe*. Proceedings of the 3rd Central European Conference in Regional Science—CERS, Košice, Slovak Republic, pp. 45–59.

Carrizales, T. (2008). Critical factors in an electronic democracy: A study of municipal managers. *The Electronic Journal of e-Government, 6*(1), 23–30.

Chourabi, H., Nam T., Walker, S., Gil-Garcia, J. R., Mellouli, S., Nahon, K., Pardo, T. A., & Scholl, H. J. (2012). *Understanding smart cities: An integrative framework*. Proceedings of the 45th Hawaii International Conference on System Science (HICSS), Honolulu, HI, USA.

Coleman, S. (2005). *Direct representation: Towards a conversational democracy*. London: Institute for Public Policy Research.

Coleman, S., & Rowe, C. (2005). *Remixing citizenship: Democracy and young people's use of the Internet*. Research report. London: Carnegie Young People Initiative.

Criado, J. I. (2008). *Public sector innovation and the Europeanization of eGovernment: An institutional approach*. Proceedings of the European Group of Public Administration Annual Conference, Rotterdam, Netherlands.

Cunha, M. A., & Miranda, P. R. M. (2013). O Uso de TIC pelos Governos: Uma proposta de agenda de pesquisa a partir da produção acadêmica e da prática nacional. *Organizações & Sociedade, 66,* 543–566.

Cunha, M. A., Duclós, L. C., Souza, G. S., & Correa S. M. (2005). *O Uso de Meios Eletrônicos no Relacionamento do Parlamentar com o Cidadão e com o Poder Executivo nos Grandes Municípios Brasileiros*. Proceedings of the XXIX Encontro Nacional da ANPAD—ENANPAD, Brasília, Brazil.

Cunha, M. A., Coelho, T. R., & Pozzebon, M. (2013). *The use of ICT in public decision-making participation*. Proceedings of the ECIS—21st European Conference on Information Systems, Utrecht, Netherlands.

Dahlberg, L. (2001). Democracy via ciberspace: Mapping the rhetorics and practices of three proeminent camps. *New Media & Society, 3*(2), 157–177.

Damodaran, L., Nicholls, J., Henney, A., Land, F., & Farbey, B. (2005). The contribution of sociotechnical systems thinking to the effective adoption of e-government and the enhancement of democracy. *The Electronic Journal of e-Government, 3*(1), 1–12.

Esteves, J., & Joseph, R. C. (2008). A comprehensive framework for the assessment of eGovernment projects. *Government Information Quarterly, 25,* 118–132.

Frey, K. (2004). *Governança Urbana e Participação Pública*. Proceedings of the XXVIII Encontro Nacional da ANPAD- EnANPAD, Curitiba, Brazil.

Gil-García, J. R., & Pardo, T. A. (2005). E-government success factors: Mapping practical tools to theoretical foundations. *Government Information Quarterly, 22*(2), 187–216.

Grönlund, Å. (2001). Democracy in an IT-framed society. *Communications of the ACM, 44*(1), 23–26.

Grönlund, Å., & Horan, T. (2005). Introducing e-GOV: History, definitions and issues. *Communications of the AIS, 15,* 713–729.

Guidi, L. (2001). E-Democracia em Bologna: a rede cívica Iperbole e como construir uma comunidade participativa online. *Revista Informática Pública, 3*(1), 49–70.

Holzer, M. (2005). *Restoring trust in government: The potential of digital citizen participation*. Proceedings of the second Sino-US international conference: Public administration in the changing world, Beijing, China.

Holzer M., & Manoharan, A. (2007). E-governance and quality of life: Associating municipal e-governance with quality of life worldwide. In C. Reddick (Ed.). *Handbook of research on strategies for local e-government adoption and implementation: Comparative studies* (pp. 408–418). Hershey: IGI Global.

ITU—International Telecommunication Unit. (2013). Measuring the information society. http://www.itu.int/en/ITU-D/Statistics/Documents/publications/mis2013/MIS2013_without_Annex_4.pdf. Accessed 5 March 2014.

Kanter, R. M., & Litow, S. S. (2009). *Informed and interconnected: A manifesto for smarter cities*. Harvard Business School General Management Unit Working Paper, 9–141. http://www.hbs.edu/faculty/Publication%20Files/09-141.pdf. Accessed 17 March 2014.

Lindskog, H. (2004). *Smart communities initiatives*. Proceedings of the 3rd ISOneWorld Conference, Las Vegas.

Macadar, M., & Reinhard, N. (2002). *Telecentros Comunitários possibilitando a inclusão digital: Um estudo de caso comparativo de iniciativas brasileiras*. Proceedings of the XXVI Encontro Nacional da ANPAD-EnANPAD, Salvador, Brazil.

Macintosh, A. (2004). *Characterizing e-participation in policy-making*. Proceedings of the 37th Annual Hawaii International Conference on System Sciences (HICSS), Honolulu, HI, USA.

Madrid. (2004). Reglamento orgánico de participación ciudadana del Ayuntamiento de Madrid. http://www.munimadrid.es/portales/munimadrid/es/Inicio/El-Ayuntamiento/Participacion-ciudadana/ANM-2004-22-Reglamento-Organico-de-Participacion-Ciudadana-del-Ayuntamiento-de-Madrid?vgnextfmt=default&vgnextoid=b5117b3cb0e4f010VgnVCM1000009b25680aRCRD&vgnextchannel=8449e3d5d3e07010VgnVCM100000dc0ca8c0RCRD&idSubPage=1. Accessed 7 March 2010.

Motta, F. C. P.(1990). *Organização e Poder: Empresa, Estado e escola*. São Paulo: Atlas.

Nam, T., & Pardo, T. A. (2011). *Conceptualizing smart city with dimensions of technology, people, and institutions*. Proceedings of the 12th Annual International Conference on Digital Government Research—Dg.o'11, College Park, MD, USA.

New Zealand. (2004). *Participation through e-government: The context*. Wellington: State Services Commission. http://www.e.govt.nz/policy/participation/participation-0305/participation-0305.pdf. Accessed 10 April 2010.

Norris, P. (2001). *Digital divide civic engagement, information poverty, and the internet worldwide*. New York: Cambridge University Press.

Odeendal, N. (2003). Information and communication technology and local governance: Understanding the difference between cities in developed and emerging economies. *Computers, Environment and Urban Systems, 27*(6), 585–607.

Paskaleva, K. (2009). Enabling the smart city: The progress of city e-governance in Europe. *International Journal of Innovation and Regional Development, 1*(4), 405–422.
Rose, J., Grönlund, A., & Andersen, K. V. (2007). What is eParticipation? In A. Avdic, K. Hedström, J. Rose, & Å. Grönlund (Eds.), *Understanding eParticipation: Contemporary PhD eParticipation studies in Europe*. Sweden: Örebro University Library.
Ruediger, M. A. (2002). *Governo Eletrônico e Democracia—Uma Análise Preliminar dos Impactos e Potencialidades na Gestão Pública*. Proceedings of the XXVI Encontro Nacional da ANPAD—ENANPAD, Salvador, Brazil.
Sæbø, Ø., Rose, J., & Flak, K. S. (2008). The shape of eParticipation: Characterizing an emerging research area. *Government Information Quarterly, 25,* 400–428.
Schaffers, H., Komninos, N., Pallot, M., Trousse, B., Nilsson, M., & Oliveira, A. (2011). Smart cities and the future internet: Towards cooperation frameworks for open innovation. In J. Domingue, et al. (Eds.). *Future internet assembly* (Vol. 6656, pp. 431–446). Cham (ZG): Springer.
Susha, I., & Grönlund, A. (2012). eParticipation research: Systematizing the field. *Government Information Quarterly, 29,* 373–382. doi:10.1016/j.giq.2011.11.005.
Thindwa, J., Monico, C., Reuben, W. (2003). Enabling environments for civic engagement in PRSP countries. *Social Development Notes N° 82*. Washington. World Bank.
United Kingdom. (2009a). *Local democracy, economic development and construction bill.* London: House of Commons, The Stationery Office Limited.
United Kingdom. (2009b). *Listening to communities: Consultation on draft statutory guidance on the duty to respond to petitions. Secretary of state for communities and local government. Department for communities and local government.* London: Communities and Local Government Publications.
United Nations. (2010). *E-government survey 2010: Leveraging e-government at a time of financial and economic crisis*. New York: UN Publishing Section.
Williamson, A. (2006). *Disruptive spaces and transformative praxis: Reclaiming community voices through electronic democracy*. Proceedings of the conference of Community Informatics Research Network, Prato, Itália.

Paulo Roberto de Mello Miranda is the municipal secretary for Information and Technology of the Government of Curitiba. He is a civil engineer and holds a master's degree in business administration from Federal University of Rio Grande do Sul (UFRGS). Former president for three mandates of the Brazilian Association of Government ICT Companies (ABEP), CEO of CELEPAR—ICT Company of the State of Paraná, director of PROCERGS—ICT Company of the State of Rio Grande do Sul, and of SERPRO—ICT Company of the Brazilian Ministry of Finance. He previously served as the executive secretary of the ICT Council of the states of Paraná and Rio Grande do Sul. He was the founder and director of the International Centre for Software Technology—CITS. Paulo Miranda has over 30 years of experience as ICT professional and consultant, leading professional projects in information and communications technology, strategic planning, and technology management in government and private organizations.

Maria Alexandra Viegas Cortez da Cunha has 30 years of experience as a professor, IT professional, and consultant. She holds a doctorate in business administration from Universidade de São Paulo (USP) and is a professor at Escola de Administração de Empresas de São Paulo (Sao Paulo Business School) of Fundação Getulio Vargas (FGV), in Master's Degree Program and Doctorate Program, where her line of research is information technology management. She has been leading academic research and professional projects in IT governance, e-government, e-governance, and e-democracy, and on models for administering computerization in governments. More recently, her research interests are IT for development and smart cities. Professor Cunha is the author of several articles, technical reports, researches, chapters and books, and is a member of the editorial board of Brazilian academic journals in business administration. She coordinated for two mandates the "Information Management" area at ANPAD (Brazilian Academy of Management).

José Maria Pugas Filho has graduated in law and in cultural and media studies, specialized in urban and consumers anthropology, and holds a specialized graduate diploma in sustainability and in business administration. For 10 years, Pugas has been engaged in innovation consulting services for national, state, and municipal governments in Brazil and South America. Pugas is a researcher and lecturer in new technologies, urban innovation, future sciences and cultural migrations in Brazilian and International research networks. Since 2014, Pugas is the innovation director in the Secretary of Information and Technology of Curitiba, developing innovation strategies to guide the city in the next decade and innovation advisor in venture capital funds dedicated to fomenting the innovative entrepreneurship community in the 3.4 million people of the Curitiba Metropolitan Area.

Technology Helps, People Make: A Smart City Governance Framework Grounded in Deliberative Democracy

Roberto Garcia Alonso and Sebastian Lippez-De Castro

Abstract In sharp contrast to perspectives which overestimate the role played by technology in promoting smart cities initiatives, this chapter explores an important dimension for a full development of smart cities, the integration of human dimensions. Based on the theoretical model of Habermasian deliberative democracy, the chapter proposes a revision of the Chourabi et al. (2012) analytical framework for smart cities initiatives, in which people are as important as technology. In particular, we state that success of smart city initiatives depends upon the capability of integrating people and communities engagement with the advantages of information and communication technologies (ICTs), within a comprehensive smart city governance framework.

Keywords Technology · Habermasian model · Governance · People and communities · Smart city · Framework

Abbreviations

ICTs Information and Communication Technologies
OECD The Organization for Economic Cooperation and Development

1 Introduction

Despite differences in our images of the future in terms of their apocalyptical, paradisiac, or fantastic character, almost all of them involve several types of advances in the use of technology. Some dream of flying cars, others envision humans on the Moon or even on Mars, and some predict some form of humanoids with artificial intelligence. In any case, nearly everyone who has envisioned our race's future an-

R. Garcia Alonso (✉)
Cll 40 No. 5—50, Ed. (9) Julio Carrizosa, Piso 2, Bogotá, Colombia
e-mail: garcia.roberto@javeriana.edu.co

S. Lippez-De Castro
Cll 40 No. 5—50, Ed. (9) Julio Carrizosa, Piso 1, Bogotá, Colombia
e mail: slippez@javeriana.edu.co

J. R. Gil-Garcia et al. (eds.), *Smarter as the New Urban Agenda A Comprehensive View of the 21st Century City,* Public Administration and Information Technology 11,
DOI 10.1007/978-3-319-17620-8_18

ticipates tremendous advances in the use of information and communication technologies (ICTs). In fact, during the past decades we have witnessed a rapid change boosted by innovation in this realm. Cellular phones, pagers, mobile devices, geographic information systems, and systems' interoperability and integration represent only a few of these advances. At the same time, our world's population has turned from living mainly in rural places to living in urban areas. Consequently, it is expected that ICTs will contribute to building up our future cities by providing support to manage the challenges associated with growing populations. In this context, the use of ICTs is frequently associated with the idea of making cities smarter.

This chapter discusses the role played by ICTs, particularly eGovernment, in walking the road toward smarter cities. In sharp contrast to perspectives which overestimate the role played by technology, we suggest that using ICTs' advantages such as those provided by eGovernment tools, m-government applications, the integration of information systems, and even the capabilities of Web 2.0 are not enough for a city to become smarter. In that vein, building up on the Habermasian deliberative democracy idea, we call for moderate expectations for smart city initiatives resting only on technology. We assert that cities aiming to become smarter should integrate ICTs' advantages with complementary and required instances and institutions (by adapting their governance structure) that, beyond those facilitated through ICTs, provide assurances for thoughtful, democratic, inclusive, and equitable citizen participation.

As a result, this chapter proposes a revision of Chourabi et al.'s (2012) analytical framework for smart city initiatives. In particular, we show that success of smart city initiatives depends upon the capability of integrating the advantages of ICTs and empowering citizens to make significant contributions, to ultimately improve social justice, equity, and sustainability. According to Chourabi et al. (2012), the influence of outer factors (governance, people and communities, natural environment, infrastructure, and economy) are in some way filtered or influenced by more influential inner factors (technology, management, and policy). Indeed, since it could heavily influence each of the other seven factors, Chorubi et al. (2012) consider technology as a meta-factor in shaping smart cities initiatives. However, we argue that, when considering an initiative as smart, people and community are as important as technology. In addition, we reconsider governance as the smart city initiatives "backdrop," instead of a single factor itself.

In doing so, we first revise the concept of smart cities related to governance and citizenship. Then, we review the role and character of citizen participation in relation to the Habermasian deliberative model of democracy. From that standpoint, we then reflect on the role played by ICTs in walking the road toward smarter cities. And finally, we rethink the Chourabi et al.'s (2012) analytical framework for smart cities initiatives and propose an alternative smart city governance framework.

To address these issues, the chapter employs an analysis of smart cities from a contextual approach to political theory and makes a literature review, focusing on interactions between human aspects and ICTs (capability, social learning, specific uses, etc.). Based on the Habermasian deliberative model of democracy we create evaluation criteria that will be relevant to revise the promises and hurdles of

ICTs in allowing citizens to better understand and participate in public affairs and decision-making. In particular, we examine whether ICTs can help to improve the characteristics of openness, rationality, and public debate which are essential to the Habermasian concept of deliberation.

Finally, we discuss problematic aspects of the smart cities clusters of factors included in the Chourabi et al. analytical framework (e.g., governance and citizenship). As a result, we suggest conceptual remedies by redefining the concept of governance and by relocating the people and communities factor in the context of an alternative smart city governance framework.

2 Smart Cities, Governance, and Citizenship

As Chourabi et al. (2012) highlight, there is a sort of "cacophony" derived from the numerous attempts to define a concept of smart cities. In fact, they identify at least six working definitions of smart cities (p. 2290) and, from that, propose an integrative framework of eight factors which contribute to understand smart city initiatives or projects, as well as, to determine their success. The eight factors are "(1) management and organization, (2) technology, (3) governance, (4) policy, (5) people and communities, (6) the economy, (7) built infrastructure, and (8) the natural environment" (p. 2291). From that cacophony, one definition rises above the noise and demands additional attention given its inclusion of governance, and people and communities, as main factors for smart city initiatives.

One such definition of smart cities overcoming the noisy environment is that drafted by Giffinger et al. (2007) which states that "A Smart City is a city well performing in a forward-looking way in these six characteristics [smart economy, smart people, smart governance, smart mobility, smart environment, and smart living], built on the "smart" combination of endowments and activities of self-decisive, independent, and aware citizens" (p. 11). In contrast, the remaining five definitions collected by Chourabi et al (2012), including those proposed by Hall (2000), the Natural Resources Defense Council, Toppeta (2010), Washburn et al. (2010), and Harrison et al. (2010), seem to neglect the role played by citizens, and instead emphasize the interconnectedness of physical and ICTs infrastructures, innovations, improving efficiency in service delivery, and the purpose of livability and sustainability. Indeed, Chourabi et al. (2012) highlight that although "addressing the topic of people and communities as part of smart cities is critical, [...] traditionally [it] has been neglected at the expense of understanding more technological and policy aspects of smart cities" (p. 2293).

Closely related to the definition showcasing citizens (Giffinger et al. 2007) is a focus on governance. There is "little literature on smart cities address[ing] issues related to governance" (according to Chourabi et al. 2012, p. 2292), however Giffinger et al. (2007) identify "smart governance" as one of the main six characteristics in which a smart city performs in a forward-looking way. Governance has been studied extensively, but little has been done relating it to smart cities. In that sense, Giffinger et al.'s (2007) definition of smart cities stands out from the others. This is not to say that "governance" has been entirely neglected, but it is not clear how "governance" is then incorporated in the way or frame in which smart city initiatives are designed or evaluated. In contrast, Giffinger et al.'s (2007) definition develop and operationalize governance as a characteristic of smart cities.

Indeed, understanding governance, particularly the new public governance, is certainly critical to better design, comprehend, and evaluate smart cities initiatives. It explains the fact that the decision-making process has become characterized by interdependent networks of actors (government, private sector, and civil society actors). As a result, the concept of governance indicates a new style of government which is a nonhierarchical control model characterized by formal and informal rules, structures, and processes in a context of a greater degree of cooperation and interaction between public and private actors (Kooiman 1993; Rhodes 1997). In the same vein, Clarke and Stewart (1998) explain that new public governance refers to community governance (Clarke and Stewart 1998). In Osborne and McLaughlin's words "the public sector is no longer defined solely in relation to the presence, or otherwise, of the government as a planner or service provider. Rather the planning, management and provision of public services is seen as something to be negotiated between a number of actors, including government, the voluntary and community sectors and the private sector" (Osborne and McLaughlin's 2002, p. 10), which defines the new understanding of new public governance. According to Prats (2003) governance is defined by actor's interactions produced by formal and informal rules, which ultimately affects governability. This perspective of governance is similar to Lynn et al.'s (2000) concept of governance identified by Chourabi et al. (2012). In that sense, technology can facilitate interactions among citizens, government, and other actors, playing an important role in improving efficiency and effectiveness in communications and public service delivery. As a result, governance cannot be seen as a factor of success for smart cities initiatives, but should be understood as the analytical tool through which new reality can be described. Therefore, this revised understanding of the concept of governance will be used to redefine the smart cities initiative framework as an alternative smart city governance framework.

This understanding of new public governance explains in part why the literature in the realm of local government (or "urban policy"/"urban politics") has had an extraordinary growth in the past decades—from the inspired elitist theories (Harding 1995), to the most modern approach to governance (Denters and Rose 2005; Pierre 2000; Kersting and Vetter 2003; Klijn 2008), or the contemporary concept of smart cities. Nowadays, the majority of empirical studies are focused on the functioning of the formal and informal modes of collaboration between public and private sectors in urban settings (Stone 1989; Mossberg 2009). The concept of "urban regime" and the analysis methodology associated with it (Urban Regime Analysis) have been the center of local network governance research. This subdiscipline has generated a sophisticated and consistent body of theories (Judge et al. 1995; Davies and Imbroscio 2009), typologies (Pierre 1999), and explanations. As a result, the concept of smart cities has entered into the analysis about local governance.

Consequently, urban performance currently depends not only on the city's endowment of hard infrastructure ("physical capital") but also on the availability and quality of communication and social infrastructure. According to Nam and Pardo (2011), the collaboration among different functional sectors, private and public organizations, and among different levels of government within a geographical region is fundamental (governance). From this standpoint, "good governance" would depend on the implementation of a smart governance infrastructure that should be accountable, responsive, and transparent. Thus, several cities have felt an increased need for better governance to manage these projects and initiatives (Mooij 2003; Johnston and Hansen 2011).

According to Nam and Pardo (2011) the smart city concept is integrated by a set of multidimensional components. The concept is aligning to three important dimensions (technology, people, and institutions): integration of infrastructures and technology-mediated services, social learning for strengthening human infrastructure, and governance for institutional improvement and citizen engagement.

In this sense, some authors suggest that governance would be a considerable challenge for democratic accountability. Why? Governance supposes a "more informal decision-making process" (Hambleton and Gross 2007, p. 9). Thus, the governance implies that decisions are taking place outside the traditional institutions of liberal democracy. That is, new public governance implies multiple actors, including people and communities, interacting, collaborating, and deliberating on public affairs. Consequently, it is not possible to understand the factor of people and communities separated from the factor of governance—they are connected.

This chapter argues that without people and communities engaged in public deliberation, acting on the local public governance arena, the smart cities initiatives would be a failure. In that sense and taking into consideration that ICTs have been pointed as tools with the potential of providing citizens to become these active actors, it is required to reflect on such ICTs'capability.

3 The Habermasian Model of Deliberative Democracy

There is a general consensus about the values that represent liberal democracy within Western societies, whereas the institutions supposed to embody these ideals show signs of crisis (Norris 2002; Putnam 2002). This paradoxical situation highlights the importance of democratic renewal, both theoretical and empirical. At a theoretical level, the deliberative democratic ideal and its communicative reformulation of the democratic vision represents an alternative to the dominant liberal tradition (Bohman and Rehg 1997; Elster 1998; Fishkin and Laslett 2003). At a practical level, ICTs have changed the organizational preconditions of democratic governance in important ways (Alexander 1998; Hacker and van Dijk 2000; Hoff et al. 2000; Kamarck and Nye 2002; see Wiklund 2005). In this context, the local arena and its institutions have become spaces of political reorganization. This new political order has been built from the basis of a participatory and deliberative agenda.

Deliberative democracy is a normative ideal of democracy. The notion of deliberative democracy in republican government appears in 1980 in the work of Joseph M. Bessette (1980). This was the democratic model of the American Constitution and marks the initial research about the model in the field of Anglo-Saxon political theory. But, the German sociologist and philosopher Jürgen Habermas was the highly influential thinker who dictated the content and limits of the academic debate in this regard, bridging continental and Anglo-American traditions of thought[1].

From a Habermasian perspective (1984, 1998), the process of democratization of modern societies goes through an increase of autonomous public spheres between participants. In the Habermasian deliberative model, the key aspect is the quality of this participation. The model is based on a robust civic society able to influence the process of decision-making. ICTs would be conceived as tools that would allow connect civil society with institutions. The deliberative model understands the civil society as a public space open to all (inclusiveness), as a "voice box" which serves to detect the problems that affect the society, filtering relevant issues that are transmitted to the formal agendas of authorities and institutions thorough informative, rational and public debate, thus it is more probable that public decisions are more rational. In this way, ICTs would be a link, a transmission belt between civil society and institutions which would be more responsive and accountable.

Habermas (1984) identifies and reconstructs universal conditions or "universal validity claims" in the structures of argumentative speech. Within communicative action, an agent must necessarily engage these claims[2] and believe that such claims can be made. These conditions which act as an "ideal speech situation," act as a regulative ideal. Habermas provides a set of criteria for measuring the quality of discourse related to structural features such as inclusiveness, equal communicative rights, and the absence of repression or manipulation. Other criteria concerned required dispositions of participants including reflexive attitudes toward one's own claims, willingness to take the demands and the counterarguments of the other seriously, and sincerity or the absence of manipulation and self-deception (Habermas 2005a).

These ideas are counterfactual in the sense that real-life discourses rarely achieved such quality or conditions—and can never empirically certify—full inclusion, noncoercion, and equality. Nevertheless, these idealizing presuppositions have an operative effect on the actual discourse providing a framework for reasonable outcomes (Habermas 2003). In this sense, this process of argumentation is a self-correcting learning process because it guarantees the rational expectation that the relevant information and reasons "put on the table" can influence the outcomes (Habermas 2005a). Other criteria concern required dispositions of participants: reflexive attitudes toward one's own claims, willingness to take the demands and the counter arguments of the other seriously, and sincerity or the absence of manipulation and self-deception (Habermas 2005b).

---

[1] The Stanford Encyclopedia of philosophy (http://plato.stanford.edu/entries/habermas/).

[2] These validity claims concern at least three dimensions of validity: (1) truthfulness, (2) rightness, and (3) truth (Habermas 1984).

Thus, in the next section we examine whether ICTs can help to improve the characteristics of openness, rationality, and public debate which are essential to the Habermasian concept of deliberation.

4 The Role Played by ICTs in Smart Cities Initiatives: Inclusiveness, Noncoercion, and Equal Participation

Taking into account the role assigned to people and communities by Giffinger et al. (2007) in smart city initiatives, the networked character of the new public governance which requires collaborations and interaction of multiple stakeholders, as well as the Habermasian deliberative model of democracy; it is relevant to revise the promises and hurdles of ICTs in allowing citizens to better understand and participate in public affairs and decision-making.

To do so, it is important to insist on the idea that "the deliberation is presented like a communication normatively constrained, communication that aims to change the content of, intensity of, or the reasons for the preferences, beliefs, actions or interpretations of one's interlocutors with respect to matters of public concern" (Neblo 2005, p. 174). In that context, decisions are legitimate because they have been taken following a procedure to ensure full inclusion, noncoercion, and equal participation of citizens in public deliberation.

4.1 Inclusiveness

Deliberative mechanisms must be designed to promote real inclusion of citizens, public debate where citizen and authorities must present arguments, listen each other's, etc. The inclusion of all citizens is an important aspect to the quality of deliberation. It helps to avoid cognitive errors and biases that can occur in situations where there is a single perspective or an interpretive framework. Bohman calls it the "epistemic value of diversity" (Bohman 2007, p. 349). In other words, an inclusive deliberation would reduce the limitations of individual reasoning, forcing us to revise the very strength of our arguments, cognitive asymmetries and then would increase the quality of the decisions (Bohman 2006, p. 188).

But, what determines that the procedure allows us to get more rational decisions? Deliberative theorists assume that the individuals would have some "deliberative capacities." A successful deliberation depends on some cognitive and moral capacities. Citizens will have or create their own beliefs and preferences, with the ability to consider alternatives and an attitude to receive favorable and serious demands and counterarguments of others, etc. In that sense, deliberative democracy implies a qualified, informed, and thoughtful citizenry.

The OECD report (2001) argues that democratic, political participation must involve the means to be informed, the mechanisms to take part in the decision-making and the ability to contribute and influence the policy agenda. In that vein, the OECD suggests three levels of participation: (1) information: a one-way relationship in which government produces and delivers information for use by citizens; (2) consultation: a two-way relationship in which citizens provide feedback to the government. Governments define the issues for consultation, set the questions and manage the process, while citizens are invited to contribute their views and opinions. And, (3) active participation: a relationship based on partnership with government in which citizens actively engage in defining the process and content of policy-making. It acknowledges equal standing for citizens in setting the agenda, although the responsibility for the final decision rests with government.

According to the OECD, the role of ICTs in the first level of participation is to provide individual citizens with enough information. The citizens can access the Internet and examine the large amount of information available. The technology may contribute to provide relevant information in a format that is both more accessible and more understandable. As noted by the OECD, ICTs could contribute to engage with citizens: E-engaging with citizens is concerned with consulting a wider audience to enable deeper contributions and support deliberative debate on policy issues. And finally, the use of technology in the third level could help to empower citizens: E-empowering citizens are concerned with supporting active participation and facilitating bottom-up ideas to influence the political agenda. From this perspective, citizens are emerging as producers rather than just as consumers of policy (Al Dalou and Shanab 2013).

Although the promise of inclusiveness seems to be satisfied because anyone could access information on public affairs, as well as make their judgments through ICTs; there are asymmetries and restrictions to access and take advantage of ICTs. One of the main restrictions is the possibility of free access to the Internet. Even if it would be possible to guarantee everyone's access to the ICTs, there are additional restrictions associated with knowledge and skills required not only to appropriately use ICTs but also to engage in a public debate. If deliberation requires some "deliberative" capacities, the "e-deliberation" demands some technological capabilities. These concerns have been associated with the concept of the so called "digital divide." The term "digital divide" has become the accepted manner for referring to "the social implication of unequal access of some sectors of community to Information and Communication Technology [ICT] and the acquisition of necessary skills" (Foster 2000, p. 445). The use of ICTs for the government must be accompanied by the ability to use this technology; without this ability some members of community cannot fully participate in economic, political, and social life. The concept of a "digital divide" is not only of technological learning but also a question of social and economic differences. The literature suggests the possibility that digital government would exclude specific groups and for those excluded the costs would become higher (Holden et al. 2003). Virtual barriers could create a "democratic divide" (Mossberger et al. 2008) for some social groups; for example persons with disabilities or people who do not speak the language of the government (e.g., Rubaii 2006, 2008).

4.2 Noncoercion and Equal Participation

Political decisions are to be made through a collective procedure of argumentation where arguing consists of exchanging reasons, oriented to the goal of rationally convincing others, instead of strategic participation oriented to impose on others personal political preferences or desires (Gutmann and Thompson 1996; Christiano 1996, pp. 53–55; Fishkin and Laslett 2003, p. 2); and this process is supposed to lead us, at least ideally, to rational consensus. In that sense, deliberative democracy not only demands some deliberative capacities but also some structural requirements as well. Participants must influence the outcome because political decisions are the result of a collective procedure of argumentation where citizens and authorities exchange arguments in order to collectively achieve rational outcomes. Thus, the promise of the ICTs would be to not only to provide enough information but also to help the decision-making process. The use of technology would be to engage and empower citizens, facilitating and supporting active participation, and facilitating bottom-up ideas to influence the political agenda.

In doing so, governments can use their websites to disseminate information either in a static manner or a dynamic one. The earliest were read-only websites while the latest involve two-way communications between citizens and public servants. In addition, websites can foster citizen participation during the decision-making process through static citizen deliberation where there is certain consultation of citizens' opinions but no real debate, or through dynamic citizen deliberation where citizens, bureaucrats, interest groups, the media, and politicians certainly debate on policy issues; indeed there is "broader and freer interaction between participants" (Holzer et al. 2004, p. 14). In the same vein, digital democracy includes "all practices to improve democratic values using information and communication technologies" (Holzer et al. 2004, p. 8). In that sense, considering that "Civic engagement consists of knowledge, discussion, interest and participation in public affairs-in government and politics, policy issues, and the community" (Mossberger and Jimenez 2009, p. 2), wherein citizens' awareness and knowledge of what government does and who does what arise as its preconditions and information constitutes a framework for the subsequent participation. Particularly, local governments can use their websites to foster citizens' involvement in the public sphere by providing information and options to participate.

In that sense, the concept of smart city cannot be understood without the technology that allows its existence: Web 2.0. And it is true that Web 2.0 refers to a new generation of Web services based on the concept of a network, which permit the creation of communities, creating networks between entities, regardless of whether they are web pages, people, photos, professional colleagues, hobbies, or any other type. These tools offer an opportunity to make things themselves and forge closer alliances and partnerships between diverse communities of interest, practice, expertise, conviction, and interdependence. Using the network as a fundamental organizational model, the individual can establish collaboration routines although they are not often aware of this collaborative process. From this point of view, the first task that cities must address is the creation of applications enabling data collection and processing, web-based collaboration, and actualization of the collective intelligence of citizens (Komninos et al. 2013).

However, improving collaboration, engagement, and active participation in the context of noncoercion and equal rights of communication by using the advantages of Web 2.0 is still a promise. The empirical evidence[3] suggests that local governments have not made Web 2.0 tools available to the public in a sufficient manner. And even when available, there is no evidence on the influence of public debate on decision-making. In addition, the promise of deliberative democracy has mainly to do with achieving more rational decisions. Indeed, Mossberger and Jimenez (2009) caution that "government websites can facilitate but not create citizens engagement" (p. 6), and Holzer et al. highlight that "there is no evidence to suggest that the quality of decision-making has improved or that decisions are more democratic given the integration of ICTs and digital-base applications" (p. 10). Thus, although citizens would tend to be more interested in participating since these applications allow them to be not only informed but also to actually debate on their concerns, as well as to influence elected officials' opinions; it is difficult to guarantee an increase of citizen participation, or more democratic and quality decision-making processes. Online citizen participation also depends on the diffusion of those alternatives before the community, it depends on the citizenry willingness to actually take advatage of such alternatives (Carter and Belanger 2004) and the appropriate use of ICTs tools by those citizens.

In that sense, improving quality of public debate and decision-making by utilizing ICTs faces the challenge of the appropriate use the tools. For instance, anonymity represents one of the most important concerns in this regard. Some researchers have found that anonymity allows for a fluidity of identity so that citizens can present themselves in varied ways, without feeling judged or constrained by conventional cultural cues (Bowker and Tuffin 2003; Kim and Jee 2006). Others argue that anonymity contributes to a lack of civility and respect in online discussion and that compelling people to use their real names encourages them to take responsibility for what they say and be more thoughtful when contributing (Polat and Pratchett 2009). Finally, the literature shows the importance of a moderator. The moderator plays a broader role in facilitating deliberation, acting as a "helper," "facilitator," and as a "filter" (Edwards 2002). Consequently, taking advantage of ICTs to enhance knowledge, communication, and deliberation, also requires us to develop civic skills and responsibility.

5 An Alternative Smart City Governance Framework

It has been argued that the promise that ICTs could contribute in providing information, engaging and empowering citizens, is not fully achieved yet. Additionally, it has also been stated that it is not clear whether the role of eGovernment and other

[3] Department of Innovation, Industry and Regional Development, State Government of Victoria, Australia (2009): Internet Content Syndication Council (2008); European Commission (2009).

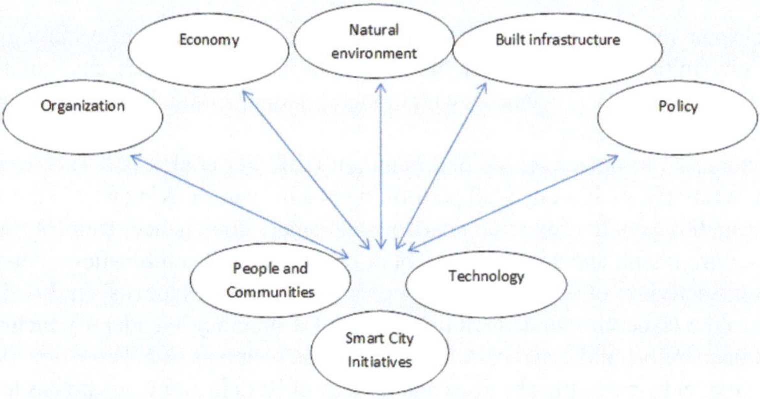

Fig. 1 Alternative smart city governance framework

ICTs' advantages in enabling public debate guarantees the quality of public decisions. In that sense and considering the main role of citizens in smart city initiatives (Giffinger et al. 2007), as well as their required informed and thoughtful participation to be involved in the complex networked local governance, the position of people and communities should be revised.

Smart cities initiatives need a new framework. Chourabi et al.'s framework is an important contribution. According to them the role of people and communities is significant, but it is only considered as an outer factor which is in some way filtered or influenced more than influential inner factors (technology, management, and policy). However, since technology acts only as an enabler itself, it cannot make cities smarter. It also requires people to take advantage of that technology, or to use other instances or institutions of the local governance network to become involved in public affairs. As a result, people and communities become as important as technology for smart city initiatives. Thus, we argue that, technology as well as people and communities should be considered as meta-factors of smart city initiatives since it could heavily influence each of the other seven factors.

In addition, and considering that governance has been characterized by the interaction and collaboration of multiple stakeholders influencing decision-making processes we argue that, given that smart city initiatives should involve technology and people and communities to affect the relationship among local actors, every smart city project ultimately becomes or affects a local governance initiative. As a result, governance should be considered as a "backdrop" of smart city initiatives instead of a factor itself (Fig. 1).

As a result, building upon Chourabi et al.'s framework, we propose the following alternative smart city governance framework:

6 Discussion

This chapter reacts to perspectives that overestimate the value of technology for smart city initiatives. In contrast, we state that successful smart city initiatives should integrate ICTs' advantages with thoughtful, democratic, inclusive, and equitable citizen participation.

To address our objective, we first highlight Giffinger et al.'s (2007) concept in which "a Smart City is a city well performing in a forward-looking way in these six characteristics [smart economy, smart people, smart governance, smart mobility, smart environment, and smart living], built on the "smart" combination of endowments and activities of self-decisive, independent, and aware citizens" (p. 11). Then, based on the Habermasian deliberative model of democracy, we identify inclusiveness, noncoercion, and equal participation as main elements of a democratic deliberative model to revise the promises and hurdles of ICTs in allowing citizens to better understand and participate in public affairs and decision-making. As a result, it should be noticed that the advantages of ICTs are still a promise. Technology helps by offering channels for improving public knowledge of government operations, for fostering communications between citizens and authorities, and among government agencies; but only people can make such communications possible because it all depends on their ability and willingness to use and take advantage of those ICTs.

Consequently, the chapter discusses Chorabi et al.'s (2012) analytical framework in which, among other factors, technology is considered as a meta-factor in shaping smart cities initiatives. We, therefore, propose a revised smart city initiatives analytical framework in which people and communities are as important as technology and we reconsider governance as the smart city initiatives "backdrop," instead of a single factor itself.

References

Al-Dalou, R., & Abu-Shanab, E. (2013). *E-participation levels and technologies*. The 6th International Conference on Information Technology (ICIT 2013), Amman, Jordan, 8–10 May (pp. 1–8). http://sce.zuj.edu.jo/icit13/images/Camera%20Ready/E-Technology/656.pdf. Accessed 29 July 2014.

Alexander, C. (Ed.). (1998). *Digital democracy*. Oxford: Oxford University Press.

Bessette, J. M. (1980). Deliberative democracy: The majority principle in republican government. In R. Goldwin & W. Schambra (Eds.), *How democratic is the constitution?* Washington, DC: American Enterprise Institute.

Bohman, J. (2006). Deliberative democracy and the epistemic benefits of diversity. *Episteme, 3*(3), 175–190.

Bohman, J. (2007). Political communication and the epistemic value of diversity: Deliberation and legitimation in media societies. *Communication Theory, 17*(4), 348–355.

Bohman, J., & Rehg, W. (Eds.). (1997). *Deliberative democracy: Essays on reason and politics*. Cambridge: MIT Press.

Bowker, N., & Tuffin, K. (2003). Dicing with deception: People with disabilities' strategies for managing safety and identity online. *Journal of Computer-Mediated Communication, 8*(2). doi:10.1111/j.1083-6101.2003.tb00209.x.

Carter, L., & Belanger, F. (2004). *Citizen adoption of electronic government initiatives*. System Sciences. Proceedings of the 37th Annual Hawaii International Conference on IEEE (p. 10). http://www.computer.org/csdl/proceedings/hicss/2004/2056/05/205650119c.pdf. Accessed 29 July 2014.

Chourabi, H., Nam, T., Walker, S., Gil-Garcia, J. R., Mellouli, S., Nahon, K., & Scholl, H. J. (2012). *Understanding smart cities: An integrative framework*. System Science (HICSS). 45th Hawaii International Conference on IEEE (pp. 2289–2297). http://www.computer.org/csdl/proceedings/hicss/2012/4525/00/4525c289.pdf. Accessed 29 July 2014.

Christiano, T. (1996). *The rule of the many*. Colorado: Westview Press.

Clarke, M., & Stewart, J. (1998). Community governance, community leadership and the new local government. YPS for the Joseph Rowntree foundation. http://www.jrf.org.uk/system/files/lgr119.pdf. Accessed 29 July 2014.

Coe, A., Paquet, G., & Roy, J. (2001). E-governance and smart communities a social learning challenge. *Social Science Computer Review, 19*(1), 80–93.

Davies, J., & Imbroscio, D. (2009). *Theories of urban governance*. London: Sage.

Denters, B., & Rose, L. E. (2005). *Comparing local governance—Trends and developments*. New York: Taylor and Francis.

Department of Innovation, Industry and Regional Development, State Government of Victoria, Australia. (2009). Web 2.0: The new tools for democratic conversations—A snapshot of initiatives in government. http://www.egov.vic.gov.au/. Accessed 29 July 2014.

Edwards, A. R. (2002). The moderator as an emerging democratic intermediary: The role of the moderator in Internet discussions about public issues. *Information Polity, 7*(1), 3–20.

Elster, J. (Ed.). (1998). *Deliberative democracy*. Cambridge: Cambridge University Press.

European Commission. (2009). Public services 2.0: The impact of social computing on public services. Luxembourg: Office for Official Publications of the European Communities. http://ftp.jrc.es/EURdoc/JRC54203.pdf. Accessed 29 July 2014.

Fishkin, J., & Laslett, R. (2003). *Debating deliberative democracy*. Oxford: Blackwell.

Foster, S. P. (2000). The digital divide: Some reflections. *International Information and Library Review, 23,* 437–451.

Giffinger, R., Kramar, H., & Haindl, G. (2007). *The role of rankings in growing city competition*. Proceedings of the 11th European Urban Research Association (EURA) Conference, Milan, Italy, October 9–11. http://publik.tuwien.ac.at/files/PubDat_167218.pdf. Accessed 29 July 2014.

Gutmann, A., & Thompson, D. (1996). *Democracy and disagreement*. Cambridge: Harvard University Press.

Habermas, J. (1984). *Theory of communicative action: Reason and the rationalization of society*. Boston: Beacon Press.

Habermas, J. (1998). *Facticidad y validez*. Madrid: Trotta.

Habermas, J. (2003). *Truth and justification*. Cambridge: MIT Press.

Habermas, J. (2005a). *Between naturalism and religion*. Malden: Polity Press.

Habermas, J. (2005b). Concluding comments on empirical approaches to deliberative politics. *Acta Política, 40,* 384–392.

Hacker, K. L., & van Dijk, J. (Eds.). (2000). *Digital democracy: Issues of theory and practice*. London: Sage.

Hall, R. E. (2000). *The vision of a smart city*. Proceedings of the 2nd International Life Extension Technology Workshop, Paris, France, September 28. http://www.osti.gov/bridge/servlets/purl/773961oyxp82/webviewable/773961.pdf. Accessed 29 July 2014.

Hambleton, R., & Gross, J. (2007). *Governing cities in a global era. Urban innovation, competition and democratic reform*. London: Palgrave.

Harding, A. (1995). Elite theory and growth machines. In D. Judge, G. Stoker, & H. Wolman (Eds.), *Theories of urban politics*. London: Sage.

Harrison, C., Eckman, B., Hamilton, R., Hartswick, P., Kalagnanam, J., Paraszczak, J., & Williams, P. (2010). Foundations for smarter cities. *IBM Journal of Research and Development, 54*(4), 1–16.

Hoff, J., Horrocks, I., & Tops, P. (2000). *Democratic governance and new technology: Technologically mediated innovations in political practice in Western Europe*. London: Routledge.

Holden, S. H., Norris, D. F., & Fletcher, P. D. (2003). Electronic government at the local level: Progress to date and future issues. *Public Performance & Management Review, 26*(4), 325–344.

Holzer, M., Melitski, J., Rho, S., & Schwester, R. (2004). Restoring trust in government: The potential of digital citizen participation. E-Government Series, IBM Center for The Business of Government. http://www.businessofgovernment.org/sites/default/files/HolzerReport.pdf. Accessed 29 July 2014.

Internet Content Syndication Council. (2008). Content creation and distribution in an expanding Internet universe: A white paper. http://internetsyndication.org. Accessed 29 July 2014.

Johnston, E. W., & Hansen, D. L. (2011). Design lessons for smart governance infrastructures. *American Governance, 3,* 197–212.

Judge, D., Stoker, G., & Wolman, H. (1995). Urban politics and theory. In D. Judge, G. Stoker, & H. Wolman (Eds.), *Theories of urban politics* (pp. 1–12). London: Sage.

Kamarck, E. C., & Nye, J. S. (Eds.). (2002). *Governance.com: Democracy in the Information Age*. Washington, DC: Brookings Institution.

Kersting, N., & Vetter, A. (Eds.). (2003). *Reforming local government in Europe: Closing the gap between democracy and efficiency.* Opladen: Springer DE, VS Verlag für Sozialwissenschaften, Leske + Budrich.

Kim, M. K., & Jee, K. Y. (2006). Characteristics of individuals influencing adoption intentions for portable Internet service. *ETRI Journal, 28*(1), 67–76.

Klijn, E. H. (2008). Governance and governance networks in Europe. An assessment of ten years of research on the theme. *Public Management Review, 10*(4), 505–525.

Kooiman, J. (Ed.). (1993). *Modern governance: New government-society interactions*. London: Sage.

Komninos, N., Pallot, M., & Schaffers, H. (2013). Special issue on smart cities and the future internet in europe. *Journal of the Knowledge Economy, 4*(2), 119–134. doi:http://dx.doi.org/10.1007/s13132-012-0083-x.

Lynn, L. E., Heinrich, C. J., & Hill, C. J. (2000). Studying governance and public management: Challenges and prospects. *Journal of Public Administration Research and Theory, 10*(2), 233–262.

Mooij, J. (2003). *Smart governance? Politics in the policy process in Andhra Pradesh, India.* (Working Paper 228). London: Overseas Development Institute. http://www.odi.org/sites/odi.org.uk/files/odi-assets/publications-opinion-files/2464.pdf.

Mossberger, K., & Jimenez, B. (2009). Can e-government promote civic engagement? A study of local government websites in Illinois and the US. Institute for Policy and Civic Engagement Research Partnership. http://www.uic.edu/cuppa/ipce/interior/egovtfinalreport2009.pdf. Accessed 29 July 2014.

Mossberger, K., Tolbert, C., & McNeal, R. S. (2008). *Digital citizenship. The internet, society, and participation*. Cambridge: MIT Press.

Nam, T., & Pardo, T. A. (2011). *Conceptualizing smart city with dimensions of technology, people, and institutions*. Proceedings of the 12th Annual International Digital Government Research Conference: Digital Government Innovation in Challenging Times (pp. 282–291). Accessed 29 July 2014.

Neblo, M. (2005). Thinking through democracy: Between the theory and practice of deliberative politics. *Acta Politica, 40*(2), 169–181.

Norris, P. (2002). *Democratic phoenix: Reinventing political activism*. Cambridge: Cambridge University Press.

OECD. (2001). *Citizens as partners. Information, consultation, and public participation in policy making*. Paris: OECD. http://www.oecd-ilibrary.org/docserver/download/4201131e.pdf?expires=1435076834&id=id&accname=id15861&checksum=2B2313630BEA54DF1FD511B4E55792A3.

Osborne, S., & McLaughlin, K. (2002). The new public management in context. In K. McLaughlin, S. Osborne, & E. Ferlie (Eds.), *New public management. Current trends and future prospects* (pp. 7–14). London: Routledge.

Pierre, J. (1999). Models of urban governance—The institutional dimension of urban politics, *Urban Affairs Review, 34*(3), 372–396.
Pierre, J. (2000). *Debating governance: Authority, steering, and democracy*. Oxford: Oxford University Press.
Polat, R. K., & Pratchett, L. (2009). E-citizenship: Reconstructing the public online. Changing local governance, changing citizens. doi:10.1332/policypress/9781847422170003.0011. Accessed 29 July 2014.
Prats, J. (2003). El concepto y el análisis de la gobernabilidad. *Revista Instituciones y Desarrollo, 14–15,* 239–269.
Putnam, R. D. (Ed.). (2002). *Democracies in flux*. Oxford: Oxford University Press.
Rhodes, R. A. W. (1997). *Understanding governance: Policy networks, governance reflexivity and accountability*. Buckingham: Open University Press.
Rubaii-Barrett, N., & Wise, L. R. (2006). Language minorities and the digital divide: A study of state e-government accessibility. *Journal of Public Management & Social Policy, 12*(2), 5–27.
Rubaii-Barrett, N., & Wise, L. R. (2008). Disability access and e-government an empirical analysis of state practices. *Journal of Disability Policy Studies, 19*(1), 52–64.
Stone, D. A. (1989). Causal stories and the formation of policy agendas. *Political Science Quarterly, 104*(2), 281. http://search.proquest.com/docview/208315275?accountid=1325.
Toppeta, D. (2010). The smart city vision: How innovation and ICT can build smart, "Livable", sustainable Cities. The Innovation Knowledge Foundation. http://www.thinkinnovation.org/file/research/23/en/Toppeta_Report_005_2010.pdf. Accessed 29 July 2014.
Washburn, D., Sindhu, U., Balaouras, S., Dines, R. A., Hayes, N. M., & Nelson, L. E. (2010). *Helping CIOs understand "Smart City" initiatives: Defining the smart city, its drivers, and the role of the CIO*. Cambridge: Forrester Research, Inc. http://public.dhe.ibm.com/partnerworld/pub/smb/smarterplanet/forr_help_cios_und_smart_city_initiatives.pdf. Accessed 29 July 2014.
Wiklund, H. (2005). A Habermasian analysis of the deliberative democratic potential of ICT-enabled services in Swedish municipalities. *New Media & Society, 7*(2), 247–270. Accessed 29 July 2014.

Roberto Garcia Alonso (Assistant Professor of Political Science)'s research interests focus on democratic theories, public opinion, and political participation. Professor Garcia Alonso earned his PhD in Political Science from University Autónoma de Madrid (Spain). He received his BA in Political Science and his Master in Democracy and Government from University Autónoma of Madrid (Spain). He was visiting scholar at Brown University, Rosario University. Professor García Alonso has also taught courses on Contemporary Political Theory, political participation and public opinion. Currently, he is conducting a research on deliberative democracy, in particular, relinking deliberative theory with empirical political science, focusing on institutional, psychological, and sociological variables that might affect deliberative quality.

Sebastian Lippez-De Castro (Assistant Professor of Political Science)'s research interests focus on governance, decentralization processes and challenges, and local government management. Professor Lippez-De Castro earned his Master in Public Administration and a Local Government Management Certificate from Binghamton University, SUNY. He received his BA in Political Science and his specialization degree in Municipal Government, from Pontificia Universidad Javeriana, Bogotá, Colombia. Currently, he is the Director of the undergraduate program in Political Science at Pontificia Universidad Javeriana, is conducting a research on potentialities of community organizations in local governance networks, and teaches on governmental technologies and processes. In addition, Professor Lippez-De Castro has also taught courses on organization theory, decentralization processes, local public management, municipalities, and local finance management in Colombia.

Privacy and Security in Smart Data Collection by Citizens

Sunil Choenni, Mortaza S. Bargh, Carmelita Roepan and Ronald F. Meijer

Abstract The question of how to make a city or government better by exploiting information and communication infrastructures, referred to as smart city, entails an emerging field of research. Large quantities of data are generated from these infrastructures and infusing these data into the physical infrastructure of a city or government may lead to better services to citizens. Collecting and processing of such data, however, may result in privacy and security issues that should be faced appropriately to create a sustainable approach for smart cities and governments. In this chapter, we focus on data collection through crowdsourcing with smart devices and identify the corresponding security and privacy issues in the context of enabling smart cities and governments. We categorize these issues in four classes. For each class, we identify a number of threats as well as solution directions for these threats.

Keywords Crowdsourcing · Privacy · Security · (Un)authorized access · (Un) authorized use · Smart devices

1 Introduction

The developments in communication and information technologies have entailed an explosive growth of data in recent years. In the context of "smart cities" and "smart governments," organizations look for opportunities to take advantage of large quantities of the available data to create a more comprehensive view of a

S. Choenni (✉) · M. S. Bargh · R. F. Meijer
Ministry of Justice & Security, WODC, P.O. Box 20301, 2500 EH, The Hague, The Netherlands
e-mail: sunilchoenni@gmail.com

M. S. Bargh
e-mail: mortaza.bargh@gmail.com

R. F. Meijer
e-mail: r.f.meijer@minvenj.nl

C. Roepan
Groene Tuin 265, 3078KG, Rotterdam, The Netherlands
e-mail: carmelita.roepan@gmail.com

J. R. Gil-Garcia et al. (eds.), *Smarter as the New Urban Agenda A Comprehensive View of the 21st Century City,* Public Administration and Information Technology 11,
DOI 10.1007/978-3-319-17620-8_19

city or government (Choenni et al. 2010; Choenni and Leertouwer 2010). Such a comprehensive view may improve policy decision-making and may lead to better services to citizens. Technological developments make it easier to involve citizens in the data collection process. This approach is regarded as data collection through crowdsourcing (Ganesan and Corner 2011; Taylor 2010). Due to the growth of smart devices, such as smartphones equipped with sensors, citizens carry measurement devices that can easily collect data about several phenomena in a city. Examples of these phenomena are street litter, deterioration of rural areas, and air pollution. These citizens may be regarded in some sense as agents of the policy-makers/local governments for data collecting. Emphasizing greater citizen involvement and participatory government, local governments stimulate active partnerships and collaborations between citizens, the private sector, and the municipality (Stembert et al. 2013). There is a wide variety of applications for mobile devices allowing citizens to collect data. For example, in the Copenhagen Wheel, some sensors are attached to city bicycles in order to report data about pollution, road conditions, congestions, etc. via such an application.

Unlike in crowdsourcing, citizens may also passively and unknowingly be involved in collecting data. Today, users download and use apps that are equipped with several tracing and logging functionalities. Users often do not know which data these applications collect and to which entities they pass the data. It also occurs that users do not change the default settings of the tracing and logging functionalities. This is partly because users do not know how to change these settings or are simply unaware of these functionalities.

Involving citizens in data collection may raise several issues concerning privacy, security, misinterpretation, or even abuse. To what extent does (extra) data collection take place without the knowledge of citizens? To what extent can the data collected by citizens be shared with other citizens and institutions? To what extent is data leakage from mobile devices acceptable? Suppose that the collected data about trees is leaked and it can be concluded that many trees in a district are ill. Then, a possible reaction of the inhabitants of the district can be to cut down these trees. Another consequence of data leakage might be that people abuse the data for their own interests. Suppose that one may conclude from the citizens' ratings that an area is indeed deteriorating. Combining this data with, for example, crime statistics that pertain to the area, someone could try to influence the prices of the houses in that area. Gutmann et al. 2008 and Kalidien et al. 2010 mention the possibility of using survey or administrative data to disclose the identity of individuals or groups to harm individuals, population subgroups, or business enterprises. An intruder might use the attributes of a small area to identify certain characteristics of individuals (e.g., ethnicity) in that area, possibly exposing them for repression or other harms. One example they mention is the use of the US Census of Population data to identify small areas with large proportions of Arab Americans after the events of September 11, 2001.

In this chapter, we provide a categorization of the security and privacy issues of crowdsourcing and accordingly present a number of guidelines to deal with such issues. To gain a sustainable value from crowdsourcing data, we need a continuation of citizens willing to collect unbiased data. The chances for having such a continuation increases whenever the security and privacy issues are handled in an adequate and transparent way and the misinterpretation and misuse of data are prevented. Therefore, we base our categorization on two criteria: whether the data access is authorized or not and whether the data use is authorized or not. Hereby, we directly relate crowdsourcing to two foundations of any trusted data collection process, namely: data privacy and data misuse. We discuss a number of mechanisms to deal with these privacy and security issues. These mechanisms include providing feedback to data subjects (i.e., those who own data) about the status of their data, reporting of aggregated data as much as possible, and developing safe and secure applications for smartphones.

The remainder of this chapter is organized as follows: We start with describing our research methodology. Subsequently, we discuss the role of smart devices and their users in collecting large volumes of data. We especially pay attention to the potentials that these devices offer with their embedded sensory capabilities for enabling a smart city/government. Furthermore, we discuss the differences between traditional data collection and data collection with smart devices. Then we discuss the privacy and security issues that may be raised if citizens are used as suppliers of crowdsourcing data. Next, we discuss a number of mechanisms and concepts that enforce privacy and security safeguards in data collection. Finally, we conclude the chapter.

2 Research Methodology

This contribution is the result of a participatory research, mainly to identify the security and privacy issues of deploying crowdsourcing with smart devices to collect data for enabling smart cities and smart governments. We formed a workgroup consisting of three researchers at the Ministry of Security and Justice of the Netherlands (a national government organization). In a period of 2 months the group had weekly brainstorming sessions to share their findings and experiences about the issues and to identify the solution directions. These brainstorming sessions were based on the knowledge acquired by the workgroup members in several projects carried out in the context of, among others, open data and mobile devices (Zuiderwijk et al. 2012; Meijer et al. 2013). Privacy and security aspects were important in these projects. Furthermore, the workgroup exchanged views with five other experts in the field of mobile applications and security in three occasions. During, for and through writing this chapter, the workgroup also conferred with the fourth coauthor who works for the Rotterdam municipality (a local government organization). Through these brainstorming instances, we followed a bottom-up approach to elucidate/elicit the hidden knowledge of the participants.

In between these meetings, the workgroup members carried out a literature study to learn from the best practices and the state of the art. This study led to identifying a number of security and privacy issues (in Sect. 4) and solution directions (in Sect. 5).

Furthermore, to gain more insight, we observed how two mobile applications Burgerschouw (Centric 2014) and Scoor Ze (Stembert et al. 2013) collect neighborhood information in some Dutch cites through crowdsourcing.

3 Smart Devices and (Local) Governments

Today, an increasing number of apps are developed to enrich the use of smart devices. These apps range from those that ease our daily life to those that adapt our contemporary society. For example, an app in the context of a hospital appointment tells you that your appointment with a specialist is postponed by an hour because his previous appointments went on longer than expected. Such an app makes our life more convenient compared to the alternative of sitting for an hour in a waiting room of a hospital. Apps that are involved in transforming a city into a smart city are typical apps to adapt our society to the contemporary developments, such as facilitating time and place independency.

Today, (local) governments exploit smartphones as an additional channel to broadcast important messages besides conventional channels such as radio and television. Many apps exploit the data of the sensors embedded in mobile devices. A sensor may be regarded as a device that measures physical quantities or signals in an environment and converts it into meaningful figures for an observer, such as location of an object and temperature. There are apps that, exploiting sensory data, remind the owners about the interesting places and events in their current neighborhood or provide some information about the object of which they are taking a photograph. These types of apps may replace the provision of information by city halls and tourist information centers.

Besides the fact that smart devices are used to provide information to their owners, smart devices also collect and pass data to servers. In some cases, device owners are (actively or passively) involved in data collection (e.g., in crowdsourcing scenarios), while in other cases device owners are not aware of it. A typical example of the latter is a mobile app that collects tracking and tracing data. This data gives rise to a number of opportunities that may be exploited by a government. For example, crowd control is such an opportunity that may rely on the movement records of people in a city obtained from their mobile devices. As a consequence, the hotspots and crowded places in a city can be located and crowd control strategies can be defined. Tracking and tracing of mobile devices may also be useful for (police) investigation purposes. A list of persons who were at a certain place within a timeframe might be interesting information if a crime was committed at that place.

Collected tracking and tracing data may also be useful to define effective and sound policies in different sectors of our society. For example in the energy sector, energy suppliers can anticipate and influence the future energy consumption by introducing apps that make citizens aware of their energy consumption. Such apps may recommend users to turn off the heating system if the app detects that nobody will be at home for a while. Furthermore, these apps can be tailored to simulate as if people are at home while they are actually on holiday. Such functionality might be useful to minimize the risk of a burglary when people are on holiday.

Citizens may also be actively involved in gathering data for the government. They may feed the government with data whether or not orchestrated by it. On request of a municipality in the Netherlands, a selected group of citizens, for example, run an app called Burgerschouw (Centric 2014) on smartphones or tablets to rate various aspects of their district, for example, the condition of trees, verges, and streets. Citizens may rate an aspect of their district as fair, high, or low. To clarify the rating criteria, the app provides users with example pictures, for example, the pictures of what should be understood as a healthy tree (good), an average healthy tree (fair), and an ill tree (low). In this case, citizens are aware of their role as data collectors and actively perform this role. Another way to feed a government with data, which is not orchestrated by a government, is to upload data via social media sites that might be relevant to the government.

In the next section, we discuss the differences between traditional and contemporary data collection methods.

3.1 Traditional Versus Smart Data Collection

Traditional data collection is grafted on privacy law and regulations. Privacy laws govern the processing of personal data, which includes all actions carried out on the data from data collection to data destruction (DPPA 2014). There are a number of guidelines and legal frameworks to handle data processing such as the Data Protection Directive of the European Union (EU 1995) and the Dutch Privacy Protection Act (DPPA 2014). From these frameworks, six principles are extracted that pertain to the processing of personal data (Chadwick 2009; OECD 2013; Cameron 2005; EU 1995; DPPA 2014):

- Finality principle, which refers to the purpose for which personal data is collected. The purpose should be explicit and the processing of the collected data must be compatible with the purpose for which the data was collected.
- Legitimacy principle, which refers to proper, careful, and legal collection and processing of the data. To this end one should take into account the context within which the data is collected and used.
- Limitation principle, which refers to having relevant, sufficient, not excessive, and correct data. This demands the data to be collected proportionally to the intended purpose (i.e., proportionality) and in ways that minimize the use of privacy sensitive data (i.e., subsidiarity).
- Transparency principle, which entitles the data subject to know when, why (i.e., for which purpose), and by whom her/his data is processed. An individual even has the right to ask a data controller about whether the data controller has his/her data within a reasonable time and expense. And if the reply is affirmative, the individual is entitled to have the data erased, rectified, completed, or amended.

- Security safeguards principle, which refers to having reasonable technical and organizational safeguards in place to protect data against various risks such as loss, unlawful and unauthorized access/use, destruction, modification, or disclosure.
- Accountability principle, which states that a data controller should be accountable for complying with the measures that materialize the principles stated above.

Data collection with smart devices (referred to as "crowdsourcing" from now on) enables us to use a multitude of easily available data sources with relatively detailed data, collected for various goals and purposes potentially by a large population. This way of data collection offers many opportunities and provokes a rise of big data driven research. As crowdsourcing is mainly based on available data sources and as it often contains personal data, the use of the collected data for big data research often changes the context of the data usage and as such it potentially conflicts with the finality, legitimacy, collection-limitation principles. For example,

- Traditional data collection emphasizes creation of original data, whereas crowdsourcing seizes the opportunities of (re)using data from existing sources.
- Researchers in traditional data collection create their own data which enables them to define and control the principles of finality, legitimacy, collection limitation, and other legal principles concerning the data collection.
- In traditional data collection, the citizens who participate in research give their explicit consent for collecting and using of their data.
- Traditional data collection is a result of a research design process whereby validity and reliability of the collected data are taken care of (e.g., degree of data details, micro or aggregated records, structured or unstructured data). Such a careful research design is not possible in crowdsourcing due to reliance on existent tools, devices and data.

Thus, problems and issues might arise as described in the following section.

4 Security and Privacy Issues

Crowdsourcing inflicts various security and privacy vulnerabilities on the citizens who participate in sensing data, the citizens over whom data is collected, and on the entities (including governments, organizations, and citizens) that consume the collected data. Wang et al. (2013) identified a number of privacy and security threats that endangered the use of smart devices for crowdsourcing. In this section, we extend and categorize these threats.

4.1 Categorization Criteria

Crowdsourcing is based on available data sources and the crowdsourcing data often contains personal data. Even when the data is aggregated, the data could be combined with other data and result in revealing of personal data (Braak et al. 2012; Kulk and van Loenen 2012). In practice, it will be infeasible to reliably predict which part of the data is privacy sensitive. Even if a part of the data is privacy sensitive, the entire data should be treated as privacy sensitive. The use of crowdsourcing data imposes several privacy and security challenges due to the character of the data and the context in which they are originally created. We argue below that as crowdsourcing has to deal with existing data and as the data often contains personal data, the focus of privacy protection and security should be on the access and use of the data. Therefore, we categorize these threats according to the way that the crowdsourcing data might be accessed and used/exploited within the context of e-government applications.

Traditionally, controlling who gets access to sensitive information has been used as an important means of protection. Access control deals with granting authorized entities and preventing unauthorized entities to access resources such as data. Therefore, the first criterion for our categorization of the security and privacy threats is whether the collected data is being accessed by authorized or unauthorized entities. After getting access to sensitive data it is often not guaranteed that the data is used (i.e., processed, stored, etc.) appropriately, for example, in the way that it is desired by the data subject or for the purposes that the data is collected. Therefore, we adopt the way that the data is used as the other categorization criterion. For this criterion we consider whether the collected data is used for an authorized or unauthorized purpose.

Considering these two criteria we identify four classes for crowdsourcing data, in being used and exploited, namely: authorized-access and authorized-use, authorized-access and unauthorized-use, unauthorized-access and authorized-use; and unauthorized-access and unauthorized-use. In the following subsections, we provide some typical threats per each of these four categories, an overview of which is given in Fig. 1.

4.2 *Authorized-Access and Authorized-Use*

Even when collected data is accessed and used according to some defined rules and policies, there is a chance that the resulting information leads to some (privacy and security) issues.

The crowdsourcing data, for example, can be inaccurate and biased. In some applications, the crowd may rely on information supplied by others to make critical decisions (e.g., to derive hazardous traffic conditions, natural disasters, human rights violations, or political unrest). In such cases, there is a possibility of incorrect or inaccurate data being reported unintentionally or in some cases maliciously (Wang et al. 2013). A source of data inaccuracy can be due to the lack of a unique mindset among the individuals who contribute to the data sensing process. For example, if crowd is supposed to rate how clean their neighborhoods are, they need to have a common understanding of cleanness and be able to rank the scenes similarly

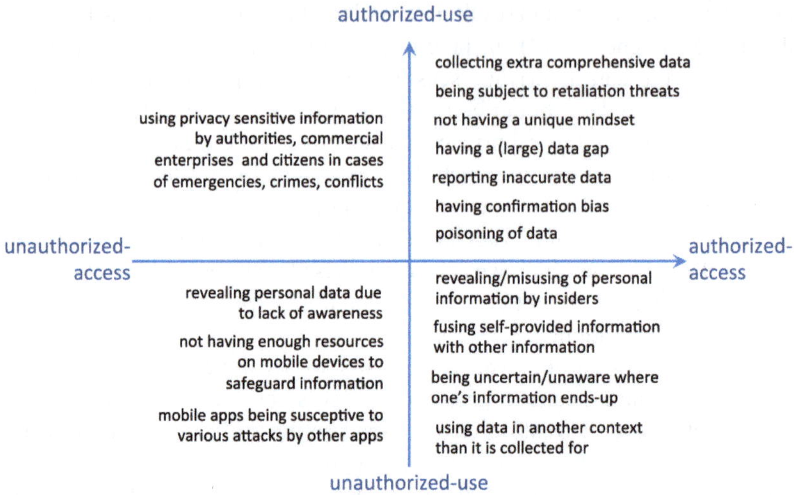

Fig. 1 Examples of the security and privacy issues within the four quadrants of (un)authorized access/use

and fairly. Having the common criteria to rate situation appropriately is a challenge. Another source of data inaccuracy can be attributed to malicious intension of sensing individuals. Through data poisoning, such individuals can inflict damages and harms on individuals and organizations. For example, a well-orchestrated malicious campaign among a number of individuals can damage the reputation of a nice neighborhood and lead to reducing of house prices there.

Collecting data through crowdsourcing is also prone to the so-called signal error when there exists a large gap of the data gathered over the phenomena under study (Zoldan 2013). One example is the use of Twitter messages to understand people's decision-making related to Hurricane Sandy, as was reported by Zoldan (2013). Based on over 20 million tweets analyzed, the tweets about storm preparations peaked the night before the storm. Interestingly, the majority of the tweets originated from Manhattan rather than the hit areas (e.g., Seaside Heights and Midland Beach) because of the high concentration of smartphone and Twitter usage in Manhattan, and power outages and low battery power levels of mobile devices in the hit areas. As such "there was a huge data-gap from communities unrepresented in the Twitter sphere" (Zoldan 2013). One can imagine how it would have been if rescue missions were decided based on those Tweets solely, without considering the context within which the data is collected (in this case, power shortage and spatial concentration of sensors in the areas). The crowdsourcing data, moreover, is subject to the so-called confirmation bias, where people tend to search data in such a way that their previous viewpoint is confirmed regardless of what the data truly conveys (Zoldan 2013). Considering the issues mentioned above, one may perceive the crowdsourcing data as less accurate and less trustworthy. Consequently, the systems and services that fully rely on such inaccurate data may inflict security and privacy threats on the users of such systems and services.

Users who are part of the crowd that contribute to the data can also be subject to some retaliation threats (Wang et al. 2013). For example, if someone reports a domestic violence to authorities, he or she may become subject to retaliation, should the suspected violator find out who reported the domestic incident. "As a result, citizens will only use the system if they trust that the system is secure and they will face no public retaliation for making reports" (Wang et al. 2013).

One goal behind data collection mechanisms should be to collect enough data that provides a view pertained to a certain purpose. One drawback of crowdsourcing based data collection is the possibility of collecting extra comprehensive/too detailed data about phenomena. Such amounts of data not only introduce extra (security and) privacy risks but also lead to information overload. Suppose that we systematically track and store timestamps of a citizen who is inspecting his or her district. If we are interested in the citizen's district ratings, the storage of timestamps and the exact route that he or she follows is irrelevant, and can therefore be regarded as a privacy breach or may introduce information overload. This is against the collection limitation principle.

4.3 *Authorized-Access and Unauthorized-Use*

When the collected data is accessed legitimately, it should still be used appropriately according to the laws, guidelines, rules, as well as the preferences of data subjects.

Crowdsourcing data may include some basic personal data from participants such as their profile data (including username, password, name, email address, and phone number), activity data (e.g., sporting, sleeping, eating, etc.), and situational data (like visited locations, adjacency to other users and objects, conversations, etc.). Such personal data must basically be accessible to a limited number of authorized entities like system administrators and specific services/systems. Authorized insiders with ill intentions (i.e., inside intruders) may reveal and misuse personal information to which they have access for their illegitimate purposes like personal satisfaction, financial gains, and political benefits. Revealing personal information makes data subjects (i.e., individuals and organizations whom the data is about) vulnerable to cyber attacks, such as identity theft, phishing, spams, and privacy breaches. Being subjected to such vulnerabilities and victimized by such attacks, the crowd may become fearful and unwilling to contribute to the data collection process. Even when voluntarily participating in crowdsourcing, users sometimes desire that their personal information should not to be shared, for instance at certain situations (like during evenings, in the weekends, and during holidays).

Integration of information systems has become a trend in the current era in various public, private, and semipublic sectors. For example, Google has merged various services like Gmail, Google+, Google Drive, and Facebook has acquired Instagram. In such cases, various databases are integrated within Google and Facebook, respectively. There are also open data initiatives to release public sector data to citizens as a measure of government transparency (Dawes 2010a, b). Such initiatives motivate combining crowdsourcing data with other data sources in order to deliver value-added services. In such cases, however, there are potential risks of privacy breaches when self-provided data of users is combined with other user data retrieved from elsewhere (Bargh and Choenni 2013). Even highly sensitive attributes may be disclosed by means of easily accessible data. Kosinski et al. (2013) show that easily accessible digital records of behavior, Facebook Likes, etc., can be used to automatically and accurately predict a range of highly sensitive personal attributes. They mention sexual orientation, ethnicity, religious and political views, personality traits, intelligence, happiness, use of addictive substances, parental separation, age, and gender (Kosinski et al. 2013). De Montjoye et al. (2013) analyzed a dataset of 15 months of human mobility data for 1.5 million individuals. They found that human mobility traces are highly unique. If the location of an individual is specified hourly, with a spatial resolution equal to that given by the carrier's antennas, four spatiotemporal points were proven to be sufficient to uniquely identify 95 % of the individuals. They also found that even rather highly aggregated datasets provide little anonymity (De Montjoye et al. 2013).

In traditional data collection, citizens may be aware of and consent to the collection of their personal data. In crowdsourcing, however, this consent may not be present. Citizens may be unaware of the extent of their contribution to the data collected and the extent of third parties using their data. Citizens may or may not consent to the data processor which uses their personal data or shares it with other organizations. Generally, in crowdsourcing it may become unclear who is using citizens' personal data and for which goals and purposes. Such an uncertainty and unawareness conflicts with the transparency principle.

Crowdsourcing data may be collected within a specific register (e.g., that of a municipality population, hospital, or judicial data) or a research context (e.g., a household or crime victimization survey, etc.). As such, data collection can be found for various purposes and on different legal domains (e.g., the health, criminal, or population register law). When data is used in another context than in which it is originally collected, a conflict may occur with the finality and legitimacy principles.

4.4 Unauthorized-Access and Authorized-Use

There are also cases possible where the collected data is accessed illegitimately while it is used legitimately. A typical example is the case where an employee accesses some data to which he has been authorized to, but he uses a colleague's credentials due to a forgotten password.

Unauthorized access for an authorized use can occur also in crowdsourcing scenarios. For example, consider the case where mobile devices of citizens are traced and stored. Based on this data, one can discover and predict the movement and travelling patterns of a citizen and determine who the co-travelers are. At first, the processing and use of a citizen's travelling and movement information may not seem interesting and relevant. Due to information overload, the data is not processed normally and therefore no privacy breaches occur. On the other hand, the timestamps and the routes that citizens follow might become very interesting data for the police if a serious crime is committed at a specific time and place around which some users were present. A citizen whose movement track coincides with the specific location and timeframe may become a suspect or may be called as a witness. Now if the police access the timestamps and route data, some questions about the unauthorized access to the data may arise, particularly whether the data use is authorized and legitimate.

4.5 Unauthorized-Access and Unauthorized-Use

The most known and acknowledged threats are those where the collected data is accessed and used illegitimately. All examples mentioned above for revealing personal information also hold for unauthorized access and unauthorized use here, where (external) intruders illegitimately get access to the systems that process and store this personal information.

Due to lack of awareness, users may reveal their personal information unwillingly. Technically such an access can be considered "authorized," because the data subject agreed on it. But this agreement is done unknowingly when the user was unaware of the impacts and consequences. As such we consider such cases as unauthorized access and unauthorized use. Often users allow applications to access their data without knowing how their data will be actually used and without realizing the risks associated with sharing their data. Users make such wrong decisions and agree with such sharing of their data due to earning immediate gain or due to lack of transparency of privacy policies. Users usually do not understand the complex terms and conditions in privacy policies.

Mobile devices are also vulnerable to security and privacy threats and attacks. This puts end users in jeopardy of losing personal data and malfunctioning of the device. Mobile devices generally have limited energy, processing power, and communication resources. This limitation makes it difficult to run protective applications on such devices against malicious programs. Therefore, compared to personal computers, mobile devices become more susceptive to the above mentioned threats and attacks. Mobile applications are also susceptive to threats such as eavesdropping, spoofing and denial of service (Chin et al. 2011). This means that crowdsourcing applications may leak information to other applications due to possible inter process communications if they are realized carelessly.

5 Frameworks and Guidelines

To handle privacy and security issues, a mixture of procedural and technical measures may be used (Hildebrandt and Koops 2010). In the following we focus on providing some solution directions and guidelines to address the privacy and security

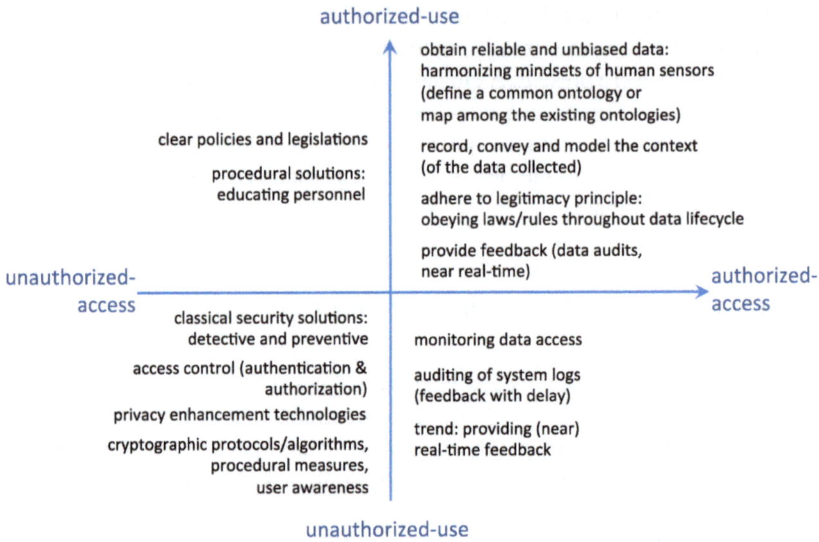

Fig. 2 Examples of the security and privacy solution directions within the four quadrants of (un) authorized access/use

challenges of crowdsourcing as described in previous sections. An overview of the discussed solution directions is given in Fig. 2.

5.1 Authorized-Access and Authorized-Use

When both data access and data use are authorized, one should be cautious and prudent in interpreting crowdsourcing data due to inaccuracy inherent in such data; after all the data is often produced for other purposes.

To make use of data objectively, a first step would be to collect reliable data as much as possible through harmonizing the mindset of (human) sensors. Those observers who judge situations and provide their perceptions as sensory data should be instructed in a way that the personal bias in situation scoring is minimized or eliminated as much as possible. This can be achieved through defining clear criteria for evaluating situations (e.g., having a limited number of categories and exemplifying each category through visual, vocal, and textual media). Having such a common set of criteria implies defining a universal ontology among all sensing units. Alternatively, one can allow each sensing domain to have its own ontology and perception and then devise a mapping function to relate these ontologies and perceptions unambiguously.

Subsequently, the context in which the data is collected should also be recorded, modeled, and conveyed to the data analysis process together with the data. The data analyzer should, in turn, take into consideration this contextual information when fusing data from various sources. Then the decision-making process that uses the fused data should be aware of the inaccuracy and uncertainty of the resulting data. In other words, any decision-making process should have an idea about the reliability of the data at hand and make an informed decision based on the level of data reliability. Hereto, the decision-maker should also consider the possibility of data poisoning by malicious individuals (e.g., suspicious data can trigger more precise investigations in order to make an operational decision). Eventually, the legitimacy principle should be respected by obeying the laws and rules throughout the whole lifecycle of data, including discarding the data properly after the use if required by law or data subjects.

Through data sharing, one increases chance of compromising privacy sensitive data. Such compromises undermine trust of data subjects (e.g., users and citizens). As information controllers (those who control the data of data subjects) are morally, ethically, and legally responsible for any misuse of the disseminated information, privacy enhancement techniques are often used to prevent unsavory disclosure of personal information. Moreover auditing the logs of data processing activities is used to control the adherence of data processors to laws and policy agreements. These audit procedures are carried out in longtime cycles and seem to be rather static. We envision that there is a need for a near real-time feedback from information processors to information controllers. When disseminating information of data subjects (e.g., citizens as carriers of smartphones) to data processors (e.g., smart city service providers), one can use feedback from data processors to data controllers to facilitate the privacy preservation process. Hereto feedback intrinsically serves as a trust enhancement mechanism by giving a good feeling directly to the data controllers (and data subject) to share data (Tsai et al. 2009). Moreover, within the context of privacy protection, feedback can enable data controllers/subjects to be in charge of revealing their data to other parties. When privacy policies cannot be defined in details beforehand, due to, for example, not knowing the information usage context, feedback can also be used to refine data privacy policies on the scene. Here, feedback works as an instrument for preventing privacy breaches. Feedback can also play a role in dealing with or preventing misuse and misinterpretation of data. This solution direction is a major step to realize the data transparency principle and as such to gain the public trust, in general, and to gain the trust of those who participate in crowdsourcing, in particular.

To prevent retaliation threats against those who contribute to crowdsourcing, the systems that collect and process the data should be trustful and have sufficient security safeguards in place such as applying data anonymization, data aggregation, and access control techniques. In this way, the identity of the contributors of crowdsourcing can be kept out of public reach.

Data aggregation can inherently help prevention of privacy breaches due to reduction of personally identifiable information. Nevertheless, one should be careful that the aggregated data does not indeed reveal any privacy sensitive information and does not contribute to deriving privacy sensitive information through fusing with other information, before and after disseminating of the data respectively (Kulk and van Loenen 2012; Gutmann et al. 2008).

5.2 Authorized-Access and Unauthorized-Use

When someone bypasses access control safeguards and gets access to sensitive information unjustifiably, one way to detect such intrusions is to use monitoring tools. We witness nowadays a surge of tools in the market that can help data custodians to monitor the way that their information resources are used. Example tools are those used for realizing a Security Operations Center for large organizations, and the one of (VDSS 2014) for small and medium size enterprises. Such tools provide (real-time) feedback for detecting security and privacy irregularities in information sharing systems. Based on this (near) real-time feedback the data custodians can respond to the detected unauthorized uses and also can take prevention measures for future unauthorized uses.

5.3 Unauthorized-Access and Authorized-Use

Using data for a legitimate purpose while the access is obtained from illegitimate ways requires solutions of a more procedural nature in our opinion. Primarily, there should be clear legislations and policies in place to define the conditions under which an access to crowdsourcing data becomes authorized without circumventing (i.e., denying) the access control process. Traditionally a court order is used to have access to crucial evidences, when it is recognized to be necessary. The procedural solutions in this area can also include campaigns to educate employees about seeking information through legitimate mechanisms.

5.4 Unauthorized-Access and Unauthorized-Use

Classical security solutions that realized functionalities such as access control (including authentication and authorization) and privacy enhancement techniques (including data anonymization, aggregation, and confidentiality) are mainly aimed at dealing with unauthorized access to and unauthorized use of sensitive data during data transit, storage, and processing. These techniques not only include technical solutions such as cryptographic protocols and algorithms but also procedural measures and user awareness enhancement campaigns and programs. While the necessity and use of technical solutions are rather well understood and acknowledged, user awareness solutions (both technical and nontechnical ones) are yet in its childhood phases. Human factors due to lack of knowledge, poor judgment, ignorance, mistake, overlooking, etc. are considered as the weakest point in the chain of defense against malicious cyber attacks.

A prerequisite for protecting crowdsourcing data against malicious attacks is to sufficiently have some security safeguards in place. This is also foreseen within the security safeguard principle of the data protection framework sketched in the previous sections. These safeguards not only include preventive measures (like access control and user awareness) but also detective measures to search for and detect those malicious activities and attacks that penetrate preventive lines of deference. For crowdsourcing data, therefore, we could use monitoring tools to detect suspicious processing of data and inform supervising authorities, data custodians, and data subjects appropriately.

Developing safe and secure applications reinforces the infrastructure that collects and distributes data from smartphones to backend servers. Important aspects to consider include limited battery energy of mobile devices and the scalability of security management among multiple devices. Furthermore, elucidating the requirements that a safe app should meet and the implementation of these requirements are some aspects that need in-depth elaboration.

6 Concluding Remarks

Technological developments make it easier to involve citizens in the data collection process. This approach is regarded as data collection through crowdsourcing. Smart devices equipped by mobile applications, that is, apps, appear to be ideally suited for this purpose. Furthermore, as discussed in this chapter, these devices contain an increasing number of apps that can ease our daily life and transform the city and government to smart entities. Besides the potentials of smart devices, data collection through crowdsourcing raises a number of privacy and security threats. We identified the privacy and security threats of deploying crowdsourcing with smart devices. To create sustainable smart cities it is necessary to take care of these threats. Otherwise, citizens will become suspicious and reluctant to use (smart cities related) apps that are intended to ease their daily life. This may have a negative impact on the economic growth of a city and its ambition to become a smart city.

We categorized the identified threats along two dimensions. The first dimension is whether the collected data is being accessed by authorized or unauthorized entities. After getting access to sensitive data it is often not guaranteed that the data is used (i.e., processed, stored, etc.) appropriately, for example, in the way that it is desired by the data subject or for the purposes that the data is collected. Therefore, we adopted the way that the data is used as the other dimension. For this dimension, we considered whether the collected data is used for an authorized or unauthorized purpose. For each class we discussed a number of threats and solutions directions.

One of these classes is authorized-access and authorized-use, where crowdsourcing data may result in, for example, an inaccurate and biased view of reality. Collecting reliable data requires, among others, defining clear criteria for evaluating situations by human sensors, unambiguous mapping of the different views of human sensors, or considering the data collection context during data analysis and decision-making processes. Another class is authorized-access and unauthorized-use, where, for example, authorized insiders with ill intentions may reveal the personal information embedded in crowdsourcing data, or data custodians may misuse crowdsourcing data without consent of data subjects. In such cases, one may use monitoring tools to detect data misuses carried out by authorized users (inside intruders). The monitoring can be done in real-time or in regular intervals (via for example auditing of data usage logs). Unauthorized-access and authorized-use is the third class where, for example, authorities and enterprises use crowdsourcing data when an emergency, crime, or conflict occurs. This way of data usage requires transparent solutions that are of a more legislative and procedural nature, whereby these authorities can obtain permission to process the data legitimately. Finally, we considered unauthorized-access and unauthorized-use, where (external) intruders illegitimately get access to crowdsourcing data and use the data for some malicious purposes. This case could stem from users who share their data and use their systems without knowing the risks involved or from using inadequate/ineffective safeguards to protect data or systems. Classical solutions such as access control and privacy enhancement technologies together with user awareness measures could be used to deal with such threats.

References

Bargh, M. S., & Choenni, S. (2013). *On preserving privacy whilst integrating data in connected information systems*. In Proceedings of the International Conference on Cloud Security Management (ICCSM'13), Seattle, US, 17–18 October.

Braak, S. W. van den, Choenni, S., Meijer, R., & Zuiderwijk, A. (2012). *Trusted third parties for secure and privacy preserving data integration and sharing in the public sector*. In Proceedings of the 13th Annual International Conference on Digital Government Research, New York, NY, USA, 2012, pp. 135–144.

Cameron, K. (2005). The laws of identity. http://www.identityblog.com/?p=352/#lawsofiden_topic3. Accessed 1 Feb 2014.

Centric. (2014). http://www.centric.nl/NL/Default/Branches/Lokale-overheid/Burgerschouw. Accessed 11 June 2015.

Chadwick, D. W. (2009). Federated identity management. In: Foundations of security analysis and design V; lecture notes in computer science (Vol. 5705), 2009, pp. 96–120.

Chin, E., Porter Felt, A., Greenwood, K., & Wagner, D. (2011). *Analyzing inter-application communication in Android*. In Proceedings of MobiSys. ACM, 2011.

Choenni, S., & Leertouwer, E. (2010). Public safety mashups to support policy makers. In Proceedings of EGOVIS 2010, Electronic Government and the Information Systems Perspective, Bilbao, Spain, August 31–September 2, 2010. LNCS 6267, pp. 234–248.

Choenni, S., van Dijk, J., & Leeuw, F., (2010). Preserving privacy whilst integrating data: Applied to criminal justice. *Information Polity, 15*(1–2), 125–138.

Dawes, S. S. (2010a). Stewardship and usefulness: Policy principles for information-based transparency. *Government Information Quarterly, 27,* 377–383.

Dawes, S. S. (2010b). Information policy meta-principles: Stewardship and usefulness. In 2010 43rd Hawaii International Conference on System Sciences (pp. 1–10). IEEE.

DPPA. (2014). Dutch data protection authority—College Bescherming Persoonsgegevens (2014), Dutch Privacy Protection Act (DPPA)—Wbp-naslag; https://cbpweb.nl/nl/over-privacy/wetten/wbp-naslag. Accessed 11 June 2015.

EU. (1995). Directive 95/46/EC of the European Parliament and of the Council of 24 October 1995 on the protection of individuals with regard to the processing of personal data and on the free movement of such data.

Ganesan, D., & Corner, M. (2011). Crowd sourcing for data collection. http://sensorlab.cs.dartmouth.edu/NSFPervasiveComputingAtScale/pdf/15693928 97.pdf. Accessed 26 Feb 2014.

Gutmann, M., Witkowski, K., Colyer, C., O'Rourke, J., & McNally, J. (2008). Providing spatial data for secondary analysis: Issues and current practices relating to confidentiality. *Population Research and Policy Review, 27*(6), 639–665.

Hildebrandt, M., & Koops, B. J. (2010). The challenges of ambient law and legal protection in the profiling era. *The Modern Law Review, 73*(3), 428–460.

Kalidien, S., Choenni, S., & Meijer, R. (2010). *Crime statistics online: Potentials and challenges*. Proceedings of the 11th Annual International Conference on Digital Government Research, New York, NY, USA, 2010, pp. 131–137.

Kosinski, M., Stillwell D., & Graepel, T. (2013). *Private traits and attributes are predictable from digital records of human behavior*. Proceedings of the National Academy of Sciences of the United States of America, 110(15) Michal Kosinski, 5802–5805. doi:10.1073/pnas.1218772110.

Kulk, S., & van Loenen, B. (2012). Brave new open data world? *International Journal of Spatial Data Infrastructures, 7,* 196–206.

Meijer, R., Choenni, S., Sheikh, A. R., & Conradie, P. (2013). *Bridging the contradictions of open data*. Proceedings of the 13th European Conference on eGovernment. Como, Italy.

Montjoye de, Y-Al., Hidalgo, C. A., Verleysen, M., & Blondel, V. D. (2013). Unique in the crowd: The privacy bounds of human mobility. *Scientific Reports, 3*(1376). doi:10.1038/srep01376.

OECD. (2013). The OECD privacy framework. http://www.oecd.org/sti/ieconomy/2013-oecd-privacy-guidelines.pdf. Accessed 21 Feb 2014.

Stembert, N., Conradie, P., Mulder, I., & Choenni, S. (2013). *Participatory data gathering for public sector reuse: Lessons learned from traditional initiatives. Electronic government* (pp. 87–98). Berlin Heidelberg: Springer.

Taylor, J. (2010). Citizens as public sensors. http://radar.oreilly.com/2010/04/crowdsourcing-the-dpw.html. Accessed 26 Feb 2014.

Tsai, J. Y., Kelley, P., Drielsma, P., Cranor, L. F., Hong, J., & Sadeh, N. (2009). *Who's viewed you? The impact of feedback in a mobile location-sharing application. Computer Human Interaction (CHI'09).* ACM Press.

VDSS Vita Data Security Systems. (2014). http://www.vdss.nl/vdss-in-your-network.ashx. Accessed 21 Feb 2014.

Wang, Y., Huang, Y., & Louis, C. (September 2013). Towards a framework for privacy-aware mobile crowdsourcing. In Social Computing (SocialCom), 2013 International Conference on Social Computing (pp. 454–459). IEEE Press.

Zoldan, A. (5 October 2013). More data, more problems: Is big data always right?. http://www.wired.com/insights/2013/05/more-data-more-problems-is-big-data-always-right/. Accessed 11 June 2015.

Zuiderwijk, A., Janssen, M., Meijer, R., Choenni, R., Charalabidis, Y., & Jeffery, K. (2012). *Issues and guiding principles for opening governmental judicial research data.* Proceedings of the 11th European Conference On Electronic Government, EGOV'12, LNCS Vol. 7443. Springer verlag, Kristiansand, Norway, September 3–6 (90–101).

Sunil Choenni holds a Ph.D. in database technology from the University of Twente and an MSc in theoretical computer science from Delft University of Technology. Currently, he heads the Department of Statistical Information Management and Policy Analysis of the Research and Documentation Centre (WODC) of the Dutch Ministry of Security and Justice. Furthermore, he is a professor of human-centered ICT at Rotterdam University of Applied Sciences in Rotterdam. Prior to joining WODC, he held several teaching and research positions at different universities and research centers. His research interests include data management, e-government, cyber security, and human-centered design.

Mortaza S. Bargh is a research scientist at WODC as well as a visiting research professor at Rotterdam University of Applied Sciences. His research interest areas include collaborative security, secure distributed systems, information fusion algorithms, and preserving privacy in big data infrastructures. His work is multidisciplinary by nature, in close collaboration with various disciplines such as administration, sociology, criminology, and economy. He has an educational background in information and communication theory. He had previously worked as a scientific researcher at Novay (former Telematica Instituut) in Enschede, the Netherlands from 1999 until 2011 in the areas of pervasive computing and information security.

Carmelita Roepan is an adviser at the Municipal Rotterdam, Centre of Youth and Family. Currently she is engaged in education for child care. Her research interests are on the application of ICT in public domains.

Ronald F. Meijer graduated in science of public administration. He worked in universities doing policy research (Amsterdam, Leiden). He is engaged in research at the Research and Documentation Centre (WODC) of the Ministry of Security and Justice of the Netherlands. His interest is focused on information management, data archiving, open data, and privacy in the domain of justice.

Innovation and Opportunities for Citizen Participation in Spanish Smart Cities

Xosé María Mahou-Lago and Enrique José Varela-Álvarez

Abstract Smart cities are a social, political, administrative, and technological phenomenon. The growth and consolidation of smart cities depends on sociotechnical conditions that implement 2.0 virtual platforms in specific sociopolitical and organizational contexts. To determine whether Spanish smart city Web portals facilitate fluid interaction between local administrations and citizens, we designed a study using heuristic test techniques to analyze Web portal usability in 20 Spanish smart cities. The objective was to identify the type and development level of electronic participation features on selected municipal Web sites.

Keywords Citizen participation · Governance · Smart cities · Spanish cities · Web portal · Usability

1 Introduction

Urban development is a hot topic today. More than half of the world's population currently lives in urban areas (EU 2011) and cities have become spaces for understanding new economic, social, political, and administrative phenomena derived from the application of information and communication technologies (ICTs).

Smart city initiatives are oriented towards different spheres of action. The 2007 report *Smart Cities—Ranking of European Medium-Sized Cities* listed experiences in areas of economy, governance, environment, people, mobility, and living. These have been the object of much study in recent years, with a view to establishing analytical frameworks that document the relationships between and among them (Schaffers et al. 2011; Chourabi et al. 2012). In this chapter, we will focus specifically on the institutional dimension, the core of smart city initiatives (Belissent 2011), by means of a Web portal usability analysis.

X. M. Mahou-Lago (✉) · E. J. Varela-Álvarez
Facultade de Ciencias Sociais e da Comunicación, University of Vigo, Campus A Xunqueira s/n, 36005 Pontevedra, Spain
e-mail: xmahou@uvigo.es

E. J. Varela-Álvarez
e-mail: evalvarez@uvigo.es

J. R. Gil-Garcia et al. (eds.), *Smarter as the New Urban Agenda A Comprehensive View of the 21st Century City,* Public Administration and Information Technology 11,
DOI 10.1007/978-3-319-17620-8_20

Institutional factors related to smart city initiatives are linked to the idea of "smart governance," which involves participation in decision-making, public and social services, and transparent government (Giffinger et al. 2007). Public and private actors and networks participate and cooperate without hierarchical structures in the formulation and implementation of public policies (Rhodes 1997).

This chapter presents the results of an analysis of online participation features hosted on the Web portals of 20 Spanish smart cities. The actual level of citizen participation is beyond the scope of this study, which instead examines the level of development of participation features that allow citizens to influence public affairs. Specifically, we were guided by three questions:

1. What types of online participation platforms do Spanish smart cities use on their Web portals in the context of e-government and specific areas of action such as environmental sustainability and mobility?
2. What is their level of development from a usability perspective?
3. Based on their level of development, do they offer a virtual environment favoring online citizen participation?

This chapter begins with a brief summary of the evolution of citizen participation as a process of improving democratic quality to its current state of e-participation in the context of e-governance. It will provide a useful framework for analyzing, by means of an expert test, the usability of online e-government participation features and smart initiatives (in this case, mobility and environmental sustainability) in 20 Spanish smart city Web portals. The final part of the chapter provides some conclusions and answers the central question of our study: do smart cities enhance opportunities for citizen participation?

2 From Citizen Participation to E-Participation in Smart Cities

If we begin by looking at participation first, we find that citizens participate in activities linked to public affairs through diverse means and degrees of intensity, according to the types of service and level of government that produces or provides the public service in question. Whether participation is effective, or whether an action produces any type of result or impact on public affairs is a very different matter (Frewer and Rowe 2005).

The outcome may involve states that cede power to their various levels of government or territorial administration, thereby reinforcing community proximity in a setting that allows new forms of citizen participation in public decision-making. This is fertile ground for both traditional presential participation and the new e-participation. The former develops along an axis from information to co-decision-making, passing through stages of consultation and comanagement. E-participation adds value to classical participation (as it is called by Ramilo and Fernández 2012), using ICTs. For Iasulaitis and Pineda (2013), the main difference between presential participation and e-participation is the "interactivity" of the latter, through new dialogic/dialogue-based information exchange processes facilitated by ITCs.

Through "upgraded" interactivity based on deliberative and dialogic practices, citizens become aware of their role as active individuals in the sphere of public policies and public services; rather than merely as clients seeking a product from a public organization. The challenge of e-participation is to link the citizen with e-government, e-democracy, and e-governance (OECD 2003). The quality of democracy is enhanced as participation reinforces ICTs and smart cities assimilate them as a transversal, integrative, and holistic dimension capable of providing services and relegitimizing public institutions.

E-participation fits within the framework of new collective action logic. New citizenship (Moro 2005, p. 112), by involving both active citizens and groups in these developments, seeks to overcome some of the unresolved classic dilemmas in democratic processes, such as the relationship between inequality and participation, the more recent "digital divide" (Siddiqi et al. 2006), or simply as another vehicle for political participation and choice (Chen et al. 2004). However, Kingston (2002) observes that "e-participation will only work, though, if the public want to participate and if they believe that their views are being listened to by the elected officials."

This new collective action logic takes place in sociopolitical and administrative innovation spaces generated by smart cities (Nam and Pardo 2011b), which can be defined as a sociotechnical outcome that occurs in a complex space combining technological, human, and institutional dimensions or factors ("smart ecosystem"). The resulting social strategy seeks to answer two key questions: "How do smart technologies change a city," and "How do traditional institutional and human factors in urban dynamics impact a smart city initiative leveraged by new technologies?" (Chourabi et al. 2012; Nam and Pardo 2011a, pp. 282–291).

Of the numerous definitions for smart cities, two in particular reflect the complexity and sociotechnical perspective that we feel should be present in their design and strategy:

- "A city performing well in a forward-looking way in economy, people, governance, mobility, environment, and living, built on the smart combination of endowments and activities of self-decisive, independent, and aware citizens" (Giffinger et al. 2007).
- "A city that gives inspiration, shares culture, knowledge, and life, a city that motivates its inhabitants to create and flourish in their own lives" (Rios 2008).

"E-citizens" are born, live, and coexist in smart cities, and digital natives are at home there (Prensky 2001). This takes us to an innovative relational scenario between public organizations and individuals who need or desire to interact, to participate in public decision-making, to function as users or as citizens. The challenge of smart cities lies principally in connecting with "smart citizens" (Hemment and Townsend 2013), enabling them to solve their own problems with the technology available through e-government practices or even with technologies created by active, committed, and community-minded citizens.

The clear link between e-participation and ICT development in local public administrations therefore requires integrating strategies and designs for social, technological, and organizational subsystems. Ramilo and Fernández (2012, pp. 5–6) propose a set of requirements for this pursuit. The first addresses "complexity" in terms of resources, time, format, or platforms. Second is the "type" of participation—open, closed, or dialogic—facilitated by each tool. Third is the "diversity of channels" for participation, such as Web sites, blogs, SMS texting, or WhatsApp.

Web portals tend to be the channel preferred by public administrations. A portal can be defined as a single access point to information, resources, and services relating to the different areas of government responsibility (Luna-Reyes et al. 2009; Granić et al. 2011; Bouzas and Mahou 2013). Portals have become the context for e-government and a tool for facilitating citizen participation and greater transparency in public management (Heeks and Foster 2013; Heeks and Stanforth 2007).

Since the start of this millennium, institutional Web usability studies have emerged as a line of research focusing on Web portal functionality and the ease with which a user interacts with the site. It seeks to measure user interaction with the Web site and ascertain levels of satisfaction for a series of factors: quality and usefulness of content, quality of service and assistance by the provider, and quality of the application design (Hassan et al. 2004). Usability studies focused on the user are abundant, diverse, and broadly interdisciplinary, including areas such as the provision of services, information management, and even participation in e-government environments (Byun and Gavin 2011; Bouzas and Mahou 2013).

3 A Usability Analysis of Online Participation Features in Spanish Smart Cities

Local governments in much of the southern European Union, including Spain, still adhere to what Parrado (2002) has classified as a Napoleonic model, with a series of sociopolitical and administrative traditions that are expressed in a mixed urban model involving a few global cities (Sassen 2001), mid-size cities, and smaller rural municipalities. The general urban–rural system in the EU (Kelder 2011, pp. 31–42) encompasses a reduced number of great cities, extensive infra-municipalism, and a series of assistance-based public policies circumscribed by an inclusive intergovernmental relations model that is only formally multilevel (Carrillo 2002; Varela 2011).

About a decade ago in Spain, e-administration, with its heterogeneous strategies, began to be adapted to asymmetrical and diverse local public management models (Ramió 2004, 2009). This diversity of e-government and public e-service models corresponds to a broad range of smart initiatives implemented by Spanish cities in various sectors, from economy (competitiveness) to persons (social and human capital), and from governance systems (participation) to mobility (transportation and ICTs), the environment (natural resources), and quality of life. Specifically, Spanish smart cities show similarities to the European environment (Capgemini Consulting and Fundación Orange 2013), with similar strategies and performance (SOCITM and IDEA 2002). However, some authors question the impact of e-participation and e-government practices in smart cities in southern Europe, especially regarding the quality of democracy (Hüller 2010) and social innovation (Calzada 2013, pp. 1–20).

In the context of administrative innovation, Spanish local governments and provincial capitals in particular are launching powerful Web portals to provide information and basic services. Using the heuristic or expert test technique, we will examine the virtual environments created for citizen participation. Heuristic tests fall within the category of "inspection methods" (Nielsen 1994, 2003) and have been applied in usability studies since the 1990s (Eschenfelder 2004; Welle et al. 2010; Sandoval-Almazán and Gil-García 2012). The heuristic technique establishes a checklist that is used by qualified personnel to examine a series of Web page dimensions. Each dimension is disaggregated into a series of quantifiable parameters that provide a way of detecting strengths and weaknesses, as well as establishing criteria for comparison with other cases. To ensure objectivity, it is common for two to five evaluators to verify the findings and use consensus to determine unclear results.

Though mindful of its subjective and qualitative limits[1] (Fernandez et al. 2011), for this study we have chosen to apply a heuristic test that allows us to gather abundant information from numerous units of analysis in a short period of time and at reduced cost.

The expert test designed for this study identifies five dimensions intended to measure the usability of online participation platforms hosted on smart city Web portals, and subsequently determine the level of interactivity between local administrations and citizens. The five dimensions include: "linkage," "citizen attention," "open government," and "e-administration services" to evaluate the "e-government" sector of local administrations (Tables 1, 2), and the "smart city initiatives" dimension to evaluate the "mobility" and "environmental sustainability" sectors (Table 3).

Each dimension contains a certain number of associated parameters consisting of the main online participation resources that Spanish citizens require from an institutional Web site (ONTSI 2013a). The maximum combined score a smart city can achieve in the e-government, mobility, and sustainability sectors is 135.5 points.

The "linkage" dimension evaluates the degree of user difficulty in accessing the citizen participation space hosted by the municipal Web site, based on easily identifiable, visible, and consistent labeling.

The "citizen attention" dimension evaluates the presence of resources located on an institutional portal (chat/virtual assistance, e-mail, and virtual mailbox for queries or consultations) that facilitate assistance and response through asynchronic communication and information exchange between administrations and citizens.

[1] This is in contrast with other methods such as the "accessibility test," for which the main automatic research applications (Cynthia Says®, eXaminator®, hera®, Truwex®, Wave®, or Taw®) allow a more quantitative and objective analysis of the *Web Content Accessibility Guidelines* (WCAG).

Table 1 E-government sector: parameters for the "linkage," "attention to citizens," and "open government" dimensions. (Source: Compiled by author based on Bouzas and Mahou 2013; Beynon-Davies and Williams 2003; Eschenfelder 2004; Fernández et al. 2011; Hassan et al. 2004; Sandoval-Almazán and Gil-García 2012)

Dimension	Parameter	Weight (A)	Values (B)	Total (A×B)
1. Linkage	1.1. Direct link to citizen participation spaces with a clear and descriptive tab, easy to identify, visible, and valid	1	1.5—Yes, with a clear heading and thorough description	
			1—Yes, with a clear heading and average description	
			0.5—Yes, with a clear heading but weak or no description	
			0—No	
	1.2. The link integrates the main contents and services for citizen participation	1	1.5—Yes	
			1—Some of them	
			0—None of them	
	Score: 3 points			
2. Attention to citizens	2.1. Availability and quality (design and functioning) of chat service/virtual assistance	2	1.5—Yes, high level of availability and quality	
			1—Yes, average level of availability and quality	
			0—No	
	2.2. E-mail availability	1	1—Yes	
			0—No	
	2.3. Virtual mailbox availability for queries/receiving information	2	1—Yes	
			0—No	
	Score: 6 points			
3. Open government	3.1. Blogs created by citizens as well as by local government	3	1.5—Yes, citizen and administration blogs	
			1—Yes, only public administration blogs	
			0—No	
	3.2. Feedback mechanisms via online questionnaires	3	1—Yes	
			0—No	
	3.3. Forums. Debate spaces, enquiries, proposals, voting, and comments by citizens and local government	3	1.5—Yes, citizen and administration forums	
			1—Yes, only administration forums	
			0—No	

Table 1 (continued)

Dimension	Parameter	Weight (A)	Values (B)	Total (A×B)
	3.4. Quality of information on citizen participation (completeness, structure, comprehensibility, and updatedness)	2	1.5—Yes, full range of information	
			1—Yes, some information	
			0—No	
	3.5. Collaborative maps	2	1—Yes	
			0—No	
	3.6. Forms for processing of complaints and suggestions on local services	3	1.5—Yes, both options	
			1—Yes, one of the two	
			0—No, neither of them	
	3.7. Open data options	3	1—Yes	
			0—No	
	3.8. Citizen participation via social network(s) and entity's own social network	3	1.5—Yes, both on the main social networks and entity's own network	
			1—Yes, only on the main social networks	
			0—No	
	3.9. Quality of information on transparency (completeness, structure, comprehensibility, and updatedness)	3	1.5—High quality (complete, updated, and comprehensible information within a well-organized structure)	
			1—Average quality (complete, updated information but with mediocre organization and not very comprehensible)	
			0.5—Low quality (incomplete information, poorly updated, incomprehensible at times, mediocre organization)	
			0—Offers no information	
	Score: 33.5 points			

Table 2 E-government sector: parameters for the "e-administration services" dimension. (Source: Compiled by author based on Beynon-Davies and Williams 2003; Bouzas and Mahou 2013; Buckley 2003; Gil-García and Martínez-Moyano (2007); Horst et al. (2007); Sandoval-Almazán and Gil-García 2012; Tan et al. 2013)

		Level of processing[a]				
		0	1	2	3	4
Requests	Activation of tele-assistance service/ home assistance service					
	Local library card					
	Residence registration					
	Home occupation license					
	License for potentially dangerous animals					
	Enrollment in vocational training centers					
	Complaint to the consumer affairs office					
	Electric appliance and furniture disposal service					
	Citizen card/public transport card					
Payments	Property tax					
	Traffic fines					
	Waste disposal tax					
Score: 48 points						

[a] Analysis depending on the procedure:
0. The service is not offered
1. Access to information
2. Offers document downloading
3. Combined online/offline procedure
4. Full online procedure

Table 3 Mobility and environmental sustainability sectors: parameters for the "smart city initiative" dimension. (Source: Compiled by author based on Bouzas and Mahou 2013; Beynon-Davies and Williams 2003; Eschenfelder 2004; Fernández et al. 2011; Hassan et al. 2004; Sandoval-Almazán and Gil-García 2012)

Parameter			Weight (A)	Values (B)	Total (A×B)
1. Entity's own Web site/Web page on the initiative			1	1—Yes	
				0—No	
2. Initiative Web site link available on the local government Web site with a tab that is clear and descriptive, easy to identify, visible, and valid			1	1.5—Yes, with a clear tab and thorough description	
				1—Yes, with a clear tab and average description	
				0.5—Yes, with a clear tab but weak or no description	
				0—No	
3. Quality of information on the initiative (completeness, structure, comprehensibility, and updatedness)			2	1.5—High quality (complete, updated, and comprehensible information within a well-organized structure)	
				1—Average quality (complete, updated, but not very comprehensible information with mediocre organization)	
				0.5—Low quality (incomplete, poorly updated, and at times incomprehensible information, with mediocre organization)	
				0—No information	
4. Specific services for mobility and environmental smart initiatives	4.1. Attention to citizens	Availability and quality (design and functioning) of chat service/virtual assistance)	2	1.5—Yes, high availability and quality	
				1—Yes, average availability and quality	
				0—No	
		E-mail availability	1	1—Yes	
				0—No	
		Virtual mailbox available for queries/requesting information	2	1—Yes	
				0—No	

Table 3 (continued)

Parameter			Weight (A)	Values (B)	Total (A×B)
	4.2. Open government	Blogs created by citizens as well as by local government	3	1.5—Yes, citizen and administration blogs	
				1—Yes, only administration blogs	
				0—No	
		Feedback mechanisms via online questionnaires	3	1—Yes	
				0—No	
		Spaces for debate, enquiries, proposals, voting, and comments by citizens and local government	3	1.5—Yes, citizen and administration forums	
				1—Yes, only administration forums	
				0—No	
		Quality of information on citizen participation (completeness, structure, comprehensibility, and updatedness)	2	1.5—Yes, full range of information	
				1—Yes, some information	
				0—No	
		Collaborative maps	2	1—Yes	
				0—No	
		Forms for processing complaints and suggestions on local services	3	1.5—Yes, both options	
				1—Yes, one of the two	
				0—No, neither of them	
		Open data options	3	1—Yes	
				0—No	
		Citizen participation via social network(s) and entity"s own social network	3	1.5—Yes, on the main social networks and the entity"s own	
				1—Yes, only on the main social networks	
				0—No	

Table 3 (continued)

Parameter			Weight (A)	Values (B)	Total (A×B)
		Quality of information on transparency (completeness, structure, comprehensibility, and updatedness)	3	1.5—High quality (complete, updated, and comprehensible information within a well-organized structure)	
				1—Average quality (complete, updated, but not very comprehensible information with mediocre organization)	
				0.5—Low quality (incomplete, poorly updated, and at times incomprehensible information, with mediocre organization)	
				0—No information	
Score: 45 points					

The "open government" dimension addresses platforms for direct involvement and citizen participation in government actions in a context of political–administrative transparency, with the aim of improving governmental efficiency and decision-making quality (Christmas 2011; Criado 2012; McDermott 2010; Meijer et al. 2012). The features included in the expert test are:

- Blogs: personal Web sites with content that can be shared with permission so that others may collaborate (Stephens 2011 in Karkin 2013). In this study, we examine links to citizen participation and debate sites created both by citizens and local administrations.
- Surveys: online questionnaires for gathering opinions.
- Forums: spaces for debate, consultation, proposals, voting, and comments launched either by citizens or local administrations.
- Information: specifically, links to information regarding citizen participation (sessions for the public, courses, association censuses, regulations, participating organs, centers, or spaces).
- Collaborative maps: tools that collect or publish collaborative data on specific facts or events taking place in the municipality.
- Complaints and suggestions: processing of complaints and suggestions regarding municipal services through online forms.
- Open data platforms: sharing of data between citizens and local administrations in standard open format that allows access and reuse (Janssen et al. 2012).
- Social networking: citizen participation and attention through social network accounts (primarily Facebook and Twitter) or city social networks.
- Transparency: quality of information (content, structure, and updatedness) on transparency.

The "e-administration" services dimension seeks to determine the level of online development of processes for certain municipal services offered to citizens (Batlle et al. 2011; Fernández et al. 2011). A catalog of services was compiled for two types of processes: requests and payments. Selection of these two services was based on an examination of Web portals from Spanish smart cities listed in the "top" and "contenders" categories in the 2012 IDC report. After examining these Web sites, we selected the most commonly offered basic services that involved some aspect of the areas of responsibility included in the *Ley de Bases del Régimen Local* (Law on Bases of the Local Regime)[2] for cities with a population greater than 50,000 inhabitants. Making use of e-government literature to verify the degree of online processing (Beynon-Davies and Williams 2003; Buckley 2003; Dhillon et al. 2008; Gil-García and Martínez-Moyano 2007; Sandoval-Almazán and Gil-García 2012; Spyros et al. 2010; Tan et al. 2013), we evaluated the services on a scale of 0 ("no service offered") to 4 ("the user can complete the entire process online").

Finally, though the smart phenomenon can actually encompass numerous interconnected fields of activity, the "smart city initiative" dimension in this study

[2] Law 7/1985, of 2 April, regulating the Bases of the Local Regime (Bases del Régimen Local-articles 15, 25, and 26).

focuses exclusively on the mobility and environmental sustainability sectors. These two areas were chosen because they appear to receive the greatest attention in cities (AMETIC 2013; CGLU 2012; Libro Blanco Smart Cities 2012).

Two objectives can be identified in this regard. The first was to determine the quality of information about the initiative and ease of access by verifying (i) the existence of a dedicated Web page; (ii) clarity and visibility of the heading from the main municipal Web page; and (iii) the completeness, expository structure, and updatedness of information. Second, this dimension was also intended to explore the citizen participation mechanisms found in the "open government" dimension, but adapted to smart city initiative Web sites. The final objective is to discover if smart city and governance sector initiatives differ when offering municipal participation services to citizens.

Smart cities were selected through representative sampling of the main Spanish municipalities, based on the degree of smart initiative development. The 2012 IDC ranking of intelligent cities was used, which assigns cities to four categories: top, contenders, players, and followers. The cities were scored according to two factors: intelligent dimensions (development and application of initiatives regarding governance, building, mobility, environment, and services) and facilitating forces (persons, economy, and ICTs that create propitious conditions for successful implementation of the initiative).

We selected the ten Spanish cities listed in the top and contenders groups (Table 4). A double filter was then applied to select ten additional cities from the 34 encountered in the broad, heterogeneous players and followers groups. Three

Table 4 Selection criteria for the Spanish smart cities. (Source: Authors' own elaboration. IDC (2012))

Development level IDC	Smart city	Population
Top	Barcelona	+ 1,000,000
	Santander	100,000–200,000
	Madrid	+ 1,000,000
	Málaga	500,000–1,000,000
	Bilbao	200,000–500,000
Contenders	Valladolid	200,000–500,000
	Zaragoza	500,000–1,000,000
	Vitoria	200,000–500,000
	San Sebastián	100,000–200,000
	Pamplona	100,000–200,000
Players	Alicante	200,000–500,000
	Burgos	100,000–200,000
	Castellón	100,000–200,000
	A Coruña	200,000–500,000
	Logroño	100,000–200,000
	Murcia	200,000–500,000
	Salamanca	100,000–200,000
Followers	Badajoz	100,000–200,000
	Palma de Mallorca	200,000–500,000
	Las Palmas GC	200,000–500,000

selection criteria were established: the candidate cities had to belong to the *Red Española de Ciudades Inteligentes*—RECI (Spanish Network of Smart Cities)[3], be provincial capitals, and house a population of 100,000–500,000 persons[4] (the most populated cities in this group). The 14 cities that passed through the first filter were then randomly sampled, and ten cities chosen for the study. The total sample size was 20 out of a sample universe of 26 Spanish smart cities that were listed in the 2012 IDC report and were also members of RECI.

Once the heuristic test had been designed and the sample cities selected, two evaluators independently explored and collected data from the 20 smart city Web portals for 7 days, using a basic computer system. The average examination time was 120 min, 30 of which were dedicated to general site inspection prior to analysis. After independent examination, the evaluators explored the Web sites and reviewed the data jointly for a week and a half. They were required to resolve discrepancies by consensus and their findings were compiled in a report.

The results of the expert test applied to 20 Spanish smart cities are presented here and visually summarized in Tables 5–10. The entire set of municipalities averaged 53.22 points out of 135.5 possible points (Table 5). Only Zaragoza scored above 100 points, (102.5), ranking above all cities in all dimensions except the "smart city initiative." Zaragoza's institutional Web site, with multiple channels for citizen participation, received the maximum possible score in "open government." Barcelona and Madrid followed, with 73 and 68.5 points, respectively. Barcelona was strongest in "citizen attention" and "e-administration services." Madrid did not stand out in any single dimension but presented relatively high averages in all dimensions, with the second highest result for "open government." Santander scored 66 points and had the best "smart city initiatives" result (27.5 points), thanks to a Web page dedicated to the exchange of ideas between citizens and the local administration.

At the opposite end of the spectrum we find the city of Murcia with 32 points. Though the "citizen attention" score was high, it presented the lowest results for "linkage" (0) and "e-administration services" accompanied by poor results in other dimensions. Logroño, Burgos, and Castellón scored below 40 points and presented the lowest "smart city initiatives" results. Castellón had the lowest scores in "citizen attention" and "e-administration services."

The average score for all cities in each dimension was below 50 % of the maximum possible on the expert test; with the exception of "e-administration services," which scored slightly above 50 %. The "smart city initiatives" average score of 11.71 points was especially negative, some 33 points below the maximum of 45 points for that dimension.

---

[3] The cities of Bilbao and San Sebastian are not members of the Spanish Network of Smart Cities (Red Española de Ciudades Inteligentes or RECI). Yet we have included them in this study due to their significant level of smart initiatives (CGLU 2012; IDC 2012).

[4] This is broken down into two categories (100,000–200,000 and 200,000–500,000 inhabitants), according to the classification of Spanish municipalities used by the Instituto Nacional de Estadística (INE).

Table 5 Total score. (Source: Authors' own elaboration)

	Link (out of 3)	Attention to citizenship (out of 6)	Open government (out of 33.5)	Smart city initiative (out of 45)	E-administration services (out of 48)	Total (out of 135.5)
A Coruña	1	2	7.5	6	19	35.5
Alicante	1.5	3	11.5	7.5	21	44.5
Barcelona	1.5	3	16.5	14	38	73
Badajoz	0.5	2	3	5	27	37.5
Bilbao	1	1	18	20.5	16	56.5
Burgos	2	2	10.5	3.5	19	37
Castellón	1	1	13	3	17	35
Las Palmas GC	0	2	4.5	16	28	50.5
Logroño	0	3	8	4	19	34
Madrid	1.5	2	21	14	30	68.5
Málaga	2.5	1	13.5	17.5	24	58.5
Murcia	0	3	7.5	7.5	14	32
Palma de Mallorca	2.5	0	15.5	7.5	22	47.5
Pamplona	0.5	1	16	8	20	45.5
Salamanca	2.5	3	10	16	24	55.5
San Sebastián	2	2	16.5	11.5	33	65
Santander	1.5	1	12	27.5	24	66
Valladolid	2.5	1	18	9	28	58.5
Vitoria	2	1	18.5	13.5	27	62
Zaragoza	3	3	33.5	23	40	102.5
Average	1.45	1.85	13.72	11.72	24.5	53.25

Table 6 Linkage dimension. (Source: Authors' own elaboration)

	Link to citizenship participation space (out of 1.5)	Contents and Services (out of 1.5)	Total (out of 3)
A Coruña	1	0	1
Alicante	0.5	1	1.5
Barcelona	0.5	1	1.5
Badajoz	0.5	0	0.5
Bilbao	1	0	1
Burgos	1	1	2
Castellón	1	0	1
Las Palmas GC	0	0	0
Logroño	0	0	0
Madrid	1.5	0	1.5
Málaga	1.5	1	2.5
Murcia	0	0	0
Palma de Mallorca	1.5	1	2.5
Pamplona	0.5	0	0.5
Salamanca	1.5	1	2.5
San Sebastián	1	1	2
Santander	1.5	0	1.5
Valladolid	1.5	1	2.5
Vitoria	1	1	2
Zaragoza	1.5	1.5	3
Average	0.92	0.52	1.45

The data for the "linkage" dimension (Table 6) indicate that Web portal users may find access difficult. Even when labels or headings for participation spaces were clear and descriptive, links presented diversity, lacked visibility due to location, and often failed to integrate all content and services. Zaragoza obtained the maximum score thanks to a prominently placed link with a clear label ("open government") on the main Web page. The links on the sites of Murcia, Logroño, and Las Palmas de Gran Canaria were entirely unusable.

In the "citizen attention" dimension (Table 7), we found that no city offered all three of the services analyzed (chat/virtual assistance, email, and virtual mailbox); and none were available on the Palma de Mallorca Web site. Though e-mail was most frequently resorted to by Spanish Web users (ONTSI 2013a), only 12 cities offered it, followed by 10 that offered a virtual mailbox. Only Zaragoza offered chat/virtual assistance and it was available for all "citizen attention" services.

The "open government" dimension (Table 8) shows a very broad distribution of scores. Over half of the municipal Web portals (55 %) offered no participation platforms, while the remainder offered high-level services including forums, blogs, social networks, and open data platforms.

A "complaints and suggestions" space was available in 17 municipalities, making this the most frequently occurring service in the "open government" dimension. Only Zaragoza offered collaborative maps. The average scores were quite modest for all services, particularly blogs, surveys/opinion, forums, and open data platforms,

Table 7 Citizen attention dimension. (Source: Authors' own elaboration)

	Chat/virtual assistant (out of 3)	E-mail (out of 1)	Virtual Mailbox (out of 2)	Total (out of 6)
A Coruña	0	0	2	2
Alicante	0	1	2	3
Barcelona	0	1	2	3
Badajoz	0	0	2	2
Bilbao	0	1	0	1
Burgos	0	0	2	2
Castellón	0	1	0	1
Las Palmas GC	0	0	2	2
Logroño	0	1	2	3
Madrid	0	0	2	2
Málaga	0	1	0	1
Murcia	0	1	2	3
Palma de Mallorca	0	0	0	0
Pamplona	0	1	0	1
Salamanca	0	1	2	3
San Sebastián	0	0	2	2
Santander	0	1	0	1
Valladolid	0	1	0	1
Vitoria	0	1	0	1
Zaragoza	3	0	0	3
Average	0.15	0.6	1.1	1.85

which did not even obtain a quarter of the maximum possible score. The exception was information on "transparency." Social networks were used extensively, and some cities even implemented their own systems. Most Web sites showed at least an average level of information regarding participation.

Zaragoza displayed the best results for each parameter, achieving the maximum of 33.5 points for this dimension. Badajoz, A Coruña, Las Palmas, Logroño, and Murcia presented scores below 10 points and had the lowest number of features.

In the "smart city initiatives" dimension, Table 9 reveals a duality between location ("Web"/"own page" and "linkage") and information parameters, and parameters related to citizen attention and open government. The data indicated that smart initiatives have their own Web pages with detailed information and a prominently located link in the municipal portal, even though online participation mechanisms are scarce.

The social network was the most common participation channel, appearing on 12 institutional webs; email and complaints/suggestions services were available on nine portals, and a virtual mailbox was available on seven sites. Other features were scarce, and chat/virtual assistance, collaborative maps, open data platforms, and information on transparency were not offered on any site. With survey/opinion, information, complaints and suggestions, and social networks as the most prominent services, Santander scored highest (27.5 points), followed by Zaragoza and Bilbao. At

Table 8 Open government dimension. (Source: Authors' own elaboration)

	Blogs (out of 4.5)	Survey (out of 3)	Forums (out of 4.5)	Information (out of 3)	Collaborative maps (out of 2)	Complaints and suggestions (out of 4.5)	Open data (out of 3)	Social networking (out of 4.5)	Transparency (out of 4.5)	Total (out of 33.5)
A Coruña	0	0	0	0	0	4.5	0	0	3	7.5
Alicante	0	0	4.5	2	0	4.5	0	0	0.5	11.5
Barcelona	3	0	0	3	0	4.5	3	3	0	16.5
Badajoz	0	0	0	0	0	0	0	3	0	3
Bilbao	0	3	0	3	0	4.5	0	3	4.5	18
Burgos	0	0	0	3	0	0	0	3	4.5	10.5
Castellón	3	0	0	2	0	4.5	0	3	0.5	13
Las Palmas GC	0	0	0	0	0	4.5	0	0	0	4.5
Logroño	0	0	0	0	0	4.5	0	3	0.5	8
Madrid	0	3	0	3	0	4.5	3	3	4.5	21
Málaga	0	0	0	3	0	3	0	3	4.5	13.5
Murcia	0	0	0	0	0	0	0	3	4.5	7.5
Palma de Mallorca	0	0	0	2	0	4.5	0	4.5	4.5	15.5
Pamplona	0	0	0	1	0	4.5	3	3	4.5	16
Salamanca	0	0	0	2.5	0	4.5	0	3	0	10
San Sebastián	3	0	0	3	0	4.5	0	3	3	16.5
Santander	0	3	0	0	0	4.5	0	0	4.5	12
Valladolid	4.5	3	0	3	0	4.5	0	3	0	18
Vitoria	0	0	4.5	2	0	4.5	0	3	4.5	18.5
Zaragoza	4.5	3	4.5	3	2	4.5	3	4.5	4.5	33.5
Average	0.9	0.75	0.67	1.77	0.1	3.75	0.6	2.55	2.62	13.71

Table 9 Smart city initiatives dimension. (Source: Authors' own elaboration)

	Own Web site/Web page (out of 1)	Link on the local government Web site (out of 1.5)	Information on initiative (out of 3)	Specific services for mobility and environmental smart initiatives (out of 39.5)	Total (out of 45)
A Coruña	1	0	2	3	6
Alicante	1	0.5	3	3	7.5
Barcelona	1	0.5	3	9.5	14
Badajoz	1	1	3	0	5
Bilbao	1	1.5	3	15	20.5
Burgos	1	0.5	2	0	3.5
Castellón	1	1	1	0	3
Las Palmas GC	1	1	3	11	16
Logroño	1	1	2	0	4
Madrid	1	0.5	3	9.5	14
Málaga	1	1.5	3	12	17.5
Murcia	1	0.5	2	4	7.5
Palma de Mallorca	1	0.5	2	4	7.5
Pamplona	1	1	2	4	8
Salamanca	1	1	2	12	16
San Sebastián	1	1.5	3	6	11.5
Santander	1	1.5	3	22	27.5
Valladolid	1	1.5	2	4.5	9
Vitoria	1	1.5	3	8	13.5
Zaragoza	1	1	3	18	23
Average	1	0.95	2.5	7.27	11.72

the opposite extreme were Badajoz, Burgos, Castellón, and Logroño, which offered no participation services on their respective smart initiative Web pages.

Finally, for the "e-administration services" dimension, Table 10 shows that services were not offered in 68 of the 240 cases analyzed; while services with varying degrees of processing were available in 172 cases. Level 4 ("full online processing") was the most frequent (91 cases), followed by 39 Level 1 cases ("user can only access information, but no processing is possible") and 39 Level 2 cases ("download documents and continue processing offline"). Level 3 (combined online–offline processing) was virtually nonexistent. From this we infer that when a city offers a service, online processing for the entire procedure is likely (in 53 % of the cases).

The services found to offer the highest level of processing were those relating to some type of payment, such as taxes (property, urban waste disposal) and traffic fines; which could be completed entirely online in most cities. All other services presented greater diversity of levels. Some institutional Web pages allowed certain services to be requested, but in most cities users simply access information (Level 1) or could download documents for subsequent offline processing (Level 2). "Residence registration" was the most frequently occurring full procedure (12 cities), while "registration in vocational training centers" was the least-offered service.

Table 10 E-administration services dimension. (Source: Authors' own elaboration)

	C	A	B	BA	BI	BU	CS	GC	LO	M	MA	MU	PM	NA	SA	SS	S	VA	VI	Z	Max.
Activation of the tele-assistance service/home assistance service	1	1	2	0	1	2	2	0	0	2	1	0	0	0	2	1	2	1	0	1	4
Local library card	4	1	4	0	2	1	0	0	2	4	4	0	0	0	0	4	2	0	0	2	4
Residence registration	1	4	3	4	4	4	4	4	1	2	2	4	2	4	2	2	4	4	4	4	4
Home occupation license	0	2	1	1	0	0	2	4	2	2	0	0	0	0	2	2	2	4	1	4	4
License for potentially dangerous animals	0	0	2	4	1	0	2	4	0	2	2	2	0	4	2	3	2	4	2	4	4
Enrollment in vocational training centers	0	0	2	4	4	0	0	0	0	0	0	0	0	0	4	4	2	4	3	4	4
Complaint to the consumer affairs office	0	0	4	4	2	0	2	4	2	1	1	0	4	0	2	0	2	4	4	4	4
Electric appliance and furniture disposal service	1	0	4	4	1	0	0	0	0	4	1	0	0	0	1	1	1	4	0	1	4
Citizen card/ public transport card	4	1	4	4	1	0	2	0	0	1	1	0	4	0	1	4	1	1	1	4	4
Property tax	4	4	4	2	0	4	1	4	4	4	4	4	4	4	4	4	1	1	4	4	4
Traffic fines	0	4	4	0	0	4	1	4	4	4	4	0	4	4	0	4	4	1	4	4	4
Waste disposal tax	4	4	4	0	0	4	1	4	4	4	4	4	4	4	4	4	1	0	4	4	4

The Zaragoza municipality had the largest number of entirely online procedures, followed by Barcelona and Las Palmas de Gran Canaria. The cities of Pamplona, Murcia, Burgos, and Palma de Mallorca offered the least number of services.

4 Conclusions

As we encountered in prior pages, for a city to be considered "smart," it must include several elements or dimensions of society, technology, and public management (Khan et al. 2010) in its development and strategy (Nam and Pardo 2011a, pp. 282–291; Lips 2012). Smart cities must articulate an integrated framework (Chourabi et al. 2012) in which citizen participation emerges as an absolute requirement.

Although processes linked to smart cities are increasingly interactive and dialogic, with more interoperable and usable ICTs, it is still uncertain *how* citizen participation will evolve, *who* the protagonists will be and especially *what* are the most appropriate means for enabling it (Schaffers et al. 2011; Batty et al. 2012, pp. 481–518). This study focused specifically on the means, seeking to contribute to the debate by examining online participation platforms in 20 Spanish smart city Web portals to discover their level of usability in areas of e-government and mobility/environmental sustainability smart initiatives.

In the e-government sector, cities offered platforms mainly directed at e-administration, with a high level of online processing for payment of taxes, fees, and fines. Other basic services such as tele-assistance and license or card issue requests were less complete. The "citizen attention" and "open government" services in smart cities offered only limited participation mechanisms: mainly social networking, e-mail, virtual mailbox, and complaint/ suggestion services.

Citizen participation opportunities were even more limited in "mobility" and "environmental sustainability" sector initiatives. Web portals mainly provided information on projects but no interactive channels for citizens to interact with its content and orientation. On the whole, these Web pages were flat and unidirectional. The citizen–administration relationship always included a presential aspect and was limited mainly to the associative sphere. Web sites that did have their own online participation platform were limited to specific e-government resources hosted by the city Web portal but managed by the smart initiative Web site.

The absence of online participation spaces observed in this study corroborates other research. The *Observatorio Nacional de Telecomunicaciones y Sociedad de la Información* (ONTSI—National Observatory for Telecommunications and the Information Society) reported in its *Estudio de la demanda y uso de Gobierno Abierto en España* (Study on the Demand and Use of Open Government in Spain 2013b) that half the population surveyed considered current public participation tools insufficient. Recommendations included improving the functionality of institutional webs, increasing the presence of public organizations on social networks, providing personalized information through mobile applications and creating online communities and forums.

If we classify Spanish smart city Web sites according to their level of usability (Table 11), Zaragoza emerges as the only city with a high usability score for online participation services. It significantly outpaces Murcia, the lowest-scoring city, by 70 points. When comparing cities in the high/medium-high and low/medium-low category groupings, the gap diminishes to 34 points.

The data presented two other points of interest. First, there was no clear correlation between population size and level of Web site usability. All city population categories were represented in the high/medium-high usability rankings. However, cities with lower population were more concentrated in the medium-low/low usability rankings (100,000–200,000/200,000–500,000).

Additionally, no defining correlation could be drawn between the level of smart city development (as listed in the 2012 IDC report) and online participation platform usability levels. The "top" and "contender" cities were generally grouped in the high/medium-high usability categories; but some "top" cities such as Málaga and Bilbao ranked in the medium usability category. Pamplona, which is listed as a "contender" city, ranked in the medium-low category. Most of the "player" and "follower" cities filled the medium-low/low categories.

Table 11 Citizen usability platforms quality. (Source: Authors' own elaboration)

Smart city	Development level IDC	Population	Total (out of 135)	Participation platforms usability
Zaragoza	Contenders	500,000–1,000,000	102.5	*High* 80–100 points
Barcelona	Top	+1,000,000	73	*Medium–high* 60–79 points
Madrid	Top	+1,000,000	68.5	
Santander	Top	100,000–200,000	66	
San Sebastián	Contenders	100,000–200,000	65	
Vitoria	Contenders	200,000–500,000	62	
Málaga	Top	500,000–1,000,000	58.5	*Medium* 50–59 points
Valladolid	Contenders	200,000–500,000	58.5	
Bilbao	Top	200,000–500,000	56.5	
Salamanca	Players	100,000–200,000	55.5	
Las Palmas GC	Followers	200,000–500,000	50.5	
Palma de Mallorca	Followers	200,000–500,000	47.5	*Medium–low* 40–49 points
Pamplona	Contenders	100,000–200,000	45.5	
Alicante	Players	200,000–500,000	44.5	
Badajoz	Followers	100,000–200,000	37.5	*Low* 20–39 points
Burgos	Players	100,000–200,000	37	
A Coruña	Players	200,000–500,000	35.5	
Castellón	Players	100,000–200,000	35	
Logroño	Players	100,000–200,000	34	
Murcia	Players	200,000–500,000	32	

We have no explanation for this, beyond a description of the Web sites analyzed. It would be interesting if future research on smart city Web site usability combined inspection methods such as the heuristic test with empirical methods focused on the user (Nielsen 1994, 2003). This could improve our understanding of *how* and *why* citizens use these platforms, and especially how they can increase citizen involvement in public decision-making.

From the dearth of information regarding these topics, and with respect to the object of this study, we conclude from the results of the heuristic test that Spanish smart city Web portals are informative but do not generate virtual environments favoring fluid interaction between local administrations and citizens. Though 92.8 % of Spanish Internet users consider citizen participation in public affairs necessary (ONTSI 2013b), these sites present deficiencies in their offer of ways for citizens to participate in designing public policies and producing or providing public goods and services.

References

AMETIC. (2013). *Smart cities 2012*. Madrid: Ministerio de Industria..

Batlle, J., Abadal, E., & Blat, J. (2011). Benchmarking del e-gobierno local: Limitaciones de los métodos de evaluación comparativa. *El Profesional de la Información, 20*(3), 251–259.

Batty, M., Axhausen, K. W., Giannotti, F., Pozdnoukhov, A., Bazzani, A., Wachowicz, M., et al. (2012). Smart cities of the future. *The European Physical Journal Special Topics, 214*(1), 481–518.

Belissent, J. (2011). *The core of a smart city must be smart governance*. Cambridge: Forrester Research, Inc.

Beynon-Davies, P., & Williams, M. D. (2003). Evaluating electronic local government in the UK. *Journal of Information Technology, 18*(2), 137–149.

Bouzas, R., & Mahou, X. (2013). The study of the functionality of government portals. A methodological proposal from the e-health. *Gestión y Política Pública, temático,* 261–308.

Buckley, J. (2003). E-service quality and the public sector. *Managing Service Quality, 13*(6), 453–462.

Byun, D., & Gavin, F. (2011). Evaluating usability, user satisfaction and intention to revisit for successful e-government websites. *Electronic Government* (An International Journal), *8*(1), 1–19

Calzada, I. (2013). Critical social innovation in the smart city era for a city-regional European Horizon 2020. *P3T, Journal of Public Policies and Territories, Social Innovation and Territory, 6,* 1–20.

Capgemini Consulting, & Fundación Orange. (2013). *Estudio Comparativo 2013 de los Servicios Públicos online en 25 ayuntamientos españoles*. Madrid: Capgemini Consulting & Fundación Orange.

Carrillo, E. (2002). Los gobiernos locales. In P. Román (coord) (Ed.), *Sistema Político Español* (pp. 315–338). Madrid: McGraw-Hill.

CGLU. (2012). SMART CITIES STUDY: Estudio internacional sobre la situación de las TIC, la innovación y el Conocimiento en las ciudades. http://www.uclg.org/es/media/noticias/nuevo-estudio-de-smart-cities-sobre-la-situaci%C3%B3n-de-las-tic-innovaci%C3%B3n-y-conocimiento. Accessed 10 Dec 2013.

Chen, D.-Y., Huang, T.-Y., & Hsiao, N. (2004). Citizen participation, e-government, and public management: A case of Taipeu city mayor's e-mail box. *International Symposium of Digital Divide and Digital Opportunity*, February 20–21, Taipei (Taiwan), 1–20.

Chourabi, H., Gil-García, J. R., Pardo, T. A., Nam, T., Melloili, S., Scholl, H. J., et al. (2012). Understanding smart cities: An integrative framework. *2012 45th Hawaii International Conference on System Sciences*. Accessed 15 Jan 2014. doi:10.1109/HICSS.2012.615.

Christmas, L. (2011). A new age of transparency and let the public know. *British Journalism Review, 22*(4), 20–22.

Criado, J. I. (2012). Redes sociales y Open Government. Hacia unas Administraciones locales en red y abiertas. *Revista Democracia y Gobierno Local,* 18–19.

Dhillon, G. S., Weerakkody, V., & Dwivedi, Y. K. (2008). Realising transformational stage e-government: A UK local authority perspective. *Electronic Government* (An International Journal)*, 5*(2), 162–180.

Eschenfelder, K. R. (2004). Behind the web site: An inside look at the production of web-based textual government information. *Government Information Quarterly, 21,* 337–358.

European Unión. (2011). *Cities of tomorrow. Challenges, visions, ways forward.* Brussels: European Commission, Directorate General for Regional Policy.

Fernandez, A., Insfran, E., & Abrahão, S. (2011). Usability evaluation methods for the web: A systematic mapping study. *Information and Software Technology, 53,* 789–817.

Fernández, Y., Gutiérrez, C., & Fernández, J. M. (2011). E-gobierno en las entidades públicas: Conocimiento, utilidad y valoración de las páginas Web de los Ayuntamientos. Un caso de estudio. *Auditoría Pública, 53,* 109–119.

Frewer, L. J., & Rowe, G. (2005). Evaluating public participation exercises: Strategic and practical issues. In C. Joanne (Ed.), *Evaluating public participation in policy making* (pp. 85–106). Paris: OECD.

Giffinger, R., Fertner, C., Kramar, H., Kalasek, R., Pichler-Milanović, N., & Meijers, E. (2007). *Smart cities ranking of European medium-sized cities*. Vienna: Centre of Regional Science, Vienna UT.

Gil-Garcia, J. R., & Martinez-Moyano, I. (2007). Understanding the evolution of e-government: The influence of systems of rules on public sector dynamics. *Government Information Quarterly, 24* (2), 266–290.

Granić, A., Mitrović y, I., & Marangunić, N. (2011). Exploring the usability of web portals: A Croatian case study. *International Journal of Information Management, 31,* 339–349.

Hassan, Y., Martín, F. J., & Iazza, G. (2004). Diseño Web Centrado en el Usuario: Usabilidad y Arquitectura de la Información. *Hipertext.net*. 2.

Heeks, R., & Foster, C. (2013). Innovation and scaling of ICT for the bottom-of-the-pyramid. *Journal of Information Technology*. Accessed 15 Jan 2014. doi:10.1057/jit.2013.19.

Heeks, R., & Stanforth, C. (2007). Understanding e-government project trajectories from an actor-network perspective. *European Journal of Information Systems, 16*(2), 165–177. Accessed 15 Jan 2014. doi:10.1057/palgrave.ejis.3000676.

Hemment, D., & Townsend, A. (2013). *Here come the smart citizens*. Manchester: FutureEverything.

Hüller, T. (2010). Playground or democratisation? New participatory procedures at the European Commission. *Swiss Political Science Review, 16,* 77–107.

Horst, M., Kuttschreuter, M., & Gutteling, J. M. (2007). Perceived usefulness, personal experiences, risk perception and trust as determinants of adoption of e-government services in The Netherlands. *Computers in Human Behavior, 23*(4), 1838–1852.

Iasulaitis, S., & Pineda, C. (2013). Una administración pública transparente, participativa y responsable: el uso de las TICs. *IV Congreso Internacional en Gobierno, Administración Pública y Políticas Públicas (GIGAPP-IUIOG-INAP)*, Madrid, 23–24 September, pp. 1–33.

IDC. (2012). Análisis de las Ciudades Inteligentes en España 2012– El Viaje a la Ciudad Inteligente. http://www.slideshare.net/laltes/idc-smartcities-analysis-spain. Accessed 15 Jan 2014.

Janssen, M., Charalabidi, Y., & Zuiderwijk, A. (2012). Benefits, adoption barriers and myths of open data and open government. *Information Systems Management, 29*(4), 258–268. Accessed 15 Jan 2014. doi:10.1080/10580530.2012.716740.

Khan, G. F., Moon, J., Park, H. W., Swar, B., & Rho, J. J. (2010). A socio-technical perspective on e-government issues in developing countries: A scientometrics approach. *Scientometrics,* 1–20. Accessed 15 Jan 2014. doi:10.1007/s11192-010-0322-5.

Karkin, N. (2013). Herramientas Web 2.0 para la participación pública a través de los sitios de Internet del gobierno. *Gestión y Política Pública, temático,* 307–331.

Kelder, T. (2011). I love it when a plan comes together: Urban agendas in EU2020. *Eixo Atlántico. Revista da Eurorrexión Galicia-Norte de Portugal, 18,* 31–42.

Kingston, R. (2002). The role of e-government and public participation in the planning process. XVI AESOP Congress, Volos (Greece). July, 10–14 (2002), 1–14.

Libro Blanco Smart Cities. (2012). Enerlis, ernst and young, Ferrovial and Madrid Network. http://www.libroblancosmartcities.com/. Accessed 11 Jan 2014.

Lips, M. (2012). E-Government is dead: Long live Public Administration 2.0. *Information Polity, 17,* 239–250.

Luna-Reyes, L. F., Hernández García, J. M., & Gil-García, R. (2009). Hacia un modelo de los determinantes de éxito de los portales de gobierno estatal en México. *Gestión y Política Pública, XVIII*(2), 307–340.

McDermott, P. (2010). Building open government. *Government Information Quarterly, 27*(4), 401–413.

Meijer, A. J., Curtin, D., & Hillebrandt, M. (2012). Open government: Connecting vision and voice. *International Review of Administrative Sciences, 78*(1), 10–29.

Moro, G. (2005). Citizen's evaluation of public participation. In OECD (Ed.), *Evaluating public participation in policy making* (pp. 109–126). Paris: OECD.

Nam, T., & Pardo, T. A. (2011a). Conceptualizing smart city with dimension of technology, people, and institutions. *The proceedings of the 12th annual international conference on digital government research,* pp. 282–291.

Nam, T., & Pardo, T. A. (2011b). Smart city as urban innovation: Focusing on management, policy, and context. *ICEGOV2011, September 26–28, Tallinn, Estonia*, pp. 185–194.

Nielsen, J. (1994). Heuristic evaluation. In J. Nielsen & R. L. Mack. (Eds.), *Usability inspection methods*. New York: Wiley.

Nielsen, J. (2003). Usability 101: Introduction to usability. http://www.useit.com/alertbox/20030825.html. Accessed 15 Jan 2014.

OECD. (2003). *Engaging citizens online for better policy making*. Paris: OECD.

ONTSI. (2013a). Servicios utilizados por los usuarios de Internet. http://www.ontsi.red.es/ontsi/es/indicador/servicios-utilizados-por-los-usuarios-de-internet-1. Accessed 15 Jan 2014.

ONTSI. (2013b). Estudio de la demanda y uso de Gobierno Abierto en España. http://www.ontsi.red.es/ontsi/es/estudios-informes/estudio-de-la-demanda-y-uso-de-gobierno-abierto-en-espa%C3%B1-edici%C3%B3n-2013. Accessed 15 Jan 2014.

Parrado, S. (2002). *Sistemas administrativos comparados*. Madrid: Tecnos y Universidat Pompeu Fabra.

Prensky, M. (2001). Digital natives, digital immigrants. *On the Horizon, 9*(5), 1–6. (MCB University Press).

Ramió, C. (2004). E-Administración y nuevos modelos de gestión pública: Una propuesta de decálogo normativo. *Revista Vasca de Administración Pública, 70,* 321–338.

Ramió, C. (2009). Evaluación de la innovación institucional y gestión de las administraciones locales. *Eixo Atlántico. Revista da Eurorrexión Galicia-Norte de Portugal, 15,* 21–52.

Ramilo, M. C., & Fernández, R. (2012). Ciudadanía y Participación en la Sociedad de la Información y el Conocimiento. *Working Papers GIGAPP-IUIOG* 1–30.

Rhodes, R. A. W. (1997). *Understanding governance. Policy networks, governance, reflexivity and accountability*. Buckingham: Open University Press.

Rios, P. (2008). Creating "the smart city". http://dspace.udmercy.edu:8080/dspace/bitstream/10429/20/1/2008_rios_smart.pdf. Accessed 5 July 2014.

Sandoval-Almazán, R., & Gil-García, J. R. (2012). Are government internet portals evolving towards more interaction, participation, and collaboration? Revisiting the rhetoric of e-government among municipalities. *Government Information Quarterly, 29*(1), 72–81.

Sassen, S. (2001). *The global city: New York* (2nd ed.). Princeton: Princeton University Press.

Schaffers, H., Komnimos, N., Pallot, M., Trousse, B., Nilsson, M., & Oliveira, A. (2011). Smart cities and future internet: Towards cooperation frameworks for open innovation. In J. Domingue, et al. (Eds.), *Future internet assembly* (pp. 431–446). Berlin: Springer-Verlag.

Siddiqi, J., Akhgar, B., Gamble, T., & Zaefarian, G. (2006). A framework for increasing participation in e-Government. *International Conference on e-Business, Enterprise Information Systems, e-Government & Outsourcing (EEE,06)*, pp. 1–7.

SOCITM, & IDEA. (2002). *Local e-government now: A worldwide view*. London: SOCITM and IDEA.

Spyros, A., Kitsios, F., & Papadopoulos, T. (2010). New service development in e-government: Identifying critical success factors. *Transforming Government: People, Process and Policy, 4*(1), 95–118.

Stephens, R. T. (2011). Web 2.0: Integration model with electronic commerce. In K. Malik & P. Choudhary (Eds.), *Business organizations and collaborative web: Practices, strategies and patterns* (pp. 18–30). Hershey: igi Global.

Tan, C., Benbasat, I., & Cenfetelli, R. T. (2013). IT-mediated customer service content and delivery in electronic governments: An empirical investigation of the antecedents of service quality. *Journal MIS Quarterly, 37*(1), 77–109.

Varela, E. 2011. Gobernanza urbana. Elementos de aprendizaje y estudios de casos. In M. Iglesias, M. Martí-Costa, J. Subirats, & M. Tomàs (Eds.), *Políticas urbanas en España* (pp. 363–386). Barcelona: Icaria.

Welle, D., DeJong, M., & Lentz, L. (2010). Usable guidelines for usable websites? An analysis of five e-government heuristics. *Government Information Quarterly, 27,* 254–263.

Xosé María Mahou Lago has a PhD in political science and administration from the University of Santiago de Compostela and is a lecturer at the Faculty of Social Sciences and Communication (University of Vigo). He has been coordinating the Master in Public Management and Institutional Leadership since 2011 (http://masterdireccionyliderazgo.com/). He has been a visiting researcher at the Georgetown University (Washington DC) and at the Internet Interdisciplinary Institute of the Universitat Oberta de Catalunya (Barcelona). His main lines of research are public policies, electronic government, web usability, and smart cities. He is also a member of the "Observatorio de Gobernanza, G³" (University of Vigo) and of the interuniversity research team "Gobernet".

Enrique José Varela Álvarez holds a bachelor's degree in Political Science and Sociology from the University Complutense of Madrid and a PhD in Governance and Public Administration by the University Research Institute Ortega y Gasset in Madrid (Spain). Currently, he teaches political science and public administration at the University of Vigo, in the bachelor's degree in public management and in the master's degree in public administration and institutional leadership courses. He is also an academic coordinator and researcher at the Centre for Governance G³ (Department of Political Science and Public Administration, University of Vigo, Spain), and a member of different teams of Spanish and Portuguese researchers (NEAPP, UMinho, Portugal; GIGAPP, IUIOG, Spain), which work on management and public services and governance at the local and cross-border level, on which he has published numerous contributions (some of them can be found on http://www.meubook.com/pg/profile/observatoriodegobernanza).

MIX
Papier aus verantwortungsvollen Quellen
Paper from responsible sources
FSC® C105338

If you have any concerns about our products,
you can contact us on
ProductSafety@springernature.com

In case Publisher is established outside the EU,
the EU authorized representative is:
Springer Nature Customer Service Center GmbH
Europaplatz 3, 69115 Heidelberg, Germany

Printed by Libri Plureos GmbH
in Hamburg, Germany